HIERARCHICAL MULTIOBJECTIVE ANALYSIS OF LARGE-SCALE SYSTEMS

HIERARCHICAL MULTIOBJECTIVE ANALYSIS OF LARGE-SCALE SYSTEMS

Yacov Y. Haimes

University of Virginia
Charlottesville, Virginia

Kyosti Tarvainen

Helsinki University of Technology
Espoo, Finland

Takashi Shima

Kanazawa Women's Junior College
Kanazawa, Japan

Jacob Thadathil

J. T. Planning and Management Software Co.
Cleveland, Ohio

HEMISPHERE PUBLISHING CORPORATION

A member of the Taylor & Francis Group

New York Washington Philadelphia London

HIERARCHICAL MULTIOBJECTIVE ANALYSIS OF LARGE-SCALE SYSTEMS

1 2 3 4 5 6 7 8 9 0 BRBR 8 9 8 7 6 5 4 3 2 1 0 9

This book was set in Times Roman by Edwards Brothers, Inc.
The editors were Todd W. Baldwin and Brenda Brienza.
Cover design by Sharon M. DePass.
Braun-Brumfield, Inc. was the printer and binder.

A CIP catalogue record for this book is available from the British Library.

Library of Congress Cataloging-in-Publication Data

Hierarchical multiobjective analysis of largescale systems / Yacov Y. Haimes . . . [et al.].
p. cm.
Includes bibliographical references.

1. Systems engineering. 2. System analysis. I. Haimes, Yacov Y.
TA168:H54 1989
620'.001'171—dc20 89-19778
CIP

ISBN 0-89116-448-0

CONTENTS

PREFACE

Systems engineering has emerged during the last three decades to play a vital role in the planning, design, construction, development, operation, maintenance, and management of engineering- and technology-based systems, creating a new body of knowledge that has expanded by leaps and bounds. In many respects, systems engineering can be viewed as philosophy—a philosophy grounded on the arts and sciences and supported by modeling concepts, frameworks, paradigms, methodologies, and tools; simulation and optimization techniques; data processing and management; and cognitive psychology, decisionmaking, and the behavioral sciences. The aim of this discipline is (a) to improve our understanding of the system under consideration and its interaction with its environment and (b) to improve a decisionmaking process that is intricate, multifaceted, dynamic, hierarchical, multiobjective, and risk based. Clearly, the complexities that characterize most large-scale systems cause their modeling, optimization, and management to be nontrivial tasks that demand both a wide, cross-disciplinary expertise from diverse fields and the development of appropriate modeling and optimization methodologies that address the hierarchical multiobjective nature of such systems. This book has been designed to provide the reader with a comprehensive treatise on the theory, methodology, and applications of modeling, optimization, and management of large-scale systems that have inherent hierarchical multiobjective characteristics.

The book draws from the advances that have been made in the field of large-scale systems over the last two decades, synthesizing and building on several doctoral dissertations on the subject that have been completed at the Systems Engineering Department at Case Western Reserve University, Cleveland, Ohio. A comprehensive book of this kind necessarily relies heavily on numerous contributions from many individuals in the field, to all of whom we are very grateful. Former colleagues from Case Western Reserve University, including Vira Chankong, Irving Lefkowitz, Leon Lasdon, Ken Loparo, Donald Macko, and Mihajlo Mesavoric, have, through many seminars and informal discussions, contributed to the ideas presented here. We are particularly in debt to Virginia Benade, who has been working with us as a technical editor on this book and its source material for over a decade now. Mary Ann Pelot, Marie Ward, and Melanie Farrish have not only provided skillful technical typing, but they have also streamlined the international communication among the authors, which has spanned three continents. The implicit and indirect contributions of many individuals and organizations are too many to enumerate here. We are grateful to Jerry Pannullo, Vijai Tulsiani, and Julie Walker for their careful and judicious proofreading of the galley proofs.

The generous support of the National Science Foundation for almost two decades has been invaluable in the development of the source material synthesized in this text. Finally, the encouragement and support of Ms. Florence Padgett, Editor, and the cooperation of Mr. Todd W. Baldwin, Production Editor, the Hemisphere Publishing Corporation, are greatly appreciated.

Yacov Y. Haimes
Kyosti Tarvainen
Takashi Shima
Jacob Thadathil

HIERARCHICAL MULTIOBJECTIVE ANALYSIS OF LARGE-SCALE SYSTEMS

CHAPTER
ONE

OVERVIEW OF HIERARCHICAL MULTIOBJECTIVE ANALYSIS

1.1 INTRODUCTION

This book reflects the evolutionary process of the field of systems analysis, a field whose origin can be traced to the 1930s and 1940s through the work of von Neumann, Koopmans, Leontief, Dantzig, Kantorovitch, and others. This early work was dominated by resource allocation and other economic problems using linear models. Two major advances in the state of the art characterized the early 1950s: the work of Bellman on dynamic programming and the work of Kuhn and Tucker on the epistemology of nonlinear programming. From the mid-1950s to the present, the field of systems analysis has advanced by leaps and bounds and has branched to such an extent that a major effort is needed to underline the remarkable achievements in the field today. Two major milestones, however, are of interest to us here: (1) the increasing dominance and centrality of the hierarchical multilevel approach in modeling and optimizing large-scale and complex systems through the use of decomposition and higher-level coordination, and (2) the maturity reached by systems analysts and decisionmakers in their realization that single-objective models are often fundamentally deficient and that multiobjective models are imperative for a viable decisionmaking process. The prolific literature on the hierarchical and multiobjective (HMO) approaches, available in the numerous books and monographs on these subjects, clearly demonstrates their importance and centrality to systems analysis and decisionmaking.

The overall goal of this book is to provide a unified framework for the integration of HMO approaches—addressing the triplets of theory, methodology, and application. For pedagogical purposes, each hierarchical concept or philosophy is first introduced in the context of a single-objective model and is subsequently extended and incorporated within a multiobjective framework.

Large-scale systems encompass a bewildering variety of aspects, components, resources, decisions, constraints, and exogenous variables, intertwined in a complex input-output causal relationship. These systems must reflect multiple goals and objectives that are often conflicting and noncommensurable, and they must respond to a hierarchy of constituencies and decisionmakers. Consequently, the art and the science of modeling such systems necessarily require the cooperative efforts of experts from multiple disciplines to untangle the intricate relationships among the system's elements.

The hierarchical approach is based on the decomposition of a large-scale, complex system and the subsequent modeling of that system into "independent" subsystems. This decentralized approach enables the systems analyst to analyze and comprehend the behavior of the subsystems at a lower level and to transmit the information obtained to fewer subsystems at a higher level.

A recent trend in systems analysis has been to consider models that have more than one objective function. This is especially important in the modeling and optimization of large-scale systems, which tend to have several conflicting and noncommensurable objectives. For example, one may want to maximize both economic efficiency, which is measured in monetary units, and environmental quality, which is measured in units of pollutant concentration. Traditionally, only one objective (economic efficiency) has been considered, with other objectives being included as constraints.

The decade of the 1970s may thus be remembered by systems engineering historians as a decade marked by the shift from the dominance of single-objective modeling toward multiple-objective modeling and optimization. This development seems natural in a field that experienced its infancy during the 1940s and 1950s, its adolescence during the 1960s and the early 1970s, and is now advancing toward maturity during the late 1970s and the 1980s.

Central to the HMO framework is the premise that large-scale systems can be best represented through a hierarchy of subsystems, each with its own set of multiple noncommensurable objectives. Assuming that the objectives of higher-level subsystems are dependent on the objectives of lower-level subsystems, this yields a hierarchy of objectives, subobjectives, sub-subobjectives, etc. The concept of higher-level coordination was developed over the years for single-objective models or structures and was then extended—both theoretically and methodologically—to accommodate multiple objectives, where Pareto optimal solutions and their associated trade-offs dominate the scene. Although other multiobjective optimization methods could be used, the surrogate worth trade-off (SWT) method and its extensions have been found to be most appropriate for higher-level coordination purposes. This is primarily because trade-offs associated with Pareto optimal solutions are generated, as a matter of course, in the SWT method. Furthermore, the SWT and its extensions are flexible in many ways—the method is applicable to models with both linear and nonlinear multiple objectives, for both static and dynamic models, and for both single and multiple decisionmakers. The Pareto optimal solutions and their respective trade-offs generated by the SWT method and its extensions provide a substantive theoretical and methodological basis for coordination within the HMO structure.

Until the mid-1970s, theoretical and methodological results of the HMO framework as well as its applications occurred at the grassroots level and were essentially preliminary. More recently, substantive theoretical grounding and procedures have been established through which generative HMO methodologies can be formed. This book presents the theoretical and methodological results at the cutting edge of the HMO framework, where applications to water resources and transport systems demonstrate the viability and usefulness of the framework.

Chapter 2, in its entirety, is devoted to acquainting the reader with the concept of multiple objectives and with the SWT method and its extensions. Chapter 3 serves a similar purpose in introducing the reader to the basic concepts of composition and higher-level coordination methodologies for single-objective models.

The various modeling and optimization methodologies discussed in this book include

- HMO coordination,
- Multiple decisionmakers,
- Dynamic HMO models,
- HMO approach for discrete dynamic models,
- Multiobjective multistage impact analysis,
- Multiobjective dynamic programming,
- Hierarchical overlapping coordination (HOC), and
- Hierarchical holographic modeling (HHM).

1.2 HIERARCHICAL MULTIOBJECTIVE COORDINATION

Higher-level coordination among the various subsystems in hierarchical structures has commonly been achieved through commensuration of the subsystem's objectives and subsequent application of single-objective optimization techniques. A superior modeling and optimization process involves the integration of the hierarchical and multiobjective approaches into a unified framework. This framework combines the strengths of both approaches. The flexibility of HMO structures makes the modeling process more realistic and understandable to the decisionmakers—the ultimate users of these models.

The integration of the HMO approaches results in a hierarchy of noncommensurable objectives at various levels. A four-level HMO structure, for example, may have four distinct levels of objectives and subobjectives (see Fig. 1.1). This type of HMO structure is indeed the actual situation in most, if not all, large-scale systems.

Coordination among several objectives at a higher level and the corresponding subobjectives at lower levels in the hierarchy represents the core of the HMO

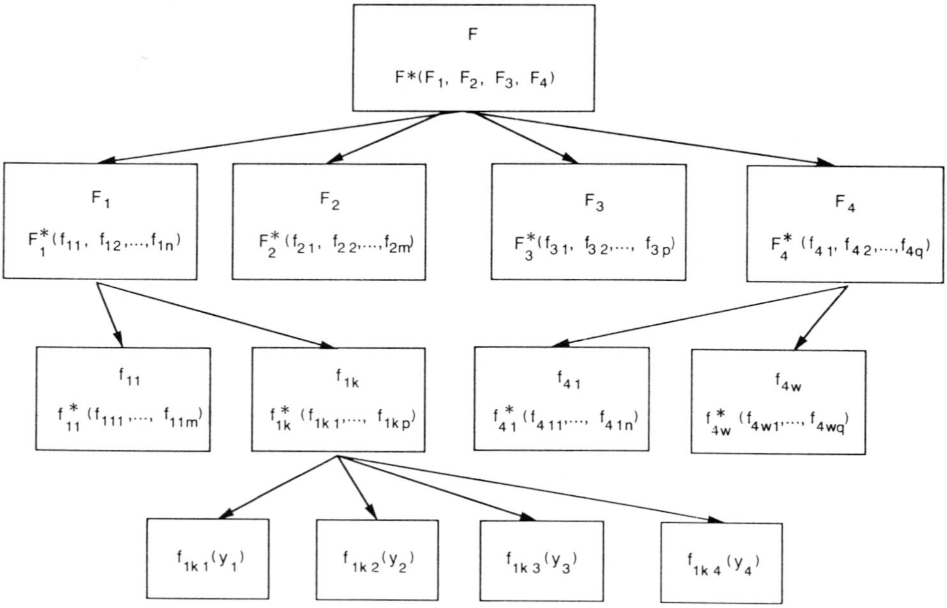

Figure 1.1 Coordination of an HMO framework.

framework. The fundamental issue here is to harmonize often competing and conflicting objectives and subobjectives at the various levels. Chapters 4 and 5 are devoted to the integration of the HMO approaches.

1.3 MULTIPLE DECISIONMAKERS

The ultimate goal in the development of HMO models and optimization methodologies is to assist decisionmakers in the complexity of the decisionmaking process. There is an irreconcilable dichotomy between the presumably more objective, analytical modeling and optimization process and the more subjective policy-making process based on human decisionmaking. The cognitive psychology and the dynamics of these stakeholders in the decisionmaking process, commonly referred to as decisionmakers, have long been the subject of study by many scholars. In this book we have devoted Chap. 6 to this fascinating area. Cases with multiple decisionmakers are addressed in many chapters, too.

1.4 DYNAMIC HMO MODELS

Obviously, the time domain plays a central and critical role in decisionmaking. In Chaps. 7 and 8, we discuss the temporal aspects of the HMO approach in two distinct though overlapping approaches: (1) discrete dynamic HMO models and (2) multiobjective dynamic programming.

1.4.1 HMO Approach for Discrete Dynamic Models

Chapter 7 addresses discrete-time dynamic hierarchical systems with multiple objectives. The methodology is an extension of the coordination schemes developed for a static HMO approach.

1.4.2 Multiobjective Dynamic Programming

In Chap. 8 we extend Bellman's principle of optimality, developed for single-objective models, to models with multiple objectives. The formulation presented is concerned with an extension to the discrete-time domain.

1.5 HIERARCHICAL OVERLAPPING COORDINATION

In 1973, hierarchical overlapping coordination (HOC) revolutionized the then-common way of thinking by suggesting (and then articulating) the efficacy of decomposing a system model into more than one decomposition. HOC is based on the premise that in decomposing a system's model into subsystems and/or levels within a hierarchical structure, more than one decomposition is possible and, most importantly, desirable. These different decompositions, each of which might be responsive to a different aspect of the system and/or data base, are coordinated through the respective couplings of the various decompositions. The coordination ultimately yields an overall system's optimum in single-objective models and preferred Pareto optimal solutions in multiple-objective models.

In water resources, for example, without HOC, modelers are implicitly forced to choose between a hydrologically or politically based model decomposition. Each decomposition represents and uncovers important aspects not demonstrable through the other. The Maumee River Basin, for example, can be decomposed into five planning subareas on the basis of political and geographic considerations, but can just as well be decomposed into eight watersheds on the basis of hydrological criteria. Yet each decomposition has its own merits, advantages, and disadvantages, and each represents a very specific data base collected and cared for by different agencies and constituencies.

A most critical concern of HOC is associated with the availability of the data base. The manipulation of data bases to serve and satisfy artificially imposed demands and constraints (imposed through the modeling process) necessitates a compromise that may cause an eventual deterioration in model credibility. For example, in the United States, data concerning streamflow, water quality, and floods are available on a hydrological basis and are collected by the U.S. Geological Survey, the U.S. Environmental Protection Agency, and the U.S. Army Corps of Engineers, respectively. On the other hand, data concerning population dynamics, employment, and other economic activities are available on political-geographic bases and are collected by other agencies, such as the U.S. Departments of Commerce, Labor, or the Treasury. HOC enables the utmost utilization of these data bases with a minimum of their manipulation or misuse. This can be achieved by resorting to two simultaneous decompositions—hydrological and political-geographic—each of which might have a different number of subsystems.

Theoretical and methodological results developed for single-objective models are then extended to models with multiple objectives. Chapters 9 through 11 are exclusively devoted to the theory, methodology, and application of the HOC and the multiobjective hierarchical overlapping coordination (MHOC).

1.6 HIERARCHICAL HOLOGRAPHIC MODELING

Hierarchical holographic modeling (HHM) is a natural evolution of state-of-the-art modeling, as derived from MHOC. HHM is grounded on the premise that large-scale systems are best modeled with more than one model—termed hierarchical holographic submodels (HHSs). Each of these HHSs can more adequately represent a certain aspect of the system associated with and/or controlled by decisionmakers, single or multiple. Here, the term holography, which is borrowed from physics, signifies the multidimensional images of the system being modeled.

In the abstract, a mathematical model may be viewed as a one-sided, limited image of the real system that it portrays. To clarify and document not only the multiple components, objectives, and constraints of a system, but also its welter of societal aspects (functional, temporal, geographic, economic, political, legal, environmental, sectoral, institutional, etc.) is quite impossible with single-model analysis and interpretation. Given this assumption and the notion that even present integrated models cannot adequately cover all of a system's aspects per se, the concept of HHM constitutes a comprehensive theoretical framework for systems modeling.

Central to the mathematical and systems basis of holographic modeling (from the perspective of theoretical constructs) is overlapping among various holographic models with respect to the objective functions, constraints, decision variables, and input-output relationships of the basic system. In this context holographic modeling may be viewed as the generalization of HOC. Alternatively, MHOC may be viewed as a degenerate case of HHM.

Common to these HHSs is a set of objective functions, constraints, decision variables, and input-output relationships. However, the existence of distinct sets of these attributes for each HHS makes HHM both a more realistic modeling approach and a more powerful decisionmaking tool. Chapter 12 is devoted to the HHM approach.

1.7 WHO SHOULD BENEFIT FROM THIS BOOK

Since this book addresses a panorama of concepts, theories, methodologies, and example problems in a unified framework, it can offer multifaceted benefits. For graduate students who have already been introduced to the basic concepts of multiple objectives, the book extends their knowledge far beyond the generation of Pareto optimal solutions and the interaction with decisionmakers that yields a preferred solution. It offers a multitude of modeling approaches and coordination procedures that add more resilience and robustness to systems analysis. For those who are familiar with decomposition techniques for single-objective models, the book extends their horizons to more realistic models that incorporate the multiple conflicting and competing objectives so fundamentally characteristic within hierarchical organizational structures. For professionals interested in problem solving, the book offers a repertoire of methodologies and techniques for modeling and optimizing large-scale systems that are applicable to a wide variety of modeling characteristics: linear/nonlinear, static/dynamic, deterministic/nondeterministic, single objective/multiple objective, and single system/hierarchy of multiple subsystems. For researchers, the book provides an excellent reference source as well as a fountainhead for new research topics. The book can also easily be adapted by university professors for use in a one-semester course on the subject.

In summary, this book is written for graduate students, professors, and systems analysts interested in large-scale systems, the hierarchical-multilevel approach, multiobjective optimization, decisionmaking, and policy analysis. The blend of and balance among theory, methodology, and applications should make this text not only a valuable reference book on the subject but also a useful textbook for a course on large-scale systems and the HMO approach.

CHAPTER

TWO

OVERVIEW OF MULTIOBJECTIVE OPTIMIZATION

2.1 INTRODUCTION

Real systems, large and small without exception, are characterized by multiple objectives and goals that often conflict and/or compete with one another. Fundamental to multiobjective analysis is the Pareto optimum concept, also known as the noninferior solution (Kuhn and Tucker, 1950; Haimes, 1977; Chankong and Haimes, 1983). Qualitatively, a noninferior solution of a multiobjective problem is one where any improvement of one objective function can be achieved only at the expense of another.

To define a noninferior solution mathematically, consider the following multiobjective function problem, also known as a vector optimization problem:

$$\min_{\mathbf{x} \in X} \{f_1(\mathbf{x}), f_2(\mathbf{x}), \ldots, f_n(\mathbf{x})\} \tag{2.1}$$

where

$\mathbf{x}$ = N-dimensional vector of decision variables
X = the set of all feasible solutions
$= \{\mathbf{x} | g_i(\mathbf{x}) \leqslant 0; \quad i = 1, 2, \ldots, \quad m\}$

Definition A decision $\mathbf{x}^*$ is said to be a noninferior solution to the problem posed by the system given in Eq. (2.1), if and only if there does not exist another

$\bar{\mathbf{x}}$ so that $f_j(\bar{\mathbf{x}}) \leq f_j(\mathbf{x}^*)$, $j = 1, 2, \ldots, n$, with strict inequality holding for at least one j.

Clearly, the solution to the multiobjective problem posed by Eq. (2.1) is not unique, and some kind of subjective judgment by the decisionmaker(s) should be added to the quantitative analysis. We will assume throughout this book that more than one decisionmaker may be involved in the selection of an acceptable and preferred solution. To avoid complexity in notation, a reference to a decision-maker denotes multiple decisionmakers unless we specify otherwise.

The various available methodologies for solving the problem given in Eq. (2.1) differ in two major ways:

1. The procedures used to generate noninferior solutions, and
2. The timing, ways, and means used to interact with the decisionmakers, and the type of information made available to them in that process (such as trade-offs).

The weighting method, also known as the parametric approach, was until recently the most common method for solving multiobjective problems. The problem given in Eq. (2.1) is essentially converted in this method into a scalar optimization:

$$\min_{\mathbf{x}} \sum_{i=1}^{n} \theta_i f_i(\mathbf{x})$$

subject to the constraints $x \in X$ and $\sum_{i=1}^{n} \theta_i = 1$, $(\theta_i \geq 0)$.

A subjective determination of the levels of the weighting coefficients, θ_i, is necessary. Subsequently, this parametric approach may yield meaningful results to the decisionmaker only when solved (parametrically) many times for different values of θ_i, $i = 1, 2, \ldots, n$. The potential existence of a duality gap is an additional important drawback to this method.

Numerous methods exist for solving multiobjective problems. They include utility functions, indifference functions, the lexicographic approach, parametric approach, ϵ-constraint approach, goal programming, goal attainment method, adaptive search approach, interactive approaches, electre method, the surrogate worth trade-off (SWT) method, and others (Haimes et al., 1975; Chankong and Haimes, 1983; Yu, 1985; Goicoechea et al., 1982). This chapter reviews the SWT method and its extensions.

2.2 EXAMPLES OF MULTIPLE ENVIRONMENTAL OBJECTIVES

The purpose of this section is to dramatize the trade-offs between the many conflicting objectives that characterize most natural systems.

2.2.1 Floor Control vs. Hydropower Generation

Consider two major objectives in the operation of reservoir systems:

1. Maximization of the hydroelectric power generation from that reservoir, and
2. Maximization of the responsiveness of the reservoir to flood control (i.e., minimization of flood damage).

Obviously, these two objectives are in conflict and competition (see Figs. 2.1 and 2.2). The higher the level of the reservoir, the more electric power generation is possible because of the high water head, yet less water storage is available for flood control purposes. Clearly, one can identify within the active storage capacity of that reservoir a Pareto optimum region whereby the enhancement of the first objective can be achieved only at the expense or degradation of the second, namely, flood control.

Also note that the units of these two objectives are noncommensurable. The first objective, which maximizes the hydroelectric power, may be measured in units of energy and not necessarily in monetary units, where the second objective can be measured in terms of acres of land, livestock, or human lives saved.

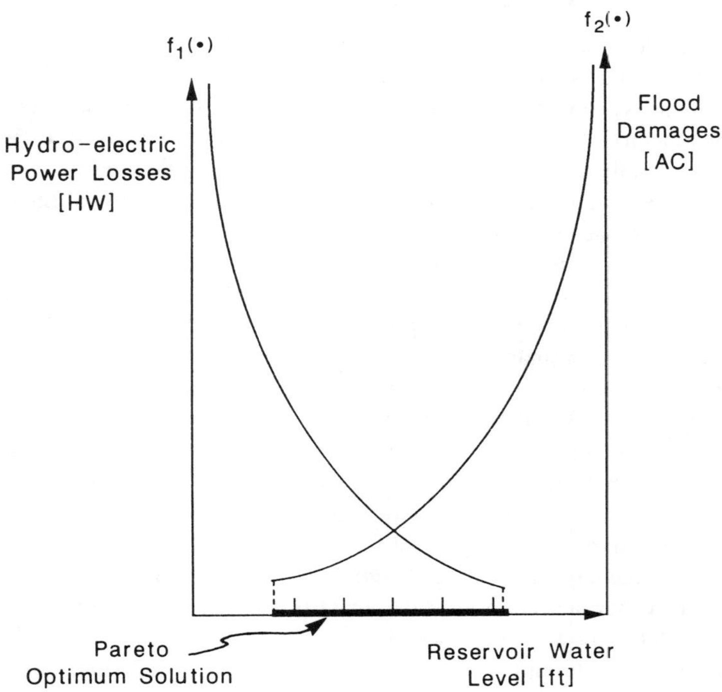

Figure 2.1 Flood damages and hydroelectric power losses in the decision space.

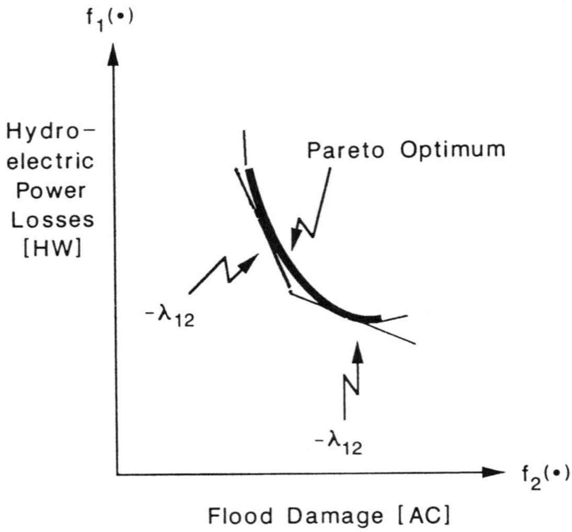

Figure 2.2 Flood damages vs. hydroelectric power losses in the functional space.

2.2.2 Water Quality vs. Cost of Treatment

Consider another example of conflicting and noncommensurable objectives in the area of water quality dealing with wastewater treatment cost and stream quality. The benefits from most wastewater treatment processes are subject to severely diminishing returns due to the exponentially increasing loss in treatment efficiency. For example, cleaning up the last 1% of pollution may be double the cost of eliminating the first 99%. The cost of achieving the known discharge goal must be viewed in terms of the sacrifices society will be compelled to make in other social demands and in terms of the large amounts of energy and natural resources that will be consumed. The elimination of the last increment of water pollution is likely to have serious, upsetting, waste-disposal impacts on air and land. These adverse impacts may be more damaging to the environment than the retention of some small amount of water pollution, particularly in areas where the self-purifying capacity of water is great and where users are not adversely affected.

Specifically, one may want to minimize the total cost of wastewater treatment on the one hand and minimize the dissolved oxygen deficiency in the stream at the outlet of the treatment plant on the other hand (see Figs. 2.3 and 2.4). Clearly, within the feasible alternatives, one may identify a Pareto optimal region where improving one objective (namely, decreasing the cost of treatment) can be achieved only at the expense of further degradation of the water quality (namely, increasing the dissolved oxygen deficiency in the stream). The argument for best available technology or zero discharge would mean that the cost of treatment would be at the very extreme of Figs. 2.3 and 2.4 with a correspondingly low dissolved oxygen deficiency.

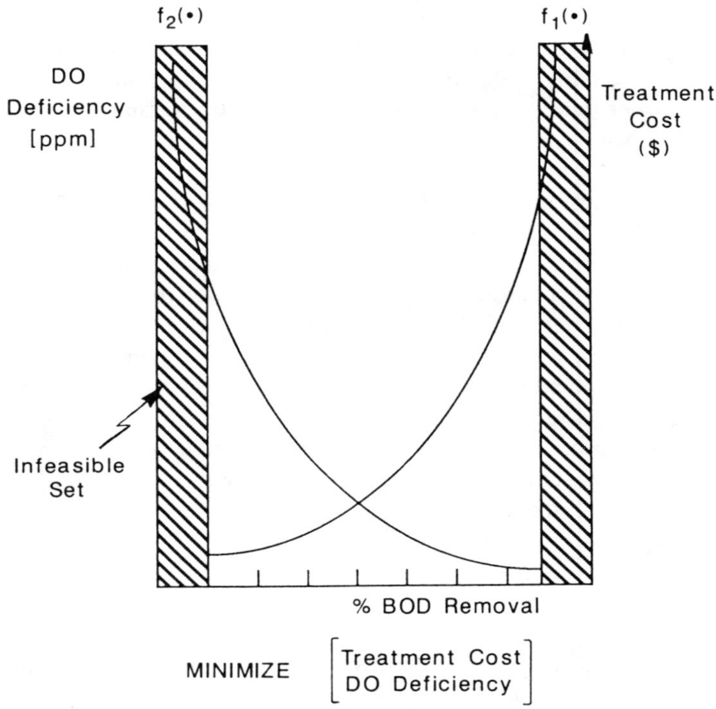

Figure 2.3 Treatment costs and dissolved oxygen (DO) deficiency in the decision space.

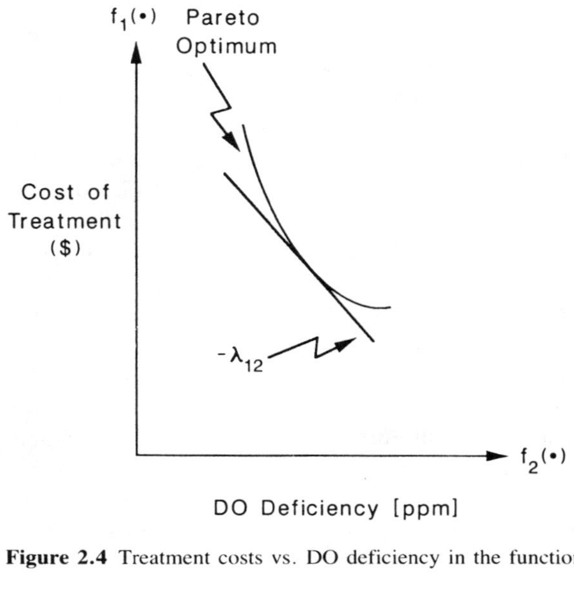

Figure 2.4 Treatment costs vs. DO deficiency in the functional space.

Using the natural capacity of water to purify itself of some kinds of wastes in limited quantities does not preclude alternatives of sequential use of the water for other purposes. Water quality standards should vary depending on the uses desired for the water. Water quality control programs should try to find the economically desirable and socially preferable uses of specific water bodies and set quality standards in relation to the preferred users and well-being of aquatic life.

Thus, depending on the purposes and on the users of that water body, different trade-offs and corresponding Pareto optimal solutions may be determined. For example, where the assimilative capacity of the stream is high and where that part of the stream is not used for recreational purposes, then a lower treatment efficiency may be feasible and even desirable while minimizing overall treatment costs. On the other hand, if the water quality is poor in a given stream, then one may wish to improve the treatment efficiency with the higher cost for the advantage of environmental quality. The point in this example is that a trade-off analysis is inevitable when considering the treatment cost vs. environmental quality. These two objectives are noncommensurable and should be dealt with through proper analysis, such as with the SWT method.

2.3 SWT METHOD

The SWT method recognizes that optimization theory is usually much more concerned with the relative value of additional increments of the various noncommensurable objectives, at a given value of each objective function, than it is with their absolute values (Haimes and Hall, 1974; Haimes et al., 1975; Hall and Haimes, 1976; Chankong and Haimes, 1983). Furthermore, given any current set of objective levels attained, it is much easier to turn to the decisionmakers to assess the relative value of the trade-off of marginal increases and decreases between any two objectives than it is to assess their absolute average values. In addition, the optimization procedure can be developed so it does no more than assess whether one more quantity of one objective is worth more or less than that lost from another at any given level of each of them. The ordinal approach can then be used with much less concern for the distortions that relative evaluation introduces into attempts to commensurate the total value of all objectives.

Since the dimension of the decision space N for most real-world problems is generally higher than the dimension of the functional space n (N decisions and n objectives; $N \gg n$), as a further simplification, one should make decisions in the functional space and only later transfer the information to the decision space.

A classic approach to the treatment of noncommensurable objectives is to select a primary or dominating objective to be optimized while constraining the decisions considered to ensure that some minimum level for all others is attained in the process. If all objectives are at or better than this minimum level of attainment with some proposed decision, such a decision can be termed satisfactory. So long as any satisfactory decision set exists, it is unnecessary to consider any decision that results in a poorer condition in any objective. Hence this approach will also help considerably in reducing the field of decisions to explore.

Let

$$\bar{f}_j = \min_{\mathbf{x}} f_j(\mathbf{x}) \qquad \mathbf{x} \in X \qquad j = 1, 2, \ldots, n \tag{2.2}$$

The ϵ-constraint approach replaces $n - 1$ objective functions by $n - 1$ constraints, as given in Eq. (2.3) (Haimes et al., 1971; Haimes, 1973):

$$\min_{\mathbf{x}} f_i(\mathbf{x}) \tag{2.3}$$

subject to

$$f_j(\mathbf{x}) \leq \epsilon_j \qquad j \neq i \qquad j = 1, 2, \ldots, n \qquad \mathbf{x} \in X$$

where ϵ_j, $j \neq i$, $j = 1, 2, \ldots, n$, are maximum tolerable levels such as $\epsilon_j = \bar{f}_j + \bar{\epsilon}_j$ and $\bar{\epsilon}_j > 0$.

The levels of satisfactory ϵ_j can be varied parametrically to evaluate the impact on the single objective function $f_i(\mathbf{x})$. Of course, the ith objective, $f_i(\mathbf{x})$, can be replaced by the jth objective, $f_j(\mathbf{x})$, and the solution procedure is then repeated. The equivalence between problems given in Eqs. (2.1) and (2.3) is well documented in the literature (Haimes et al., 1971; Haimes, 1973). The ϵ-constraint approach facilitates the generation of noninferior solutions as well as the trade-off functions, as we discuss later.

By considering one objective function as primary and all others at minimum satisfying levels as constraints, the Lagrange multipliers related to the $(n - 1)$ objectives as constraints will be zero or nonzero. If nonzero, that particular constraint does limit the optimum. It will be shown that nonzero Lagrange multipliers correspond to the noninferior set of solutions. Furthermore, the set of nonzero Lagrange multipliers represents the set of trade-off ratios between the principal objective and each of the constraining objectives respectively. Clearly, these Lagrange multipliers are functions of the optimal level attained by the principal objective function, as well as the level of all other objectives satisfied as equality (binding) constraints. Consequently, these Lagrange multipliers form a matrix of trade-off functions.

The question of the worth ratios still remains after the matrix of trade-off functions has been computed. The worth ratios are essentially achieved via an interaction with the decisionmaker. However, since the worth ratio need only represent relative worth, not the absolute level of worth of the objectives, any surrogate ratio that varies monotonically with the correct one will suffice.

2.3.1 Trade-Off Function

The following development shows that the trade-off functions can be found from the values of the dual variables associated with the constraints in a reformulated problem. Reformulate the problem given in Eq. (2.1) as follows:

$$\min_{\mathbf{x}} f_1(\mathbf{x}) \tag{2.4}$$

subject to

$$\mathbf{x} \in X \qquad f_j(\mathbf{x}) \leq \epsilon_j \qquad j = 2, 3, \ldots, n$$

where $\epsilon_j = \bar{f}_j + \bar{\epsilon}_j$, $\bar{\epsilon}_j > 0$, $j = 2, 3, \ldots, n$, $\bar{f}_j$ are as defined in Eq. (2.2) and $\bar{\epsilon}_j$ will vary parametrically in the process of constructing the trade-off function.

Form the generalized Lagrangian L to the system given in Eq. (2.4):

$$L = f_1(\mathbf{x}) + \sum_{j=2}^{n} \lambda_{1j}[f_j(\mathbf{x}) - \epsilon_j] \tag{2.5}$$

where λ_{1j} ($j = 2, 3, \ldots, n$) are generalized Lagrange multipliers. The subscript $1j$ in λ denotes that λ is the Lagrange multiplier associated (in the ϵ-constraint vector optimization problem) with the jth constraint, where the objective function is $f_1(\mathbf{x})$. λ_{1j} will be subsequently generalized to associate with the ith objective function and the jth constraint λ_{ij}. Denote by $\hat{X}$ the set of all x_i ($i = 1, 2, \ldots, N$), and denote by Ω the set of all λ_{ij} ($j = 2, 3, \ldots, n$) that satisfy the Kuhn-Tucker condition for the problem given in Eq. (2.5). The conditions of interest to our analysis are

$$\lambda_{1j}[f_j(\mathbf{x}) - \epsilon_j] = 0 \qquad \lambda_{1j} \geqslant 0 \qquad j = 2, 3, \ldots, n \tag{2.6}$$

Note that if $f_j(\mathbf{x}) < \epsilon_j$ for any $j = 2, 3, \ldots, n$ (i.e., the constraint is not binding), then the corresponding Lagrange multiplier $\lambda_{1j} = 0$.

The value of λ_{1j} ($j = 2, 3, \ldots, n$) corresponding to the binding constraints is of special interest since it indicates the marginal benefit (cost) of the objective function $f_1(\mathbf{x})$ due to an additional unit of ϵ_j. From Eq. (2.5), assuming that the solution is global, the following results can be derived:

$$\lambda_{1j}(\epsilon_j) = -\frac{\partial L}{\partial \epsilon_j} \qquad j = 2, 3, \ldots, n \tag{2.7}$$

Note, however, that for $\mathbf{x} \in X$, $\mathbf{x} \in \hat{X}$, $\lambda_{1j} \in \Omega$ for all j

$$f_1(\mathbf{x}) = L \tag{2.8}$$

Thus,

$$\lambda_{1j}(\epsilon_j) = -\frac{\partial f_1(\cdot)}{\partial \epsilon_j} \qquad j = 2, 3, \ldots, n \tag{2.9}$$

In the derivation of the trade-off functions in the SWT method, only those $\lambda_{ij} > 0$ corresponding to $f_j(\mathbf{x}) = \epsilon_j$ are of interest (since they correspond to the noninferior solution). Thus, for $f_j(\mathbf{x}) = \epsilon_j$, Eq. (2.9) can be replaced by Eq. (2.10):

$$\lambda_{1j}(\epsilon_j) = -\frac{\partial f_1(\cdot)}{\partial f_j(\cdot)} \qquad j = 2, 3, \ldots, n \tag{2.10}$$

Clearly, Eq. (2.10) can be generalized where the index of performance is the ith objective function of the system given in Eq. (2.1) rather than the first one. In this case the index i should replace the index 1 in λ_{1j}, yielding λ_{ij}. Accordingly,

$$\lambda_{ij}(\epsilon_j) = \frac{\partial f_i(\cdot)}{\partial f_j(\cdot)} \qquad i \neq j \qquad i, j = 1, 2, \ldots, n \tag{2.11}$$

For the rest of this section, only $\lambda_{ij}(\epsilon_j) > 0$, which correspond to binding constraints, are considered since there exists a direct correspondence between λ_{ij} associated with the binding constraints and the noninferior set given in Eq. (2.1).

The possible existence of a duality gap and its effect on the SWT method is discussed in detail in Everett (1963) and Haimes et al. (1975). Note that if a duality gap does exist, the ϵ-constraint method still generates all needed noninferior solutions. However, a given value of the trade-off function λ_{ij} may correspond to more than one noninferior solution.

Definition *Indifference band* is defined to be a subset of the noninferior set where the improvement of one objective function is equivalent (in the mind of the decisionmaker) to the degradation of another.

Definition *Optimum solution (or preferred solution)* is defined to be any noninferior feasible solution that belongs to the indifference band.

We now demonstrate the computational derivation of the trade-off function λ_{ij} through the derivation of λ_{ij}. The system given by Eq. (2.4) is solved for K values of ϵ_2, say $\epsilon_2^1, \epsilon_2^2, \ldots, \epsilon_2^K$, where all other ϵ_j ($j = 3, 4, \ldots, n$), are held fixed at some level ϵ_j^0. Only those $\lambda_{12}^k > 0$ that correspond to the active constraints $f_2^k(\mathbf{x}) = \epsilon_2^k$ ($k = 1, 2, \ldots, K$) are of interest, since they belong to the noninferior solution.

Assume that for ϵ_2^1, $\lambda_{12}^1 > 0$ with the corresponding $\mathbf{x}^1$. Then $f_2(\mathbf{x}^1) = \epsilon_2^1$. Clearly, not all other λ_{ij} ($j = 3, 4, \ldots, n$) corresponding to this solution ($\mathbf{x}^1$) are positive. Thus, the following problem is solved:

$$\min f_1(\mathbf{x}) \qquad \mathbf{x} \in X \tag{2.12}$$

subject to

$$f_j(\mathbf{x}) \leq f_j(\mathbf{x}^1), \qquad j = 2, 3, \ldots, n$$

where ϵ_j^0 were replaced by $f_j(\mathbf{x}^1)$ ($j = 3, 4, \ldots, n$). A small variation δ_j from $f_j(\mathbf{x}^1)$ may be needed to ensure positive λ_{1j} ($j = 3, 4, \ldots, n$) in the computational procedure. The trade-off λ_{12} is a function of all ϵ_j ($j = 2, 3, \ldots, n$) [i.e., $\lambda_{12} = \lambda_{12}(\epsilon_2, \ldots, \epsilon_n)$].

We will show in subsequent discussions that the trade-off function $\lambda_{ij}(\cdot)$ may need to be constructed (via multiple regression) in the vicinity of the indifference band, namely, the vicinity of the solution to the problem given in Eq. (2.1).

Similarly, the trade-off function λ_{13} can be generated, where again the prime objective function is $f_1(\mathbf{x})$, and the system given in Eq. (2.4) is solved for K' different values of ϵ_3^k ($k = 1, 2, \ldots, K'$) with a fixed level of $\epsilon_2^0, \epsilon_4^0, \ldots, \epsilon_n^0$. Similarly, the trade-off functions λ_{1j} can be generated for $j = 4, 5, \ldots, n$. Once all trade-off functions λ_{1j} ($j = 2, 3, \ldots, n$) have been generated, the prime objective may be changed to the ith and thus all trade-off functions λ_{ij}, $i \neq j$, $i, j = 1, 2, \ldots, n$ can be generated. It can be shown, however, that not all λ_{ij} need be generated computationally, since the following relationships hold:

$$\lambda_{ij} = \lambda_{ik} \cdot \lambda_{kj} \qquad i \neq j \qquad i, j = 1, 2, \ldots, n \tag{2.13}$$

In addition, the relationship $\lambda_{ij} = 1/\lambda_{ji}$ for $\lambda_{ji} \neq 0$ can also be used (Haimes et al., 1975).

2.3.2 Surrogate Worth Function

The surrogate worth function provides the interface between the decisionmaker and the mathematical model. The value of the surrogate worth function W_{ij} is an assessment of how much (on an ordinal scale, say from -10 to $+10$ with zero signifying equal preference) the decisionmaker prefers trading λ_{ij} marginal units of f_i for one marginal unit of f_j, given the values of all the objectives $f_l, \ldots, f_n$ corresponding to λ_{ij}. Note that $W_{ij} > 0$ means the decisionmaker does prefer making such a trade; $W_{ij} < 0$ means he/she does not; and $W_{ij} = 0$ implies indifference. The following is a formal definition of W_{ij}.

$W_{ij} > 0$ when λ_{ij} marginal units of $f_i(\mathbf{x})$ are preferred over one marginal unit of $f_j(\mathbf{x})$ given the satisfaction of all objectives at level ϵ_k, $k = 1, 2, \ldots, n$

$= 0$ when λ_{ij} marginal units of $f_i(\mathbf{x})$ are equivalent to one marginal unit of $f_j(\mathbf{x})$ given the satisfaction of all objectives at level ϵ_k, $k = 1, 2, \ldots, n$

< 0 when λ_{ij} marginal units of $f_i(\mathbf{x})$ are not preferred over one marginal unit of $f_j(\mathbf{x})$ given the satisfaction of all objectives at level ϵ_k, $k = 1, 2, \ldots, n$

It is important to note here that the decisionmaker(s) is provided with the trade-off value (via the trade-off function) of any two objective functions at a given level of attainment of the other objective functions. Furthermore, all trade-off values generated from the trade-off function are associated with the noninferior set. Thus, it is evident that any procedure that can generate a surrogate worth function, which in turn can provide the band of indifference of λ_{ij}, $i \neq j$, $i, j = 1, 2, \ldots n$, will solve the multiobjective problem [the computational procedure for generating the overall solution to the system given in Eq. (2.1) is discussed subsequently]. In this respect, much of the experience developed and gained in the field of decision theory and team theory can be utilized in the SWT method (Raiffa, 1968; Marschak, 1955; Fishburn, 1970).

The band of indifference can be determined as follows: the decisionmaker is asked whether λ_{ij} units of $f_i(\mathbf{x})$ is $\{\gtreqless\}$ one unit of $f_j(\mathbf{x})$ for two distinct values of λ_{ij}. A linear combination of the corresponding two answers $W_{ij}(\lambda_{ij})$ obtained from the decisionmaker in ordinal scale can be made (see Fig. 2.5). Then the value of $\lambda_{ij} = \lambda_{ij}^*$ is chosen so that $W_{ij}(\lambda_{ij}^*) = 0$ on the line segment fitting the two values of λ_{ij}. With λ_{ij}^* determined, the indifference band is assumed to exist within the neighborhood of λ_{ij}^*. Additional questions to the decisionmaker can be asked in the neighborhood of λ_{ij}^* to improve the accuracy of λ_{ij}^* and the band of indifference.

The surrogate worth function assigns a scalar value (on an ordinal scale) to any given noninferior (efficient, Pareto optimal) solution. There are three ways of specifying a noninferior solution:

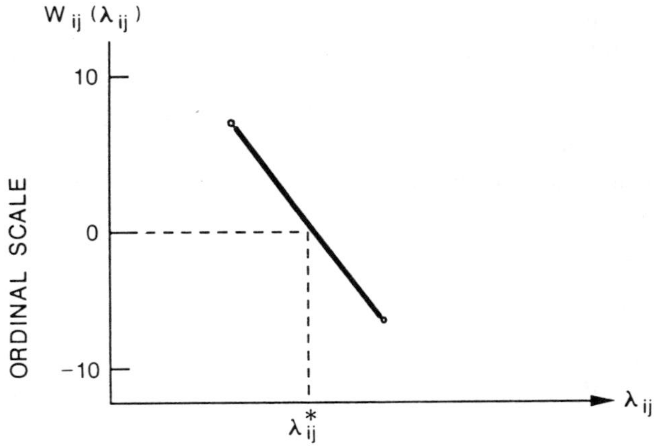

Figure 2.5 Determination of the indifference band at λ^*_{ij}.

1. By the values of its decision variables, $x_1, \ldots, x_N$,
2. By the trade-off functions $\lambda_{i1}, \ldots, \lambda_{in}$, or
3. By its objective function values $f_1, \ldots, f_n$.

Hence, we can have $W_{ij}(x_1, \ldots, x_N)$ or $W_{ij}(\lambda_{i1}, \ldots, \lambda_{in})$ or $W_{ij}(f_1, \ldots, f_n)$. The first is generally ruled out by the inefficiencies of decision space manipulations. The second may suffer from problems when discontinuities or nonconvexities occur in the functional space, but can be used in other problems. The third, or objective function space approach, appears to be the best.

As an example of how the method works, consider a three-objective problem. Several noninferior points, $(f_2, f_3)_0, \ldots, (f_2, f_3)_k$, and their trade-offs, $(\lambda_{12}, \lambda_{13})_0, \ldots, (\lambda_{12}, \lambda_{13})_k$, are determined, for example, via the ϵ-constraint method. The decisionmaker is then questioned in order to get values $W_{12}(f_2, f_3)_0 \ldots W_{12}(f_2, f_3)_k$ and $W_{13}(f_2, f_3)_0 \ldots W_{13}(f_2, f_3)_k$. (It can be shown that the other W_{ij} need not be determined.) Now, since generally none of these will be zero, we must determine more noninferior solutions and their trade-offs than before, and ask more questions of the decisionmaker until we find an $(f_2, f_3)^*$ so that $W_{12}(f_2, f_3)^*$ and $W_{13}(f_2, f_3)^*$ both equal zero. A functional relation (via regression or interpolation) for $W_{12}(f_2, f_3)$ and $W_{13}(f_2, f_3)$ can be used as an approximation when setting new constraint levels in determining new noninferior solutions.

Since the worth is only evaluated at known noninferior points, it is guaranteed that $(f_2, f_3)^*$ will give rise to a feasible solution when put into the overall mathematical model. The same guarantee holds when $W_{ij}(\lambda_{i1}, \ldots, \lambda_{in})$ is used.

What happens if there cannot be found a pair of $(f_2, f_3)^*$ whose worth functions are both zero? In that case, we can take the one whose worth functions are closest to zero as an approximate preferred solution. Note that the noninferior solutions whose surrogate worth functions are all zero are the maximum utility solutions. The noninferior solution whose worth functions are closest to zero will be the one closest to the maximum utility solution. There is a close relation between the surrogate worth function W_{ij} and the partial derivatives of the utility function.

In multiobjective analysis it is assumed implicitly that the decisionmaker maximizes his or her utility, which is a monotonic decreasing function of the various criteria. Given a decision $\mathbf{x}$ and the associated consequences $\mathbf{f}(\mathbf{x})$, the utility is given by

$$u = u(f_1(\mathbf{x}), \ldots, f_n(\mathbf{x})) \tag{2.14}$$

For a small change in f_i one can linearize Eq. (2.14).

$$\Delta u = \frac{\partial u}{\partial f_1} \Delta f_1 + \frac{\partial u}{\partial f_2} \Delta f_2 + \cdots + \frac{\partial u}{\partial f_n} \Delta f_n \tag{2.15}$$

However, for noninferior points $f_1 = f_1(f_2, \ldots, f_n)$, or

$$\Delta f_1 = \sum_{i=2}^{n} \frac{\partial f_1}{\partial f_i} \Delta f_i = -\sum_{i=2}^{n} \lambda_{1i} \Delta f_i \tag{2.16}$$

Eliminating Δf_1 [Eq. (2.16)] from Eq. (2.15) yields

$$\Delta u = \sum_{i=2}^{n} \left(\frac{\partial u}{\partial f_i} - \frac{\partial u}{\partial f_1} \lambda_{1i} \right) \Delta f_i = \sum_{i=2}^{n} \Delta u_{1i} \Delta f_i$$

where

$$\Delta u_{1i} = \frac{\partial u}{\partial f_i} - \frac{\partial u}{\partial f_1} \lambda_{1i}$$

Let

$$a_i = \frac{\partial u}{\partial f_i}$$

Then

$$\Delta u_{1i} = a_i - a_1 \lambda_{1i}$$

The surrogate worth function W_{1i} is a monotonic function of Δu_{1i} with the property that $W_{1i} = 0 \leftrightarrow \Delta u_{1i} = 0$, and can therefore be written as

$$W_{1i} = h_i(a_i - a_1 \lambda_{1i})$$

where h_i is some monotonic increasing function of its argument, with a range of -10 to $+10$, and with the property that $h_i(0) = 0$. If a_i is considered constant or varies only slightly with f_i, $i = 1, \ldots, n$, then it is possible to assume that W_{1i} depends only on λ_{1i}.

Finally, one may question whether an interaction with the decisionmaker in the function space should always yield a $W_{12}(\lambda_{12}) = 0$, that is, an indifference solution. Two cases may be identified here:

1. The decisionmaker's response is always on one side of the W_{12} scale for all λ_{12} corresponding to the Pareto optimum solutions. That is, the decisionmaker's answers are either all on the positive or all on the negative scale of W_{12}. This really means that the decisionmaker is always willing to improve—say objective 1—at the expense of degrading objective 2 in the entire Pareto optimal space. This case, while it may actually happen, is of no particular interest here, since it reduces the multiobjective problem to a single-objective optimization problem.
2. Should the decisionmaker's response in the function space be on the positive scale of W_{12} for some values of λ_{12}, and negative for other sets of values of λ_{12}, then (assuming consistency in the decisionmaker's response and continuity in λ_{12}), it can be guaranteed that a value of $W_{12} = 0$ exists that corresponds to an indifference solution with λ_{12}^*, i.e., $W_{12}(\lambda_{12}^*) = 0$.

2.3.3 Transformation to the Decision Space

Once the indifference bands have been determined for λ_{ij}^*, the next and final step in the SWT method is to determine an $\mathbf{x}^*$ that corresponds to all λ_{ij}^*. To each λ_{ij}^* determined from the surrogate worth function via the interaction with the decisionmaker, there corresponds $f_j^*(\mathbf{x})$, $j = 1, 2, \ldots, n$, $j \neq i$. These $f_j^*(\mathbf{x})$ are the values of the functions $f_j(\mathbf{x})$ at the equality constraints ϵ_j so that $\lambda_{ij}^*[f_j^*(\mathbf{x}) - \epsilon_j] = 0$. Accordingly, the optimal value of decisions, $\mathbf{x}^*$, can be obtained by simply solving the following optimization problem:

$$\min_{\mathbf{x} \in X} f_i(\mathbf{x}) \tag{2.17}$$

subject to

$$f_j(\mathbf{x}) \leq f_j^*(\mathbf{x}) \qquad j = 1, 2, \ldots, n \qquad j \neq i$$

The system given in Eq. (2.17) is a common optimization problem with a single objective function. The solution of Eq. (2.17) yields the desired $\mathbf{x}^*$ for the total vector optimization problem posed by Eq. (2.1).

The consistency of the decisionmaker should not be always assumed. The decisionmaker may show irrational behavior or provide misinformation at times. The SWT method safeguards against this by cross-checking the resulting λ_{ij}^*. Haimes et al. (1975) have shown that one set of $\lambda_{11}, \ldots, \lambda_{1n}$ will suffice for solving the multiobjective problem posed previously. It is always possible, however, to generate, for example, λ_{12}^*, λ_{23}^*, and λ_{13}^* (via an interaction with the decisionmaker), and to check that indeed the following relation holds: $\lambda_{13}^* = \lambda_{12}^* \cdot \lambda_{23}^*$ (i.e., satisfies the general relationship $\lambda_{ij} = \lambda_{ik} \cdot \lambda_{kj}$).

Theorem Given the multiobjective problem defined and referred to as problem 1, then for any feasible set of λ_{ij}^*, there exists a corresponding feasible set of decisions $\mathbf{x}^*$.

PROOF Rewrite problem 1 as follows:

$$\min_{\mathbf{x} \in X} \left[f_i(\mathbf{x}) + \sum_{j \neq i} \lambda_{ij}^* f_j(\mathbf{x}) \right]$$

If all $f_k(\mathbf{x})$ $(k = 1, 2, \ldots, n)$ are continuous and the solution set X is compact (a set X is said to be compact if it is both closed and bounded, that is, if it is closed and is contained within some sphere of finite radius), then this problem must have a solution (by Weierstrass theorem, see Luenberger, 1973).

These assumptions are very mild. Compactness of X can be guaranteed by imposing finite upper and lower bounds on each component of the decision vector $\mathbf{x}$, assuming all the constraint functions $g_i(\mathbf{x})$ are continuous. Continuity assumption of all $f_j(\mathbf{x})$ and $g_i(\mathbf{x})$ is common in mathematical programming.

Let $\mathbf{x}^*$ be a solution for a given λ_{ij}^*. Then λ_{ij}^* are the optimal trade-off values (Lagrange multipliers) for the problem:

$$\min_{\mathbf{x} \in X} f_i(\mathbf{x})$$

subject to the constraints

$$f_j(\mathbf{x}) \leq f_j(\mathbf{x}^*) \equiv \epsilon_j \qquad j = 2, \ldots, n$$

Thus $\mathbf{x}^*$ is in X and λ_{ij}^* are the desired Lagrange multipliers.

The feasibility of a solution $\mathbf{x}^*$ corresponding to λ_{ij}^* can also be shown on the basis of the lambda theorem by Everett.

It is helpful to summarize the three major steps in the SWT method:

1. Identify and generate noninferior (Pareto optimum) solutions, along with the trade-off functions λ_{ij} between any two objective functions $f_i(\mathbf{x})$ and $f_j(\mathbf{x})$, $i \neq j$. It can be shown that under certain mild conditions, one set of n trade-off functions, $\lambda_{11}, \ldots, \lambda_{1n}$, will suffice to generate all other λ_{ij}, $i \neq j$, $i, j = 1, 2, \ldots, n$
2. Interact with the decisionmaker to assess the indifference band where the surrogate worth function $W_{ij}(\lambda_{ij}^*) = 0$. It was shown that, under certain mild conditions, W_{ij} depends only on λ_{ij}
3. Determine the optimal decision set $\mathbf{x}^*$ using the optimal trade-off values λ_{ij}^*.

2.3.4 SWT Method with Multiple Decisionmakers

Large-scale systems are characterized by multiple decisionmakers at the various levels of the decisionmaking process. This is true for both planning and management purposes. For example, the planning board of the Maumee River Basin consists of eight members from federal, state, and regional agencies. The board is in charge of developing a basin-wide comprehensive plan that is responsive to environmental, economic, social, legal, political, and institutional needs. However, members of the board, as decisionmakers, exercise their mandate to be responsive simultaneously to their professional judgment, their agency's stand, and the public preferences as voiced by various public hearings and other media. Clearly, in applying the SWT method, different indifference bands may result by interacting with each planning board member. The key question is how to modify the SWT method to handle this situation.

Three major cases of multiobjective optimization problems with multiple decisionmakers may be identified. These are the direct group decisionmaking systems, the representative decisionmaking systems, and the political decision simulations. These three classifications are not discussed here, but rather we assume a more general case for simplicity.

Consider the multiobjective optimization problem posed by Eq. (2.1), where an interaction with the decisionmakers for assessing the corresponding trade-offs and preferences that lead to $W_{ij} = 0$ takes place. Two cases will be identified here: the ideal and the probable.

2.3.4.1 Ideal case In assessing the trade-offs and preferences with the decisionmakers, it is assumed in the ideal case that the indifference bands generated by all the decisionmakers for all W_{ij}, $i \neq j$, $i,j = 1, 2, \ldots, n$, have a common indifference band Δ, as depicted in Fig. 2.6. This situation is unlikely to happen; however, it provides a medium for understanding the probable case. All the indifference bands in Fig. 2.6 correspond, of course, to $W_{ij} = 0$; however, they are plotted at different levels on the W_{ij} scale in order to distinguish among the indifference bands of the various decisionmakers.

2.3.4.2 Probable case In the probable case, no common indifference band can be found for all the decisionmakers. This case is depicted in Fig. 2.7. The SWT method provides an explicit and quantitative mechanism in simulating the decisionmakers' preferences with respect to the trade-offs between any two objective functions. The identification of the differences in their preferences is a first step in closing these gaps through the inevitable process of negotiation and compromise. These negotiations may take different forms and are expected to lead to an agreeable decision (depending on the rules of the game: whether a simple majority, absolute majority, consensus, or other is needed for an agreed-upon decision).

2.3.5 Summary

The SWT method can be used to analyze and optimize the multiobjective planning problem. The following is a brief summary of the SWT method.

1. It is capable of generating all needed noninferior solutions to a vector optimization problem.
2. The method generates the trade-offs between any two objective functions on the basis of duality theory in nonlinear programming. The trade-off function

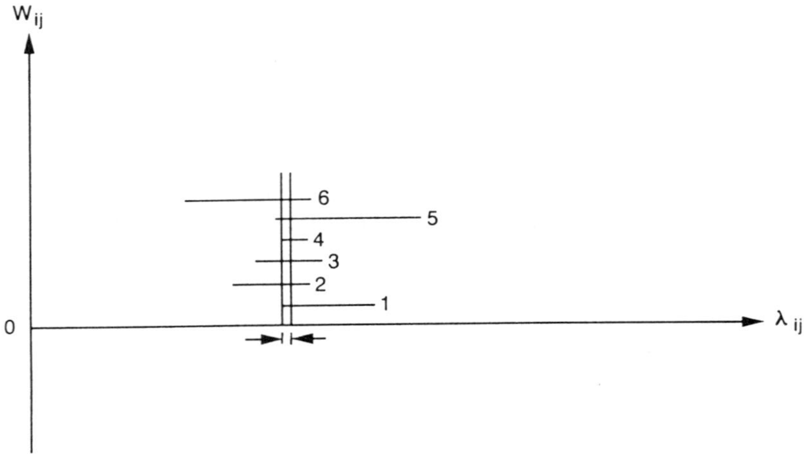

Figure 2.6 Common indifference band in the ideal case.

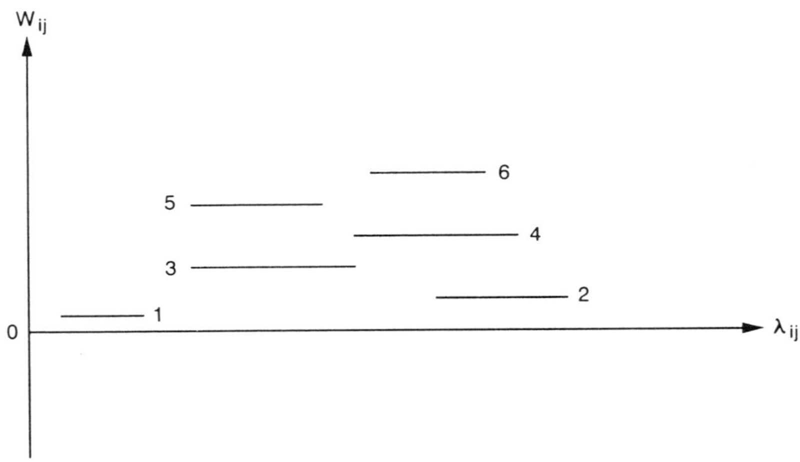

Figure 2.7 Indifference bands in the probable case.

between the *i*th and *j*th objective functions, λ_{ij}, is explicitly evaluated and is equivalent to $-\partial f_i/\partial f_j$.

3. The decisionmaker interacts with the systems analyst and the mathematical model at a general and very moderate level. This is done via the generation of the surrogate worth functions, which relate the decisionmakers' preferences to the noninferior solutions through the trade-off functions. These preferences are constructed in the objective function space (more familiar and meaningful to the decisionmakers) and only then are transferred to the decision space. This is particularly important, since the dimensionality of the objective function

space is often smaller than that of the decision space. These preferences yield to an indifference band where the decisionmaker is indifferent to any further trade-off among the objectives.

4. The SWT method provides for the quantitative analysis of noncommensurable objective functions.
5. The method is well suited for the analysis and optimization of multiobjective functions with multiple decisionmakers.
6. The method has an appreciable computational advantage over all other existing methods when the number of objective functions is three or more.

2.4 SWT METHOD AND UTILITY FUNCTION APPROACH

The theoretical background behind the SWT method may be better understood by examining the utility function approach of multiobjective optimization (Kaplan, 1975; Kaplan and Haimes, 1975). We will show that the optimality conditions of the SWT method may be derived from the optimality conditions of the utility function approach. This discussion is strictly theoretical, however, since it is difficult or impossible to implement the utility function approach in practice.

The vector optimization problem given in Eq. (2.1) may be viewed as a scalar optimization problem, where the objective function is a scalar function of the objectives. Ideally this scalar objective, known as the utility function, should be maximized, subject to a number of constraints. The decisionmaker may represent a consumer, while the various objectives represent goods he or she desires (Intrilligator, 1971). The utility function is thus a scalar function of the objectives and indicates the value of various combinations of the objectives to the decisionmaker.

2.4.1 Utility Function

Define a scalar function of the objective functions, $U(f_1(\mathbf{x}), \ldots, f_n(\mathbf{x}))$, where the objective functions are given by $f_i(\mathbf{x})$, $i = 1, 2, \ldots, n$. This scalar function may be referred to as a utility function and has the following properties:

1. If $f_i(\mathbf{x}^1) \leqslant f_i(\mathbf{x}^2)$, for $i = 1, 2, \ldots, n$, then $U(f_1(\mathbf{x}^1), f_2(\mathbf{x}^1), \ldots, f_n(\mathbf{x}^1)) \geqslant U(f_1(\mathbf{x}^2), f_2(\mathbf{x}^2), \ldots, f_n(\mathbf{x}^2))$.
2. $U(f_1(\mathbf{x}^1), f_2(\mathbf{x}^1), \ldots, f_n(\mathbf{x}^1)) \geqslant U(f_1(\mathbf{x})^2, f_2(\mathbf{x}^2), \ldots, f_n(\mathbf{x}^2))$ implies that the combination of objectives $(f_1(\mathbf{x}^1), f_2(\mathbf{x}^1), \ldots, f_n(\mathbf{x}^1))$ is preferred to the combination of objectives $(f_1(\mathbf{x}^2), f_2(\mathbf{x}^2), \ldots, f_n(\mathbf{x}^2))$.
3. If $U(f_1(\mathbf{x}^1), f_2(\mathbf{x}^1), \ldots, f_n(\mathbf{x}^1)) = U(f_1(\mathbf{x}^2), f_2(\mathbf{x}^2), \ldots, f_n(\mathbf{x}^2))$, then the decisionmaker is indifferent to the combinations $(f_1(\mathbf{x}^1), f_2(\mathbf{x}^1), \ldots, f_n(\mathbf{x}^1))$ and $(f_1(\mathbf{x}^2), f_2(\mathbf{x}^2), \ldots, f_n(\mathbf{x}^2))$. In other words, given the choice, the decisionmaker would not prefer or be able to choose between the two combinations.

While it is extremely difficult or impossible to actually determine the decisionmaker's utility function—that is, assigning numerical utilities to the various combinations of the objectives—the following theoretical development will be useful in developing the optimality conditions of the SWT method. This discussion should serve to motivate further development of the SWT method.

The contours of the utility function, $U(f_1(\mathbf{x}), f_2(\mathbf{x}), \ldots, f_n(\mathbf{x})) = c$, are called indifference curves because the decisionmaker is indifferent to any pair of combinations along a given curve. However, if $c^1 > c^2$, then all combinations along the curve $U(f_1(\mathbf{x}), f_2(\mathbf{x}), \ldots, f_n(\mathbf{x})) = c^1$ are preferred to any combination along the curve $U(f_1(\mathbf{x}), f_2(\mathbf{x}), \ldots, f_n(\mathbf{x})) = c^2$. Again, it may be difficult or impossible to determine these curves.

Let the solution of the ϵ-constraint problem given in Eq. (2.3), where $i = 1$, be represented by the vector $\mathbf{x}^*$ $(\epsilon_2, \epsilon_3, \ldots, \epsilon_n)$, where ϵ_j is specified for $j = 2, 3, \ldots, n$. The value of the primary objective attained, given ϵ_j, $j = 2, 3, \ldots, n$, is then $f_1(\mathbf{x}^*(\epsilon_2, \epsilon_3, \ldots, \epsilon_n))$, if we assume that all constraints are binding at $\mathbf{x}^*(\epsilon_2, \epsilon_3, \ldots, \epsilon_n)$. The solutions of Eq. (2.3) for various values of ϵ_j, $j = 2, 3, \ldots, n$, specify a trade-off surface, $f_1(\mathbf{x}^*(\epsilon_2, \epsilon_3, \ldots, \epsilon_n))$, a function of the satisfactory levels of the secondary objectives, ϵ_j, $j = 2, 3, \ldots, n$. Thus, the combinations $[f_1(\mathbf{x}^*(\epsilon_2, \epsilon_3, \ldots, \epsilon_n)), \epsilon_2, \epsilon_3, \ldots, \epsilon_n]$ form the set of noninferior solutions, since $f_j(\mathbf{x}^*(\epsilon_2, \epsilon_3, \ldots, \epsilon_n)) = \epsilon_j$, $j = 2, 3, \ldots, n$.

The vector optimization problem given in Eq. (2.1) may now be written as a scalar maximization problem, where the new objective, $U(\cdot)$, is a function of the original objectives. The constraints remain unchanged. Therefore,

$$\max_{\mathbf{x} \in X} U(f_1(\mathbf{x}), f_2(\mathbf{x}), \ldots, f_n(\mathbf{x})) \tag{2.18}$$

where $X = \{\mathbf{x} | g_i(\mathbf{x}) \leq 0, i = 1, 2, \ldots, m\}$.

The utility function is constructed, however, so that noninferior solutions are preferred to inferior ones. Therefore, only noninferior solutions must be examined in Eq. (2.18). The solution of the ϵ-constraint problem given in Eq. (2.3) generates the noninferior region. Restricting the utility maximization problem to the noninferior region simplifies the approach considerably. The decision variables are now the desired levels of the objectives, ϵ_j, $j = 2, 3, \ldots, n$, rather than the original decision variables, $\mathbf{x}$. The optimization is carried on in the objective function space, E^{n-1}, not in the decision variable space, E^N. As mentioned before, in most realistic problems, $N >> n$. Only $n - 1$ objective values must be specified, since the primary objective, $f_1(\mathbf{x}^*(\epsilon_2, \epsilon_3, \ldots, \epsilon_n))$, is specified by the solution of Eq. (2.3).

Substituting the optimal values of the original decision variables, $\mathbf{x}^*(\epsilon_2, \epsilon_3, \ldots, \epsilon_n)$, given desired levels of the secondary objectives ϵ_j, $j = 2, 3, \ldots, n$, the utility maximization problem may be restated as follows:

$$\max_{\epsilon_2, \epsilon_3, \ldots, \epsilon_n} U(f_1(\mathbf{x}^*(\epsilon_2, \epsilon_3, \ldots, \epsilon_n)) \tag{2.19}$$

No constraints are involved in Eq. (2.19), since all constraints were considered in the solution of the ϵ-constraint problem given in Eq. (2.3). Again, the decision variables in the utility maximization problem are now the desired levels of the secondary objectives ϵ_j, $j = 2, 3, \ldots, n$. The original decision variables $\mathbf{x}$

are ignored at this stage, having been employed to determine the trade-off surface, $f_1(\mathbf{x}^*(\epsilon_2, \epsilon_3, \ldots, \epsilon_n))$, by repeated solution of Eq. (2.3), with various values of the secondary objectives, ϵ_j, $j = 2, 3, \ldots, n$. Once the optimal values of the objectives are determined, ϵ_j^*, $j = 2, 3, \ldots, n$, Eq. (2.3) will be solved once more, to find the optimal values of the decision variables, $\mathbf{x}^*(\epsilon_2^*, \epsilon_3^*, \ldots, \epsilon_n^*)$. The optimal values of the objectives are found by solving Eq. (2.19).

Since Eq. (2.19) involves unconstrained optimization, the first-order necessary conditions for a stationary point $(\epsilon_2^*, \epsilon_3^*, \ldots, \epsilon_n^*)$ are as follows:

$$\frac{\partial U(\cdot)}{\partial \epsilon_j} = 0 \qquad j = 2, 3, \ldots, n \tag{2.20}$$

Applying the chain rule to Eq. (2.20) yields

$$\frac{\partial U(\cdot)}{\partial f_1(\cdot)} \frac{\partial f_1(\cdot)}{\partial \epsilon_2} + \frac{\partial U(\cdot)}{\partial \epsilon_2} = 0 \tag{2.21a}$$

$$\frac{\partial U(\cdot)}{\partial f_1(\cdot)} \frac{\partial f_1(\cdot)}{\partial \epsilon_3} + \frac{\partial U(\cdot)}{\partial \epsilon_3} = 0 \tag{2.21b}$$

$$\vdots \qquad\qquad \vdots$$

$$\frac{\partial U(\cdot)}{\partial f_1(\cdot)} \frac{\partial f_1(\cdot)}{\partial \epsilon_n} + \frac{\partial U(\cdot)}{\partial \epsilon_n} = 0 \tag{2.21n}$$

Equations (2.21a)–(2.21n) yield

$$\frac{\partial U(\cdot)}{\partial f_1(\cdot)} = -\frac{\partial U(\cdot)/\partial \epsilon_2}{\partial f_1(\cdot)/\partial \epsilon_2} \tag{2.22a}$$

$$\frac{\partial U(\cdot)}{\partial f_1(\cdot)} = -\frac{\partial U(\cdot)/\partial \epsilon_3}{\partial f_1(\cdot)/\partial \epsilon_3} \tag{2.22b}$$

$$\vdots \qquad\qquad \vdots \qquad\qquad \vdots$$

$$\frac{\partial U(\cdot)}{\partial f_1(\cdot)} = -\frac{\partial U(\cdot)/\partial \epsilon_n}{\partial f_1(\cdot)/\partial \epsilon_n} \tag{2.22n}$$

Equating Eqs. (2.22a)–(2.22n) yields

$$-\frac{\partial U(\cdot)/\partial \epsilon_2}{\partial f_1(\cdot)/\partial \epsilon_2} = -\frac{\partial U(\cdot)/\partial \epsilon_3}{\partial f_1(\cdot)/\partial \epsilon_3} = \cdots = -\frac{\partial U(\cdot)/\partial \epsilon_n}{\partial f_1(\cdot)/\partial \epsilon_n} = \frac{\partial U(\cdot)}{\partial f_1(\cdot)} \tag{2.23}$$

Since utility is measured in arbitrary units, we may assign

$$\frac{\partial U(\cdot)}{\partial f_1(\cdot)} = 1 \tag{2.24}$$

In this way, utility is now measured in units commensurable with the units of the objective $f_1(\mathbf{x})$.

2.4.2 Trade-Offs and Marginal Rate of Substitution

Since the trade-off surface, $f_1(\mathbf{x}^*(\epsilon_2, \epsilon_3, \ldots, \epsilon_n))$, was determined by repeated solution of Eq. (2.3) for various values ϵ_j, $j = 2, 3, \ldots, n$, the tradeoffs, λ_{1j}, $j = 2, 3, \ldots, n$, are known as well. Rewriting Eq. (2.9) as

$$\lambda_{1j}(\epsilon_2, \epsilon_3, \ldots, \epsilon_n) = -\frac{\partial f_1(\cdot)}{\partial \epsilon_j} \qquad j = 2, 3, \ldots, n \tag{2.25}$$

and combining it with Eqs. (2.23) and (2.24) yields

$$\frac{\partial U(\cdot)/\partial \epsilon_2}{\lambda_{12}(\epsilon_2, \epsilon_3, \ldots, \epsilon_n)} = \frac{\partial U(\cdot)/\partial \epsilon_3}{\lambda_{13}(\epsilon_2, \epsilon_3, \ldots, \epsilon_n)} = \cdots = \frac{\partial U(\cdot)/\partial \epsilon_n}{\lambda_{1n}(\epsilon_2, \epsilon_3, \ldots, \epsilon_n)} = 1 \tag{2.26}$$

Therefore, at the optimum

$$\begin{aligned} \lambda_{12}(\epsilon_2^*, \epsilon_3^*, \ldots, \epsilon_n^*) &= \frac{\partial U(\cdot)}{\partial \epsilon_2} \\ \lambda_{13}(\epsilon_2^*, \epsilon_3^*, \ldots, \epsilon_n^*) &= \frac{\partial U(\cdot)}{\partial \epsilon_3} \\ &\vdots \\ \lambda_{1n}(\epsilon_2^*, \epsilon_3^*, \ldots, \epsilon_n^*) &= \frac{\partial U(\cdot)}{\partial \epsilon_n} \end{aligned} \tag{2.27}$$

where $\epsilon_2^*, \epsilon_3^*, \ldots, \epsilon_n^*$ represents the optimal values of the secondary objectives ϵ_j, $j = 2, 3, \ldots, n$

Define a new function

$$m_{1j}(\epsilon_2, \epsilon_3, \ldots, \epsilon_n) = \frac{\partial U(\cdot)}{\partial \epsilon_j} \qquad j = 2, 3, \ldots, n \tag{2.28}$$

This new function, $m_{1j}(\epsilon_2, \epsilon_3, \ldots, \epsilon_n)$, represents the marginal rate of substitution of objective $f_1(\mathbf{x})$ with respect to objective $f_j(\mathbf{x})$, given levels of the remaining objectives, $f_i(\mathbf{x})$, $i = 2, 3, \ldots, n$, $i \neq j$. That is, these rates indicate the trade-offs that the decisionmaker is willing to make. The trade-off that must be made to remain on the trade-off surface is given by $\lambda_{1j}(\epsilon_2, \epsilon_3, \ldots, \epsilon_n)$. Given $f_1(\mathbf{x}^*(\epsilon_2, \epsilon_3, \ldots, \epsilon_n))$, $\epsilon_2, \epsilon_3, \ldots, \epsilon_n$, it would cost the decisionmaker $\lambda_{1j}(\epsilon_2, \epsilon_3, \ldots, \epsilon_n)$ units of $f_1(\cdot)$ to reduce ϵ_j by one unit, while he or she would be willing to spend $m_{1j}(\epsilon_2, \epsilon_3, \ldots, \epsilon_n)$ units of $f_1(\cdot)$ to make the same reduction in ϵ_j.

At the optimum, therefore, the marginal rates of substitution must be equal to the corresponding trade-offs for $f_1(\cdot)$ with respect to ϵ_j, $j = 2, 3, \ldots, n$. The utility function is then tangent to the trade-off surface. Analogously, this procedure (utility maximization) finds the highest indifference curve tangent to the trade-off surface.

The optimality condition for Eq. (2.19) is then

$$m_{1j}(\epsilon_2^*, \epsilon_3^*, \ldots, \epsilon_n^*) = \lambda_{1j}(\epsilon_2^*, \epsilon_3^*, \ldots, \epsilon_n^*) \qquad j = 2, 3, \ldots, n \tag{2.29}$$

where ϵ_j^* represents the optimal desired value of the jth objective, $j = 2, 3, \ldots, n$. Of course, it may be impossible to satisfy these conditions exactly. It is sufficient to determine the range of values for which the optimality conditions are approximately satisfied (within some specified tolerance). This range indicates the indifference band, and any solution within the indifference band will be satisfactory. The λ_{1j}'s are determined by the solution of the ϵ-constraint problem given in Eq. (2.3) for various values of ϵ_j, $j = 2, 3, \ldots, n$. The m_{1j}'s, however, are specified by the decisionmakers, through their objective interpretation of the utility function, relating their preferences among the competing multiple objectives.

The first phase in solving a multiobjective problem is to determine the trade-offs, $\lambda_{1j}(\epsilon_2, \epsilon_3, \ldots, \epsilon_n)$, $j = 2, 3, \ldots, n$, and the trade-off surface, $f_1(\mathbf{x}^*(\epsilon_2, \epsilon_3, \ldots, \epsilon_n))$, for various levels of the objectives satisfied as constraints ϵ_j, $j = 2, 3, \ldots, n$. This phase is concerned with optimization in the decision variable space E^N, choosing optimal values of the decision variables $\mathbf{x}^*(\epsilon_2, \epsilon_3, \ldots, \epsilon_n)$. The entire noninferior region may be found in this phase. Several approaches are available for this phase, including the ϵ-constraint method and the weighting approach.

The second phase involves interaction with the decisionmaker to determine the desired levels of the multiple objectives ϵ_j^*, $j = 2, 3, \ldots, n$. While the decisionmakers may not actually know their utility function, the development of the utility maximization problem provides a worthwhile motivation for formulating the SWT method. This phase requires the satisfaction of the optimality conditions for the utility maximization problem given in Eq. (2.29); however, decisionmakers may not have enough information available to determine the m_{1j}'s. Again, several alternative approaches are presented. Decisionmakers are questioned about their preferences among the multiple objectives. From the noninferior solutions, the indifference band is determined. An optimal solution is then chosen from the indifference band.

2.4.3 Interactive Procedures

Several types of interaction with the decisionmaker may be possible, depending on the complexity of the information required. The SWT method requires the least information of the proposed schemes. While the decisionmaker may not actually know the utility function, it may be possible to infer information concerning its shape, through the interaction process. An interactive scheme proposed by Brandl and modified here for notational purposes, is given below (Brandl, 1968; Chankong and Haimes, 1983; Tarvainen, 1984).

The decisionmaker is asked, "At what point(s) along the trade-off surface would you be indifferent to changes in either direction of ϵ_j, given levels of ϵ_i, $i = 2, 3, \ldots, n$, $i \neq j$?" This question would be repeated for all ϵ_j, $j = 2, 3, \ldots,$

n. The optimal solution occurs at the point, if it exists, where the decisionmaker is simultaneously indifferent to moves in any direction.

Another interactive scheme also proposed by Brandl involves asking the decisionmaker, "Given levels of objectives $f_1(\mathbf{x}), f_2(\mathbf{x}), \ldots, f_n(\mathbf{x})$ satisfied as constraints, $\epsilon_2, \epsilon_3, \ldots, \epsilon_n$, how much would you be willing to spend to reduce ϵ_j by one unit?" This scheme attempts to determine the marginal rates of substitution. If no point is found at which the trade-offs exactly match the marginal rates of substitution, a linear multiple regression analysis of the differences between the trade-offs and the corresponding marginal rates of substitution vs. the various levels of the multiple objectives would be required. These linear equations could then be solved for the point at which all the differences simultaneously equal zero.

The scheme proposed by the SWT method involves an ordinal ranking of the trade-offs, as compared to the marginal rates of substitution. The decisionmaker would be asked, "Given levels of objectives $f_1(\cdot), f_2(\cdot), \ldots, f_n(\cdot)$, would you be willing to spend (1) much more, (2) more, (3) about the same, (4) less, or (5) much less than $\lambda_{1j}(\epsilon_2, \epsilon_3, \ldots, \epsilon_n)$ units of $f_1(\cdot)$ to reduce ϵ_j by one unit?" The surrogate worth function, $W_{1j}(\epsilon_2, \epsilon_3, \ldots, \epsilon_n)$, would be based on an ordinal scale, where -10 might indicate that λ_{1n} units of $f_1(\cdot)$ are very much less worthwhile than one marginal unit of ϵ_j, and $+10$ would indicate the opposite extreme, while 0 would signify that the exchange is an even trade, that is, the solution belongs to the indifference band. The optimum is found at the point where all surrogate worth functions are simultaneously equal to zero. By questioning the decisionmaker and determining the surrogate worth function, the shape of the utility function may be inferred. The surrogate worth function tends to make the objectives commensurable.

Several algorithms for computational implementation of the SWT method are available (Haimes et al., 1975; Chankong and Haimes, 1978). In general, the surrogate worth function may take any form so that $W_{1j}(\epsilon_2^*, \epsilon_3^*, \ldots, \epsilon_n^*) = 0$, $j = 2, 3, \ldots, n$, implies that the optimality conditions given in Eq. (2.29) are satisfied. For example, alternative forms of $W_{1j}(\cdot)$ may be

$$W_{1j}(\epsilon_2, \epsilon_3, \ldots, \epsilon_n) = m_{1j}(\epsilon_2, \epsilon_3, \ldots, \epsilon_n) - \lambda_{1j}(\epsilon_2, \epsilon_3, \ldots, \epsilon_n) \tag{2.30}$$

or

$$W_{1j}(\epsilon_2, \epsilon_3, \ldots, \epsilon_n) = \log \frac{m_{1j}(\epsilon_2, \epsilon_3, \ldots, \epsilon_n)}{\lambda_{1j}(\epsilon_2, \epsilon_3, \ldots, \epsilon_n)} \tag{2.31}$$

Obviously, if either of the above forms of the surrogate worth function is employed, the conditions $W_{1j}(\epsilon_2^*, \epsilon_3^*, \ldots, \epsilon_n^*) = 0$ do indeed imply that the objective values $\epsilon_2^*, \epsilon_3^*, \ldots, \epsilon_n^*$ are optimal.

Any surrogate worth function may be used as long as

$$\begin{aligned} W_{1j}(\epsilon_2, \epsilon_3, \ldots, \epsilon_n) &> 0 \text{ implies } m_{1j}(\epsilon_2, \epsilon_3, \ldots, \epsilon_n) > \lambda_{1j}(\epsilon_2, \epsilon_3, \ldots, \epsilon_n) \\ &< 0 \text{ implies } m_{1j}(\epsilon_2, \epsilon_3, \ldots, \epsilon_n) < \lambda_{1j}(\epsilon_2, \epsilon_3, \ldots, \epsilon_n) \\ &= 0 \text{ implies } m_{1j}(\epsilon_2, \epsilon_3, \ldots, \epsilon_n) = \lambda_{1j}(\epsilon_2, \epsilon_3, \ldots, \epsilon_n) \end{aligned}$$

An ordinal ranking will suffice if not enough information is available to actually assign numerical values to the m_{1j}'s.

The SWT method does not depend on the utility function, but only on an ordinal ranking of trade-offs and marginal rates of substitution along the trade-off surface, that is, in the noninferior region. The decisionmaker must compare the values of the objective functions and their associated trade-offs that he or she is actually willing to make. The optimality conditions of the utility maximization problem are employed to formulate a surrogate worth function, which may be determined with less information than is required for the utility maximization approach.

Once the optimal values of the multiple objectives are determined, ϵ_j^*, $j = 2, 3, \ldots, n$, the final phase of the planning process involves solving the ϵ-constraint problem given in Eq. (2.3), with the optimal objective values as on the right-hand side. The optimal decision variables are given by $\mathbf{x}^*(\epsilon_2^*, \epsilon_3^*, \ldots, \epsilon_n^*)$, and the solution of the multiobjective optimization problem is complete.

In summary, then, the first phase of solving the problem given in Eq. (2.1) involves the generation of the noninferior region by the solution of the ϵ-constraint problem with a number of different right-hand sides. The results of this phase include the trade-offs and the trade-off surface. Next, interaction with the decisionmaker helps determine the surrogate worth function. The optimality conditions are then satisfied, yielding the optimal values of the objective functions. Finally, the optimal objective values are substituted in the ϵ-constraint problem, resulting in the optimal decision variables.

2.5 EXAMPLE PROBLEMS

Two example problems are presented here, mainly for pedagogical purposes. There are two objective functions and two decision variables in Example 2.1, and three objective functions and two decision variables in Example 2.2. The corresponding solutions are relatively simple; therefore, they do not necessarily demonstrate the actual computational procedures involved in large-scale problems.

Example 2.1 Solve the following multiobjective optimization problem via the SWT method:

$$\min_{x_1, x_2} \begin{Bmatrix} f_1(x_1, x_2) = (x_1 - 2)^2 + (x_2 - 4)^2 + 5 \\ f_1(x_1, x_2) = (x_1 - 6)^2 + (x_2 - 10)^2 + 6 \end{Bmatrix} \tag{2.32}$$

A solution to Eq. (2.32) necessitates the existence of a decisionmaker who selects a preferred solution from the noninferior solutions. For simplicity, no constraints are introduced in this example problem.

SOLUTION The first phase in applying the SWT method is converting Eq. (2.32) into the ϵ-constraint form presented by Eq. (2.33):

$$\min_{x_1, x_2} f_1(x_1, x_2) \tag{2.33}$$

subject to

$$f_2(x_1, x_2) \leqslant \epsilon_2$$

Form the Lagrangian function $L(x_1, x_2, \lambda_{12})$:

$$L(x_1, x_2, \lambda_{12}) = f_1(x_1, x_2) + \lambda_{12}[f_2(x_1, x_2) - \epsilon_2] \tag{2.34}$$

Substituting Eq. (2.32) into Eq. (2.34) yields

$$L(x_1, x_2, \lambda_{12}) = (x_1 - 2)^2 + (x_2 - 4)^2 + 5$$
$$+ \lambda_{12}[(x_1 - 6)^2 + (x_2 - 10)^2 + 6 - \epsilon_2] \tag{2.35}$$

Note that the Kuhn-Tucker necessary conditions for stationarity are simplified here since there are no constraints on x_1 and x_2. These conditions are reduced to Eqs. (2.36)–(2.40):

$$\frac{\partial L(\cdot)}{\partial x_1} = 2(x_1 - 2) + 2\lambda_{12}(x_1 - 6) = 0 \tag{2.36}$$

$$\frac{\partial L(\cdot)}{\partial x_2} = 2(x_2 - 4) + 2\lambda_{12}(x_2 - 10) = 0 \tag{2.37}$$

$$\frac{\partial \mathrm{L}(\cdot)}{\partial \lambda_{12}} = (x_1 - 6)^2 + (x_2 - 10)^2 + 6 - \epsilon_2 \leqslant 0 \tag{2.38}$$

$$\lambda_{12}[(x_1 - 6)^2 + (x_2 - 10)^2 + 6 - \epsilon_2] = 0 \tag{2.39}$$

$$\lambda_{12} \geqslant 0 \tag{2.40}$$

Eq. (2.36) yields

$$\lambda_{12} = \frac{x_1 - 2}{6 - x_1} \tag{2.41}$$

Eq. (2.37) yields

$$\lambda_{12} = \frac{x_2 - 4}{10 - x_2} \tag{2.42}$$

Since $\lambda_{12} > 0$ guarantees a noninferior solution, then Eqs. (2.38)–(2.40) are reduced to Eqs. (2.43)–(2.44):

$$(x_1 - 6)^2 + (x_2 - 10)^2 + 6 - \epsilon_2 = 0 \tag{2.43}$$

$$\lambda_{12} > 0 \tag{2.44}$$

Note that both Eqs. (2.41) and (2.42) should be satisfied. Therefore, equating Eqs. (2.41) and (2.42) yields

$$\lambda_{12} = \frac{x_1 - 2}{6 - x_1} = \frac{x_2 - 4}{10 - x_2} \tag{2.45}$$

Upper and lower limits on x_1 and x_2 may easily be derived by satisfying Eqs. (2.41), (2.42), and (2.44):

$$2 < x_1 < 6 \tag{2.46}$$

$$4 < x_2 < 10 \tag{2.47}$$

The boundary points 2 and 6 for x_1 and 4 and 10 for x_2 result in either $\lambda_{12} = 0$ or $\lambda_{12} = \infty$.

Solving Eq. (2.45) simplifies the generation of noninferior points, as presented in Table 2.1:

$$x_2 = \tfrac{3}{2} x_1 + 1 \tag{2.48}$$

Figures 2.8–2.10 depict the noninferior solution in the decision space x_1 and x_2, the noninferior solution in the functional space $f_1(x_1, x_2)$ and $f_2(x_1, x_2)$, and trade-off function $f_2(x_1, x_2)$ vs. $\lambda_{12}(f_2)$, respectively. Assuming that an interaction with a decisionmaker does take place, resulting in a selection of an indifference level of trade-off, λ_{12}^*, then the corresponding preferred solution x_1^* and x_2^* can be obtained either directly from Table 2.1, or by solving Eq. (2.45) with $\lambda_{12} = \lambda_{12}^*$.

Note that noninferior solutions and their corresponding trade-off values were not generated by varying ϵ_2, as is suggested by the SWT method, because a closed form and direct solution was obtained instead. In larger-scale problems with decision variables exceeding four or five, the above closed form will not be computationally feasible and noninferior solutions will be generated by varying the ϵ's. This explanation also applies to Example 2.2, discussed in the next section.

Example 2.2 Solve the following multiobjective optimization problem via the SWT method:

$$\begin{aligned} \min_{x_1, x_2} f_1(x_1, x_2) &= (x_1 - 2)^2 + (x_2 - 4)^2 + 5 \\ f_2(x_1, x_2) &= (x_1 - 6)^2 + (x_2 - 10)^2 + 6 \\ f_3(x_1, x_2) &= (x_1 - 10)^2 + (x_2 - 15)^2 + 10 \end{aligned} \tag{2.49}$$

Table 2.1 Noninferior solutions and trade-off values for Example 2.1

x_1	x_2	$f_1(x_1, x_2)$	$f_2(x_1, x_2)$	λ_{12}
2.00	4.00	5.00	58.00	0
2.50	4.75	5.81	45.81	0.14
3.00	5.50	8.25	35.25	0.33
3.50	6.25	12.31	26.31	0.60
4.00	7.00	18.00	19.00	1.00
4.50	7.75	25.31	13.31	1.67
5.00	8.50	34.25	9.25	3.00
5.50	9.25	44.81	6.81	7.00
6.00	10.00	57.00	6.00	∞

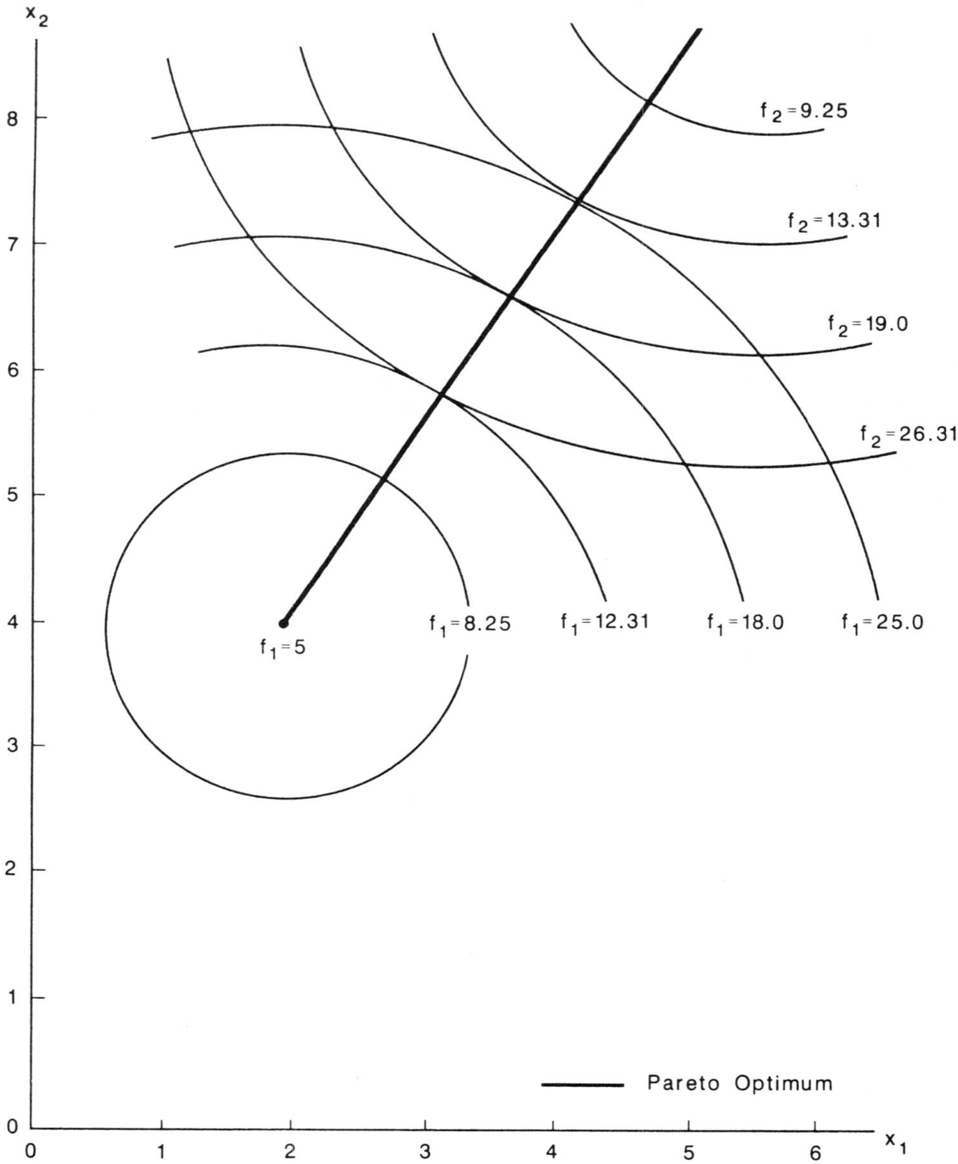

Figure 2.8 Noninferior solution in the decision space.

Solution Rewrite the problem given in Eq. (2.49) into the ϵ-constraint form:

$$\min_{x_1, x_2} f_1(x_1, x_2) \tag{2.50}$$

subject to the constraints $f_2(x_1, x_2) \leqslant \epsilon_2$ and $f_3(x_1, x_2) \leqslant \epsilon_3$.

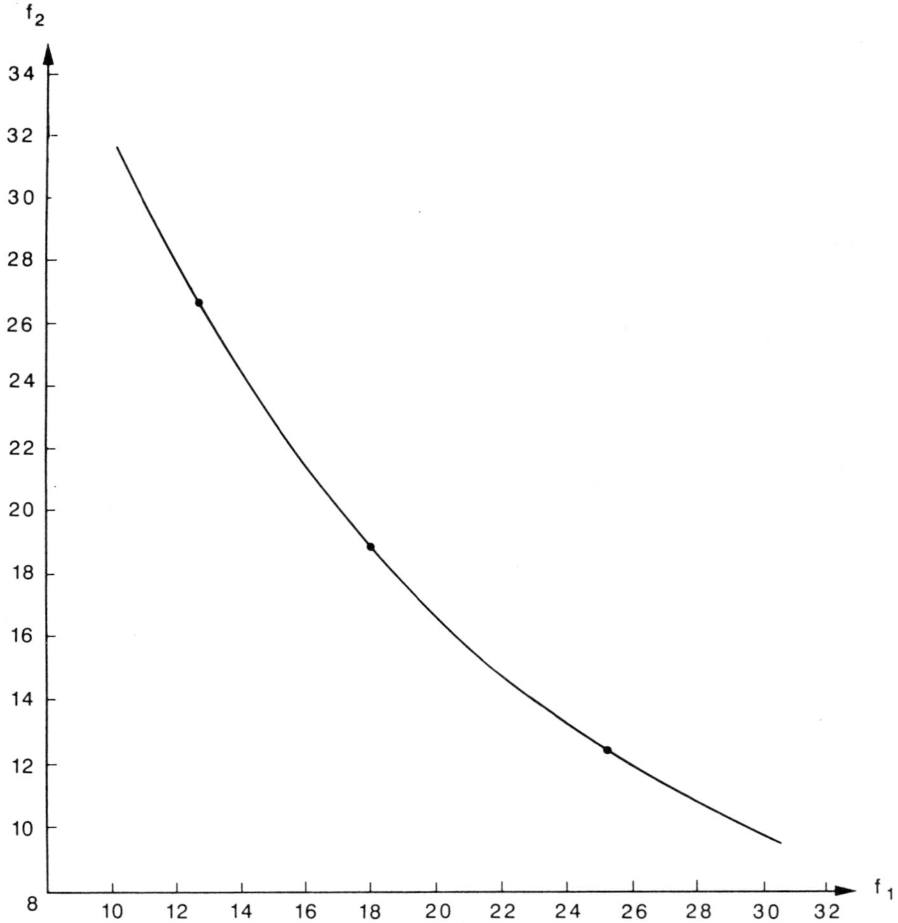

Figure 2.9 Noninferior solution in the functional space.

Form the Lagrangian $L_1(\cdot)$ for Eq. (2.50):

$$L_1(x_1, x_2, \lambda_{12}, \lambda_{13}) = f_1(x_1, x_2) + \lambda_{12}[f_2(x_1, x_2) - \epsilon_2] + \lambda_{13}[f_3(x_1, x_2) - \epsilon_3] \quad (2.51)$$

Substituting the values of $f_1(\cdot)$, $f_2(\cdot)$, and $f_3(\cdot)$ from Eq. (2.49) into Eq. (2.51), and solving the Kuhn-Tucker necessary conditions (similar to Example 2.1) yields

$$\lambda_{12} = \frac{11x_1 - 8x_2 + 10}{-5x_1 + 4x_2 - 10} \quad (2.52)$$

$$\lambda_{13} = \frac{-6x_1 + 4x_2 - 4}{-5x_1 + 4x_2 - 10} \quad (2.53)$$

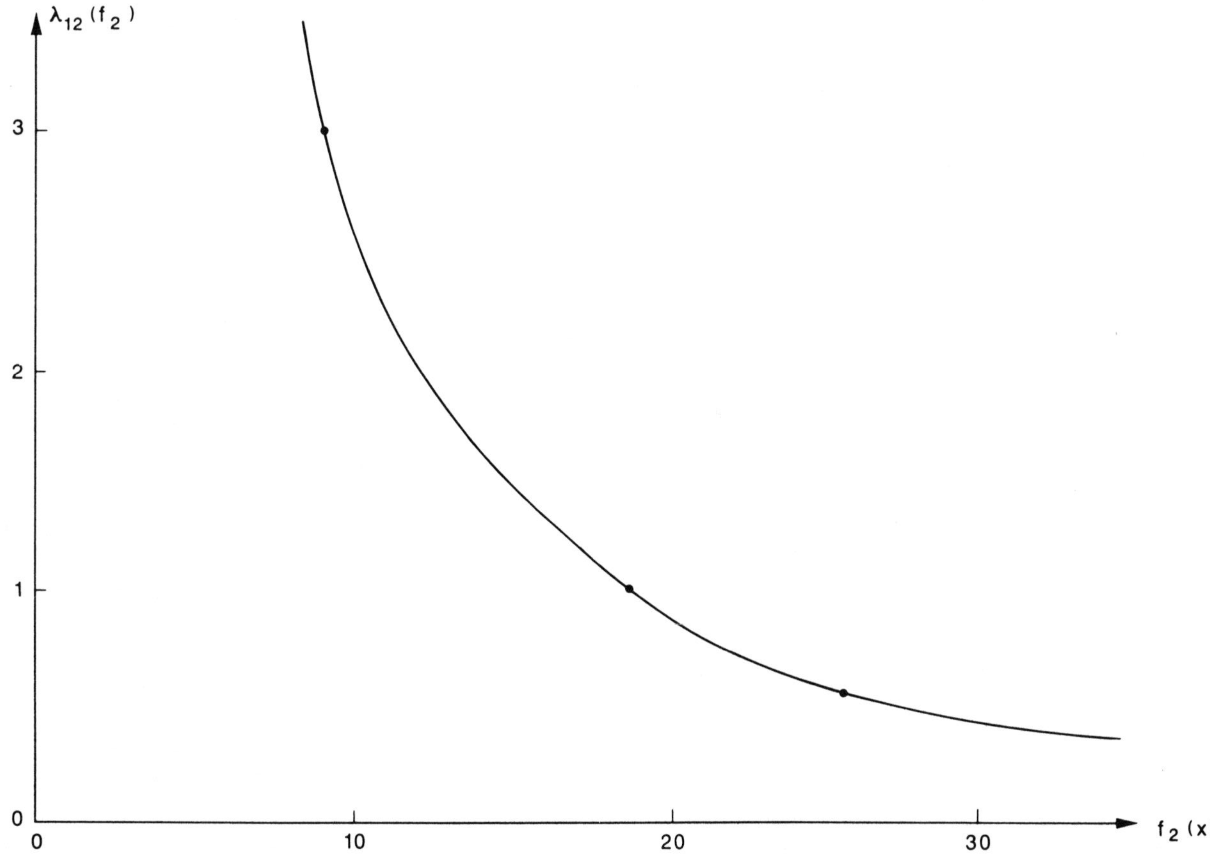

Figure 2.10 Trade-off function $\lambda_{12}(f_2)$ vs. $f_2(x_1, x_2)$.

Note that there is no requirement for $f_1(x_1, x_2)$ to be the primary objective function with $f_2(x_1, x_2)$ and $f_3(x_1, x_2)$ as constraints. The multiobjective optimization problem given in Eq. (2.49) can be alternatively written in the ϵ-constraint form as follows:

$$\min_{x_1, x_2} f_2(x_1, x_2) \tag{2.54}$$

subject to the constraints $f_1(x_1, x_2) \leq \epsilon_1$ and $f_3(x_1, x_2) \leq \epsilon_3$.

Form the Lagrangian $L_2(\cdot)$ for Eq. (2.54):

$$L_2(x_1, x_2, \lambda_{21}, \lambda_{23}) = f_2(x_1, x_2) + \lambda_{21}[f_1(x_1, x_2) - \epsilon_1] + \lambda_{23}[f_3(x_1, x_2) - \epsilon_3]. \tag{2.55}$$

Again, substituting the values of $f_1(\cdot)$, $f_2(\cdot)$, and $f_3(\cdot)$ from Eq. (2.49) into Eq. (2.55) and solving the Kuhn-Tucker necessary conditions yields

$$\lambda_{21} = \frac{-5x_1 + 4x_2 - 10}{11x_1 - 8x_2 + 10} \tag{2.56}$$

$$\lambda_{23} = \frac{-6x_1 + 4x_2 - 4}{11x_1 - 8x_2 + 10} \tag{2.57}$$

Note that Eqs. (2.52), (2.53), (2.56), and (2.57) satisfy Eq. (2.13), which is rewritten here for convenience:

$$\lambda_{ij} = \lambda_{ik} \cdot \lambda_{kj} \tag{2.58}$$

for positive λ's, and $i \neq j \neq k$, and

$$\lambda_{ij} = \frac{1}{\lambda_{ji}} \tag{2.59}$$

for $i \neq j$, $\lambda_{ji} > 0$.

Similar to Example 2.1, Table 2.2 summarizes several noninferior solutions with the corresponding trade-off values, and Fig. 2.11 depicts the noninferior solution in the decision space x_1 and x_2. Assuming that an interaction with a decisionmaker takes place and that the values of the trade-offs λ_{12}^* and λ_{13}^* corresponding to the surrogate worth functions at $W_{12} = 0$ and $W_{13} = 0$, respectively,

Table 2.2 Noninferior solutions and trade-off values for Example 2.2

x_1	x_2	$f_1(x_1, x_2)$	$f_2(x_1, x_2)$	$f_3(x_1, x_2)$	λ_{12}	λ_{13}
4	6.88	17.29	19.73	111.93	0.42	0.19
5	8.25	32.06	10.06	80.56	0.50	0.50
6	9.63	52.70	6.14	54.84	0.70	1.00
7	11.00	79.00	8.00	35.00	1.00	2.00
8	12.38	111.22	15.66	20.86	2.17	5.17

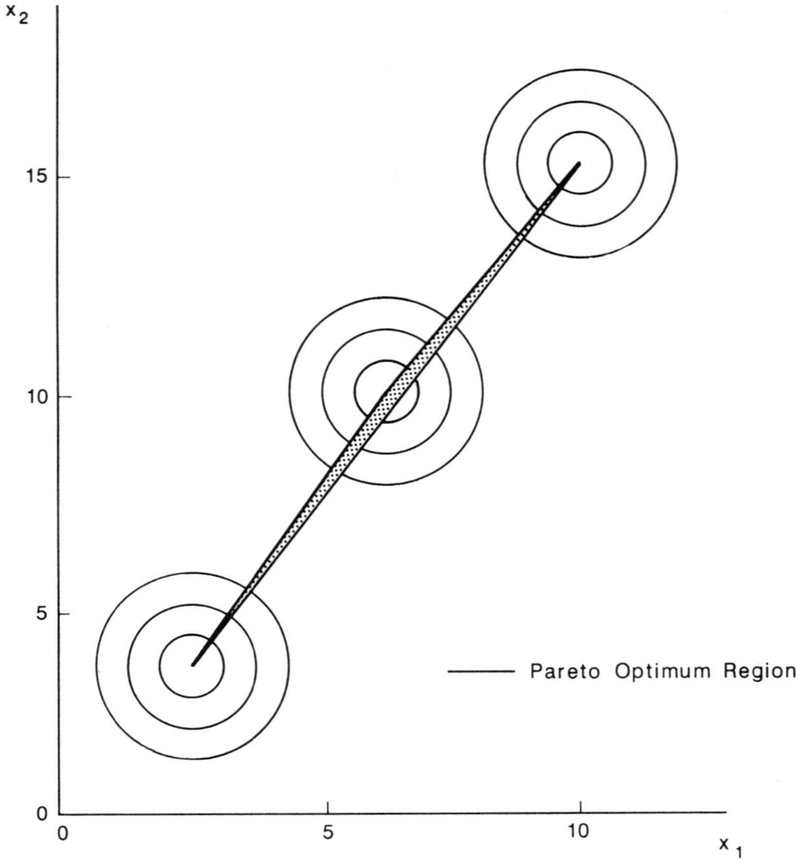

Figure 2.11 Noninferior solution in the decision space.

are obtained, then the preferred solution can be generated by substituting the values of λ_{12}^* and λ_{13}^* in Eqs. (2.52) and (2.53) and solving for x_1^* and x_2^*.

The reader is again reminded that for larger problems, a closed-form solution cannot be obtained, as in this example, and that the generation of noninferior solutions and their corresponding trade-off values should be then obtained by varying the ϵ's.

REFERENCES

Brandl, J. E., On the Treatment of Incommensurables in Cost-Benefit Analysis, *Land Econ.* vol. 44, no. 4, pp. 523–525, 1968.

Chankong, V., and Haimes, Y. Y., The Interactive Surrogate Worth Trade-off (ISWT) Method for Multiobjective Decisionmaking, in *Multicriteria Problem Solving*, ed. S. Zionts, Springer, New York, 1978.

Chankong, V., and Haimes, Y. Y., *Multiobjective Decison Making: Theory and Methodology*, Elsevier-North Holland, New York, 1983.

Everett III, H., Generalized Lagrange Multiplier Method for Solving Problems of Optimum Allocation of Resources, *Operations Res.*, vol. 11, pp. 399–418, 1963.

Fishburn, P. C., *Utility Theory for Decision-Making*, Wiley, New York, 1970.

Goicoechea, A., Hansen, D. R., and Duckstein, L., *Multiobjective Decision Analysis with Engineering and Business Applications*, Wiley, New York, 1982.

Haimes, Y. Y., Integrated System Identification and Optimization, in *Advances in Control Systems Theory and Applications*, ed. C. T. Leondes, vol. 10, pp. 435–518, Academic Press, New York, 1973.

Haimes, Y. Y., *Hierarchical Analyses of Water Resources Systems: Modeling and Optimization of Large-Scale Systems*, McGraw-Hill, New York, 1977.

Haimes, Y. Y., and Hall, W. A., Multiobjectives in Water Resources Systems Analysis: The Surrogate Worth Trade-off Method, *Water Resources Res.*, vol. 10, no. 4, pp. 615–624, 1974.

Haimes, Y. Y., Hall, W. A., and Freedman, H. T., *Multiobjective Optimization in Water Resources Systems: The Surrogate Worth Trade-off Method*, Elsevier, Amsterdam, 1975.

Haimes, Y. Y., Lasdon, L. S., and Wismer, D. A., On the Bicriterion Formulation of the Integrated System Identification and Systems Optimization, *IEEE Trans. Syst. Man Cybern.*, vol. SMC-1, pp. 296–297, 1971.

Hall, W. A., and Haimes, Y. Y., The Surrogate Worth Trade-off Method with Multiple Decision-Makers, in *Multiple Criteria Decision Making: Kyoto 1975*, ed. M. Zeleny, Springer-Verlag, New York, 1976.

Intrilligator, M. D., *Mathematical Optimization of Economic Theory*, Prentice-Hall, Englewood Cliffs, N.J., 1971.

Kaplan, M. A., *Multiobjective Analysis in Water Resources Planning and Management: A Framework for Policy Evaluation*, Ph.D. dissertation, Systems Engineering Department, Case Western Reserve University, Cleveland, Ohio, 1975.

Kaplan, M.A., and Haimes, Y. Y., Multiobjective Analysis in Water Resource Planning and Management: A Framework for Policy Evaluation, Report No. SED-WRG-75-7, submitted to National Science Foundation, Research Applied to National Needs, by Case Western Reserve University, Cleveland, Ohio, 1975.

Kuhn, H. W., and Tucker, A. W., Nonlinear Programming, *Proceedings Second Berkeley Symposium on Mathematical Statistics and Probability*, pp. 481–492, University of California Press, Berkeley, Calif., 1950.

Luenberger, D. G., *Introduction to Linear and Nonlinear Programming*, Addison-Wesley, Reading, Mass., 1973.

Marschak. J., Elements for a Theory of Teams, *Management Sci.*, vol. 1, no. 2, pp. 127–137, 1955.

Raiffa, H., *Decision Analysis Introductory Lectures on Choices Under Uncertainty*, Addison-Wesley, Reading, Mass., 1968.

Tarvainen, K., On the Implementations of the Interactive Surrogate Worth Trade-off (ISWT) Method, in *Interactive Decision Analysis*, ed. M. Grauer and A. P. Wierzbick, Springer-Verlag, New York, 1984.

Yu, P.-L., *Multiple-Criteria Decision Making*, Plenum Press, New York, 1985.

CHAPTER
THREE

OVERVIEW OF HIERARCHICAL ANALYSIS

3.1 INTRODUCTION

In studying large-scale systems with technological, societal, and environmental variables, problems of modeling and optimization are magnified and often overwhelm the analysis. This is due to the high dimensionality (very large number of variables) and complexity (nonlinearity in the coupling and interactions among the variables) of the resulting models.

The following dilemma is often faced. A detailed and comprehensive model that closely represents the real system is desired, yet such realistic models are generally too complicated, so that the application of solution strategies—namely, optimization methodologies—is difficult, if not prohibitive. This dilemma has often been resolved by grossly simplifying the system models and using linearization, to allow the use of linear programming and the simplex method. During the last decade, an alternative method has evolved with the establishment of the theoretical bases and the development of the computational tools of decomposition and multilevel optimization.

The concept of the hierarchical-multilevel approach is based on the decomposition of large-scale and complex systems and the subsequent modeling of the systems into "independent" subsystems. This decentralized approach utilizes the concepts of strata, layers, and echelons, enabling the systems analyst to analyze and comprehend the behavior of the subsystems at a lower level and to transmit the information obtained to fewer subsystems at a higher level. The reader is referred to the work of Mesarovic et al. (1970), Lefkowitz (1966), Lasdon (1964,

1970), Bauman (1966), Wismer (1971), Haimes (1973, 1975), Haimes and Li (1988), Haimes and Macko (1973), Sage (1977), Singh and Titli (1979), and others.

Whenever more decentralization is needed, the system is further decomposed. This decomposition is accomplished by introducing into the system new variables, which are called pseudo-variables. Then each subsystem is separately and independently optimized, with perhaps different optimization techniques being applied, based on the nature of the subsystem models as well as on the objectives and constraints of the subsystems. This is termed a first-level solution. The subsystems are joined by coupling variables that are manipulated at a second or higher level in order to arrive at the optimal solution of the whole system. This is termed the second- or higher-level solution. One way to achieve subsystem "independence" is by first relaxing one or more of the necessary conditions for optimality, and then satisfying these conditions at the second level.

In order to illustrate the hierarchical-multilevel approach, consider a company with N departments. Each department has a manager who reports to the president of the company. The sole objective of the ith manager is to maximize the performance of his or her department. The sole objective of the president is to maximize the performance of the whole company by imposing internal prices for transactions between the department products or services. The company's overall optimal policy is thus achieved through an iterative procedure. The separate optimization of each department does not necessarily imply the overall maximization of the company's performance, unless the performance of each department is coordinated at a higher level (i.e., by the president).

3.2 ATTRIBUTES OF DECOMPOSITION AND MULTILEVEL OPTIMIZATION

Decomposition and multilevel optimization have several significant advantages in the solution of large-scale complex optimization problems over conventional optimization methods. The following subsections summarize the attributes that are explored throughout this book.

3.2.1 Conceptual Simplification of Complex Systems

By decomposing the problem into several subproblems (subsystems), a conceptual simplification of a complex system is achieved. This is especially important for highly coupled systems, where the outputs of one subsystem are the inputs to others.

3.2.2 Reduction in Dimensionality

The decomposition yields a reduction in the dimensionality of the problem at hand at the expense of having to solve several subproblems of lower dimensions. Often this reduction in dimensionality enables the modeling and optimization of problems not otherwise solvable.

3.2.3 Simple Programming and Computational Procedures

The reduction in dimensionality in turn reduces the computational effort involved, such as problem formulation time, programming effort, and debugging effort. The trade-off between computer and manpower time has not yet been established for simple problems. It is clear, however, that for large-scale and complex systems, decomposition and multilevel optimization yield an appreciable reduction in both computer time and manpower efforts.

3.2.4 More Realistic System Models

A significant advantage of the hierarchical-multilevel approach is that none of the system model functions needs to be linear, and thus more flexible mathematical models can be constructed to represent the real system. Note that a major shortcoming and deficiency of classical systems engineering practices is that they often result in an unbalance between system modeling and system optimization. This is reflected by the vast number of linearized models in the literature that take advantage of the simplex methods and their extensions. By applying the decomposition and multilevel optimization techniques, no such costly sacrifice of realism in modeling is needed, as more representative and sophisticated nonlinear mathematical models can be constructed.

3.2.5 Interactions among Subsystems Are Permissible

Interactions among subsystems can be handled, since at the lower levels the subsystems' "independences" are achieved via pseudo-variables. This is a very important attribute to the approach since all real-world systems are highly coupled; thus, the interactions among their various components cannot be ignored. Water resource systems can probably serve as an excellent example in this respect. The benefits resulting from the construction of a dam cannot be properly assessed unless the impacts on water quality, land use, sheet erosion, flood control, socioeconomic conditions, and other factors are considered. These couplings are well handled via the hierarchical-multilevel approach.

3.2.6 Applicability to Both Static and Dynamic Systems

Decomposition and multilevel optimization are applicable to both static and dynamic systems. The time domain, which plays an important role in large-scale systems, need not be imbedded or ignored in the analyses (as in the case of static models). Therefore, large-scale systems can be modeled by both static algebraic equations and dynamic differential equations. Thus, the trade-off between system modeling and system optimization is further minimized by the applicability of the approach to both static and dynamic systems.

3.2.7 Different Optimization Techniques for Subsystems' Solution

Each subsystem in the hierarchy can be handled with different optimization techniques, depending on the nature of the subsystem. Very often dynamic programming is applied to one or more subsystems where linear programming, generalized reduced gradient (GRG), or Fletcher-Powell methods are applied to other subsystems. This attribute provides an extremely important flexibility in the solution procedures and enables the analyst to utilize the state of the art in optimization techniques for solving the problem at hand.

3.2.8 Use of Existing Models

Well-documented simulation and mathematical models that are aimed at investigating specific and selected aspects of large-scale systems are available in the literature. The hierarchical-multilevel approach enables the analyst to identify the appropriate existing models in the literature, then modify and incorporate these models as subsystems in the overall hierarchical structure for modeling and solving the total system under consideration.

3.2.9 Economic Interpretation of the Variables

Central to the computational procedure in multilevel optimization is the duality theory of nonlinear programming. In particular, the economic interpretation of the Lagrange multipliers as shadow prices often sheds more light on the understanding of the system's behavior.

3.2.10 Applicable to Multiobjective Analyses

In a hierarchical structure, different objective functions may be associated with each subsystem. The resulting overall system coordination becomes a multiobjective optimization problem. In particular, when these objective functions are noncommensurable, then the surrogate worth trade-off (SWT) method can be successfully applied.

3.3 GENERAL HIERARCHICAL STRUCTURES

Hierarchical systems or structures are families of subsystems arranged in a hierarchical fashion. Each subsystem is concerned with a specific aspect of the overall system and occupies a specific level in the hierarchy. Operation of a subsystem on any given level of the hierarchy is directly and explicitly influenced by higher-level subsystems. Influence of higher-level subsystems is binding on the lower-level subsystems and reflects a priority of importance in the actions and goals of the higher-level subsystems. Note that fulfillment of the higher-level objectives depends on the actions and performance of the lower-level subsystems. Three kinds of hierarchical-multilevel structures can emerge in dealing with large complex systems.

3.3.1 Multistrata Hierarchy

The first kind of hierarchy, a multistrata hierarchy, arises in attempting to resolve the dilemma in describing (modeling) a large complex system. The conflict is between simplicity (a requisite for understanding) in modeling, and consequently in the solution strategy applied, and competence in accounting for the large complex system's numerous behavioral aspects. The levels of this kind of hierarchy as descriptive levels are referred to as strata. Lower strata involve more detailed and specialized descriptions of the system than the higher strata. Each stratum has its own concepts and principles and may deal with different aspects of the system.

3.3.2 Multilayer Hierarchy

A multilayer hierarchy appears in resolving complex decisionmaking situations. Two trivial but important necessities of almost all real-life decision situations are (1) to act without delay and thereby avoid decision by default, and (2) to take time to better understand the situation.

Layers are essentially levels of decisionmaking complexity. Parameters in lower-layer problems are fixed by solutions of higher-layer problems, and a solution of the original problem is obtained when all subproblems in the sequence are solved. In this way a complex decisionmaking hierarchy naturally emerges in reference to three essential phases of decisionmaking:

1. Determining operational objectives and constraints,
2. Gathering information and reducing uncertainties, and
3. Selecting a preferred course of action.

3.3.3 Multiechelon Hierarchy

The third kind of hierarchy—multiechelon hierarchy—arises when a large complex system is explicitly viewed as consisting of many interacting subsystems. These deal with the mutual relationships between the various subsystems comprising a large complex system. Each subsystem is a goal-seeking system aiming either at optimizing an objective function or at satisfying an aspiration level.

Conflicts between subsystems in one echelon are resolved by higher-echelon subsystems. Coordination, the resolution of conflicts, is accomplished by intervention which is permitted by including certain variables in the subsystem's objectives that are free to be manipulated by higher-echelon subsystems. Intervention may be of the following kinds:

1. Goal intervention, which affects goal-related factors,
2. Information, which affects outcome expectations, and
3. Constraint intervention, which affects available alternative actions.

The assumption that the overall system goals will be met when all subsystems operate according to their own objective and within their prescribed operating constraints is of particular importance.

3.3.4 Summary

All three kinds of hierarchies may emerge simultaneously in dealing with a large complex system. The different hierarchies serve different purposes and may be imbedded within one another: the multistrata hierarchies are for modeling purposes; the multilayer hierarchies are for resolving the complexities of decision-making; the multiechelon hierarchies are for organization purposes. Features common to all three kinds of hierarchies are that (1) higher-level subsystems portray or are concerned with a larger portion or broader aspects of the overall system; (2) they have longer decision horizons—they are concerned with longer-range behavior; and (3) they have priority of action over lower-level subsystems.

3.4 HIERARCHICAL MODELING

Because one part of the problem is tackled while knowledge of the other parts is assumed, these submodels are often one-sided and usually do not represent the overall system behavior. There are many examples of such submodels in the literature, such as capacity expansion, planning for operation or multireservoir systems, identification of watershed and aquifer parameters, groundwater and surface water management, and water quality management. The reason for their simplification in system modeling can be explained by the difficulties of solving large-scale and complex systems by means of conventional systems engineering tools and methodologies. This is particularly true for nonlinear, dynamic large-scale systems.

Several hierarchical-multilevel structures for the planning and management of a regional water resources system can be developed to account for the intersystem and intrasystem coupling and interaction. Four major descriptions are identified in water resources systems: (1) temporal, (2) physical-hydrological, (3) political-geographical, and (4) goal or functional. All or several of these four descriptions of the system can be incorporated through a hierarchical modeling and overlapping coordination.

3.4.1 Temporal Description

A planning time horizon for water supply projects often spans 30–50 years. This long-term planning is usually supplemented by an intermediate term of 10–15 years, often referred to as planning for operation, and by a short term of 2–5 years. Clearly the short, intermediate, and long terms have to be compatible and coordinated with one another, since they relate to the same system. Planning horizons of water resources for crop and related land use can be on the order of 1–2 years. However, when a crop has been selected and the water for its seasonal growth has been allocated, horizons of decisions with respect to the periodic irrigation within the season are on the order of weeks or days.

3.4.2 Physical-Hydrological Description

A river basin is, by definition, a hydrologically self-contained region, separated from adjacent basins by a ridge or other topographic divider. Often a water resources management system covers a region consisting of a complex of several river basins (North Atlantic region, NAR; see Haimes 1977). Thus, a region can be divided into several subregions (six in the NAR study), further divided into several river basins (21 in the NAR study), and further divided into several subbasins (50 in the NAR study), and even further divided into smaller subbasins (as was done in the NAR study).

3.4.3 Political-Geographic Description

A regional water resources system often includes different local governments such as metropolitan areas, counties, and states. Modeling for water resource planning and management may consider the political-geographic description as a criterion for decomposing the regional area into subregions.

3.4.4 Goal or Functional Description

Most water resources systems have been analyzed with respect to their economic and functional goals. Various models following this pattern are available in the literature, such as demand and supply models, models for hydroelectric power generation, irrigation, industrial and municipal use, and recreation.

3.4.5 Overlapping Decomposition

The decomposition of a regional area into subregions depends on the viewpoint and aims of the analyst. One decomposition may be with respect to the region's hydrology. The region would be decomposed into subregions, such as river basins and subbasins, having topographic divisions as their boundaries. A second decomposition may be with respect to political boundaries. The regional area would be decomposed into political subregions such as townships, municipalities, and counties. A third decomposition may be with respect to regional goals and functions; a fourth may be subregions having distinct goals or purposes related to functions such as irrigation, navigation, hydroelectric power generation, and recreation. In regional water resource management, the major aspects of the regional area cannot be divorced from one another.

The decompositions just cited overlap one another. Hydrological subregions can easily overlap or span political boundaries. Hydroelectric generating stations may be dispersed through a region, not confined to any one political or hydrological subregion. Indeed, the subregional boundaries in one decomposition generally do not coincide with those in another. The problem considered here is how overlapping decompositions of a regional area can be coordinated.

3.5 GENERAL PROBLEM FORMULATION

Consider a regional area that has been decomposed from two different viewpoints so that the subregions of one decomposition overlap those of the other. To make our discussion more concrete, we assume one decomposition is from a hydrological viewpoint and the other is from a goal or functional viewpoint. Each subregion of the latter decomposition is concerned with a particular function such as irrigation, navigation, or recreation.

It is convenient to define a system optimization problem for a region R and use it as a vehicle in explaining the overlapping decomposition to water resources systems.

Let the overall optimization problem for a given region R be represented by

$$\max_{\mathbf{m}} f(\mathbf{y}, \mathbf{u}, \mathbf{m}, \boldsymbol{\alpha}) \tag{3.1}$$

subject to

$$\mathbf{g}(\mathbf{y}, \mathbf{m}, \mathbf{u}, \boldsymbol{\alpha}) \leq \mathbf{0} \qquad \mathbf{y} = \mathbf{H}(\mathbf{u}, \mathbf{m}, \boldsymbol{\alpha})$$

where

$\mathbf{y}$ = vector of regional outputs
$\mathbf{u}$ = vector of regional inputs (not manipulatable)
$\mathbf{m}$ = vector of manipulatable variables (i.e., decision variables)
$\boldsymbol{\alpha}$ = vector of model parameters
$\mathbf{g}$ = vector representing the regional operating constraints

It is an elementary extension to also include equality constraints in the development. Substituting the output relations gives the overall system optimization problem as

$$\max_{\mathbf{m}} f(\mathbf{u}, \mathbf{m}, \boldsymbol{\alpha}) \tag{3.2}$$

subject to

$$\mathbf{g}(\mathbf{u}, \mathbf{m}, \boldsymbol{\alpha}) \leq \mathbf{0}$$

where $f(\mathbf{u}, \mathbf{m}, \boldsymbol{\alpha})$ is a scalar-valued function whose value is an indication of the overall system performance. The performance of most water resource systems is characterized by a vector of goals and objectives. Introducing multiobjective functions at this stage will only complicate our discussion on hierarchical-multilevel structures. (The extension to multiobjective functions is discussed in the following chapters.)

We will assume hereafter in this section that the region R is decomposed into N subregions. Generally, the ith subregion R_i will have an objective function $\mathbf{f}_i$ $(\mathbf{x}_i, \mathbf{u}_i, \mathbf{m}_i, \boldsymbol{\alpha}_i, \boldsymbol{\sigma})$ and operating constraints $\mathbf{g}_i\ (\mathbf{x}_i, \mathbf{u}_i, \mathbf{m}_i, \boldsymbol{\alpha}_i, \boldsymbol{\sigma}) \leq \mathbf{0}$; $\boldsymbol{\sigma}$ is a vector of coordination (pseudo-) variables enabling the decomposition. More will be said about $\boldsymbol{\sigma}$ in the ensuing sections. The vectors $\mathbf{u}_i$, $\mathbf{m}_i$, and $\boldsymbol{\alpha}_i$ are subvectors of $\mathbf{u}$, $\mathbf{m}$, and $\boldsymbol{\alpha}$, respectively, and $\mathbf{x}_i$ is a vector of inputs to subregion R_i from other subregions. In general, the subregions are coupled via their inputs and outputs:

$$\mathbf{x}_i = \sum_{j=1}^{N} C_{ij}\mathbf{y}_j \qquad i = 1, \ldots, N$$
$$\mathbf{y}_i = \mathbf{H}_i(\mathbf{x}_i, \mathbf{u}_i, \mathbf{m}_i, \boldsymbol{\alpha}_i) \qquad i = 1, \ldots, N \tag{3.3}$$

where $\mathbf{y}_i$ is a vector of outputs of subregion R_i, and the C_{ij} are coupling matrices.

In general, the ith subregional optimization problems assume the form

$$\max f_i(\mathbf{x}_i, \mathbf{u}_i, \mathbf{m}_i, \boldsymbol{\alpha}_i, \boldsymbol{\sigma}) \tag{3.4}$$

subject to

$$\mathbf{g}_i(\mathbf{x}_i, \mathbf{u}_i, \mathbf{m}_i, \boldsymbol{\alpha}_i, \boldsymbol{\sigma}) \leq \mathbf{0}$$

The specifics of the subregional optimization problems depend on the specific decomposition and coordination methods used.

Each of the decomposition and coordination methods discussed here requires that the overall regional optimization problem be expressible in a form that is separable in the subregions' variables. Therefore, let us assume the overall regional problem can be rewritten as

$$\max \sum_{i=1}^{N} f_i(\mathbf{x}_i, \mathbf{u}_i, \mathbf{m}_i, \boldsymbol{\alpha}_i) \tag{3.5}$$

subject to

$$\mathbf{g}_i(\mathbf{x}_i, \mathbf{u}_i, \mathbf{m}_i, \boldsymbol{\alpha}_i) \leq 0 \qquad \mathbf{y}_i = H_i(\mathbf{x}_i, \mathbf{u}_i, \mathbf{m}_i, \boldsymbol{\alpha}_i) \qquad \mathbf{x}_i = \sum_{j=1}^{N} C_{ij}\mathbf{y}_j$$

where $i = 1, \ldots, N$.

3.5.1 Nonfeasible or Interaction Balance Methods

Nonfeasible or interaction balance methods form independent subregional optimization problems by decoupling the subregions and attaching prices to the coupling variables. The vector $\boldsymbol{\sigma}$ of coordination variables is a vector of prices entering the subregions' objective functions. In particular, the ith subregional optimization problem is then

$$\max_{\mathbf{m}_i, \mathbf{x}_i} \mathbf{f}_i(\mathbf{x}_i, \mathbf{u}_i, \mathbf{m}_i, \boldsymbol{\alpha}_i) + \left(\sum_{j=1}^{N} \boldsymbol{\sigma}_j C_{ji} \right) \mathbf{y}_i - \boldsymbol{\sigma}_i \mathbf{x}_i \tag{3.6}$$

subject to

$$\mathbf{g}_i(\mathbf{x}_i, \mathbf{u}_i, \mathbf{m}_i, \boldsymbol{\alpha}_i) \leq \mathbf{0} \qquad \mathbf{y}_i = \mathbf{H}_i(\mathbf{x}_i, \mathbf{u}_i, \mathbf{m}_i, \boldsymbol{\alpha}_i)$$

The inputs $\mathbf{x}_i$ entering the subregion from other subregions are treated as if they could be freely manipulated.

The subregions are coordinated via the prices entering their objective functions. The coordinating prices admit optimal subregion solutions that satisfy the coupling equations (3.3) (i.e., the interactions balance). If the optimal subregion solutions satisfy the coupling equations, they will be optimal for the overall region. Therefore, the coordinator's objective is to drive the coupling errors to zero.

Barring duality gaps, the coordinating prices in fact minimize the dual overall objective function, which is the sum of the optimal values of the subregion's objectives in Eq. (3.6). This function is generally not available to the coordinator. However, it is always convex with respect to prices, and the optimal subregion solutions readily provide its subgradients with respect to the prices. In particular, the errors in the coupling equations (3.3) are its subgradients. Consequently, the coordinator can use the coupling errors to obtain directions in which to adjust the prices. On the other hand, the coordinator may use the coupling errors to build up a tangential approximation (outer linearization) of the dual overall objective function. The coordinator can then obtain new prices by minimizing the approximate function.

3.5.2 Dantzig-Wolfe Extensions

Extensions of the Dantzig-Wolfe decompositions provide a different approach to coordinating the subregions. See Dantzig and Wolfe (1960) and Lasdon (1970). The subregional optimization problems remain precisely the same as they are for the nonfeasible or interaction balance methods. The coordinator, however, obtains new prices by solving a Dantzig-Wolfe master program. This approach requires the overall regional optimization problem to have a concave objective function with convex inequality constraints and linear equality constraints.

The Dantzig-Wolfe master program is an inner linearization of the overall regional optimization problem. The coordinator uses previous subregional solutions to build up the linear approximation. The solution of the Dantzig-Wolfe master program is a convex combination of previous subregional solutions, and the prices supplied to the subregions are the corresponding optimal multipliers. Although the subregional solutions generally do not satisfy the coupling equations, their convex combinations given by the Dantzig-Wolfe master program do satisfy the coupling equations. Moreover, these convex combinations give increasing values of the overall regional objective function as more and more subregional solutions become available.

3.5.3 Feasible Methods

The feasible methods form independent subregional optimization problems by fixing the subregion coupling variables. The vector $\boldsymbol{\sigma}$ of coordination variables is a vector of outputs transferred between the subregions. In particular, the ith subregional optimization problem becomes

$$\max_{\mathbf{m}_i} \mathbf{f}_i(\mathbf{x}_i, \mathbf{u}_i, \mathbf{m}_i, \boldsymbol{\alpha}_i) \tag{3.7}$$

subject to

$$\mathbf{g}_i(\mathbf{x}_i, \mathbf{u}_i, \mathbf{m}_i, \boldsymbol{\alpha}_i) \leq \mathbf{0} \qquad \boldsymbol{\sigma}_i - \mathbf{H}_i(\mathbf{x}_i \, \mathbf{u}_i, \mathbf{m}_i, \boldsymbol{\alpha}_i) = \mathbf{0} \qquad \sum_{j=1}^{N} C_{ij}\, \boldsymbol{\sigma}_j - \mathbf{x}_i = \mathbf{0}$$

The subregions are coordinated via the outputs transferred between the subregions.

The coordinating values of $\boldsymbol{\sigma}$ are feasible outputs which maximize the total of the subregions' optimal returns, namely, the sum of the optimal values of the subregions' objective functions in Eq. (3.7). This function is generally not available to the coordinator. However, necessary conditions for $\boldsymbol{\sigma}$ to be coordinating are that the specified outputs be feasible and the subregions' marginal returns all be equal. Care must be exercised to specify feasible output transfers. This requires the subregions to provide the coordinator with sufficient information to ensure that this will be the case.

3.5.4 Interaction Prediction Methods

The interaction prediction methods supply the subregions with "predicted" values of their inputs and prices on their outputs (see Mesarovic et al., 1970; Macko 1967). The subregions assume these predictions are correct, that is, the outcomes will be as predicted. Vector $\boldsymbol{\sigma}$ of coordination variables consists of two subvectors: $\boldsymbol{\sigma}_x$, predicted values of the subregional inputs, and $\boldsymbol{\sigma}_y$, prices of the subregional outputs. In particular, the ith subregional optimization problem is

$$\max_{\mathbf{m}_i} f_i(\mathbf{x}_i, \mathbf{u}_i, \mathbf{m}_i, \boldsymbol{\alpha}_i) + \boldsymbol{\sigma}_{iy}\mathbf{y}_i \tag{3.8}$$

subject to

$$\mathbf{g}_i(\mathbf{x}_i, \mathbf{u}_i, \mathbf{m}_i, \boldsymbol{\alpha}_i) \leq \mathbf{0} \qquad \boldsymbol{\sigma}_{ix} - \mathbf{x}_i = \mathbf{0} \qquad \mathbf{y}_i = \mathbf{H}_i(\mathbf{x}_i, \mathbf{u}_i, \mathbf{m}_i, \boldsymbol{\alpha}_i)$$

The subregions are coordinated via the predicted values of the subregional inputs and prices of the subregional outputs. The conditions that $\boldsymbol{\sigma}$ must satisfy in order to coordinate the subregions are based on the necessary conditions for the overall regional optimum. In particular, the predictions must be correct. That is, the subregional optimal solutions must satisfy the coupling equations (3.3), and the prices assigned to the outputs together with the subregional marginal returns with respect to the predicted input values must satisfy the dual coupling constraints, that is,

$$\boldsymbol{\sigma}_{iy} = -\sum_{j=1}^{N} \boldsymbol{\lambda}_j C_{ji} \tag{3.9}$$

where $\boldsymbol{\lambda}_j$ is the marginal return of the jth subregion with respect to the predicted input value $\boldsymbol{\sigma}_{jx}$. Mesarovic et al. (1970) proposed that the errors in the subregional coupling equations (3.3) and the dual coupling equations (3.9) be used by the coordinator to update its predictions.

3.6 GENERAL FORMULATION OF AN OVERLAPPING DECOMPOSITION

Suppose the overall regional area R has been decomposed into N hydrological subregions indexed, as previously done, by subscripts. The overall regional optimization problem expressed in the subregional variables of this decomposition is given in Eq. (3.5). In the same way, suppose a different decomposition has been obtained from a goal or functional viewpoint and consists of M subregions indexed by superscripts. Suppose the overall regional optimization problem expressed in the subregional variables of the second decomposition is

$$\max \sum_{k=1}^{M} f^k(\mathbf{x}^k, \mathbf{u}^k, \mathbf{m}^k, \boldsymbol{\alpha}^k) \tag{3.10}$$

subject to

$$\mathbf{g}^k(\mathbf{x}^k, \mathbf{u}^k, \mathbf{m}^k, \boldsymbol{\alpha}^k) \leq \mathbf{0} \qquad \mathbf{y}^k = \mathbf{H}^k(\mathbf{x}^k, \mathbf{u}^k, \mathbf{m}^k, \boldsymbol{\alpha}^k) \qquad \mathbf{x}^k = \sum_{j=1}^{M} C^{kj}\mathbf{y}^j$$

where $k = 1, \ldots, m$. The variables in Eq. (3.10) have precisely the same interpretation as their counterparts in Eq. (3.5).

We make the assumption that interactions between the hydrological and goal-functional aspects of the region enter the subregional models via $\mathbf{u}_i$ and $\mathbf{u}^k$ as given by Eq. (3.11). In other words, we assume $\mathbf{u}_i$ contains the goal-function inputs pertinent to the ith hydrological subregion, and we make a similar assumption for $\mathbf{u}^k$. In particular, we assume the coupling between the two decompositions is given by

$$\begin{aligned} \mathbf{u}_i &= \sum_{k=1}^{M} B^{ik}\mathbf{y}^k \qquad i = 1, \ldots, N \\ \mathbf{u}^k &= \sum_{i=1}^{N} B_{ki}\mathbf{y}_i \qquad k = 1, \ldots, M \end{aligned} \tag{3.11}$$

The matrices B^{ik} and B_{ki} are coupling matrices indicating which subregional outputs of one decomposition affect the subregions of the other decomposition.

The two decompositions form two distinct hierarchical multilevel structures. The subregions of each decomposition with their objectives and constraints form the first level of the hierarchies. The coordinator of the subregions with a given decomposition occupies the second level of the hierarchy. It coordinates the subregions via the variable $\boldsymbol{\sigma}$. On top of the subregion coordinators is another variable that coordinates the two overlapping decompositions via the variable $\boldsymbol{\lambda}$. It joins the two overlapping hierarchies and so occupies the third level.

3.7 FEASIBLE AND NONFEASIBLE DECOMPOSITION

Section 3.5 presented a general formulation of the feasible and the nonfeasible decompositions (also known as the interaction balance method). The two decompositions can be viewed as dual to each other. In the feasible decomposition the iterations are on the state variables where the constraints are always satisfied. In the nonfeasible decomposition, the iterations are on the Lagrange multipliers where the constraints are not satisfied until convergence is achieved (see Bauman, 1966).

For pedagogical purposes, a simple unconstrained optimization problem will be used as a vehicle to explain the mechanics of the two decomposition methods. Extension of the methodologies to constrained optimization problems will follow. Furthermore, several second-level coordination methods will be introduced and related to the decomposed system.

Consider the following problem:

$$\min_{x_1,x_2} [f(x_1, x_2) = (x_1 - 2)^2 + x_1x_2 + (x_2 - 1)^2] \tag{3.12}$$

The simple optimization problem (3.12) has two decision variables x_1 and x_2, which are coupled through the term x_1x_2. For example, x_1 and x_2 may be vectors of N and M dimensions, respectively, each associated with a different department in a company. Without the coupling term x_1x_2, the problem would be separable, and thus no decomposition would be needed.

In order to uncouple the system, a pseudo-variable σ will be introduced:

$$x_1 = \sigma \tag{3.13}$$

The variable x_1 will be replaced by σ whenever it appears in a coupling term. The unconstrained problem (3.12) can be replaced by the following:

$$\min_{x_1,x_2} [f(x_1, x_2, \sigma) = (x_1 - 2)^2 + \sigma x_2 + (x_2 - 1)^2 \tag{3.14}$$

subject to $x_1 = \sigma$.

System (3.14) is a constrained optimization problem that can be solved by forming the Lagrangian function $L(\cdot)$ as follows:

$$L(x_1, x_2, \sigma, \lambda) = (x_1 - 2)^2 + \sigma x_2 + (x_2 - 1)^2 + \lambda(x_1 - \sigma) \tag{3.15}$$

Both the feasible and nonfeasible decompositions will be applied to solve the system (3.15). The solution to the original problem (3.12) can be readily obtained via calculus to be

$$x_1^* = 2 \qquad x_2^* = 0 \qquad f(x_1^*, x_2^*) = 1$$

3.7.1 Feasible Decomposition

In the feasible decomposition, the pseudo-variable σ is determined at the second level of the two-level optimization problem. The Lagrangian function presented in Eq. (3.15) is decomposed into two sub-Lagrangians L_1 and L_2:

$$L(x_1, x_2, \sigma, \lambda) = L_1(x_1, \lambda; \sigma) + L_2(x_2; \sigma) \tag{3.16}$$

where

$$L_1(x_1, \lambda; \sigma) = (x_1 - 2)^2 + \lambda(x_1 - \sigma) \tag{3.17}$$

$$L_2(x_2; \sigma) = (x_2 - 1)^2 + \sigma x_2 \tag{3.18}$$

The pseudo-variable σ, as an argument in the sub-Lagrangians, is preceded by a semicolon (;) to indicate that it should be viewed as a known parameter, at the first-level optimization. Figure 3.1 depicts the transfer of information between the first and second levels.

3.7.1.1 First-level optimization There are two subsystems in the first level of the hierarchical structure. In subsystem 1, there seem to be two independent decision variables x_1 and λ. However, this is not the case, as will be shown. Necessary conditions for stationarity for L_1 and L_2 are

Subsystem 1:

$$\frac{\partial L_1}{\partial x_1} = 2(x_1 - 2) + \lambda = 0 \tag{3.19}$$

or

$$\lambda = 2(2 - x_1) \tag{3.20}$$

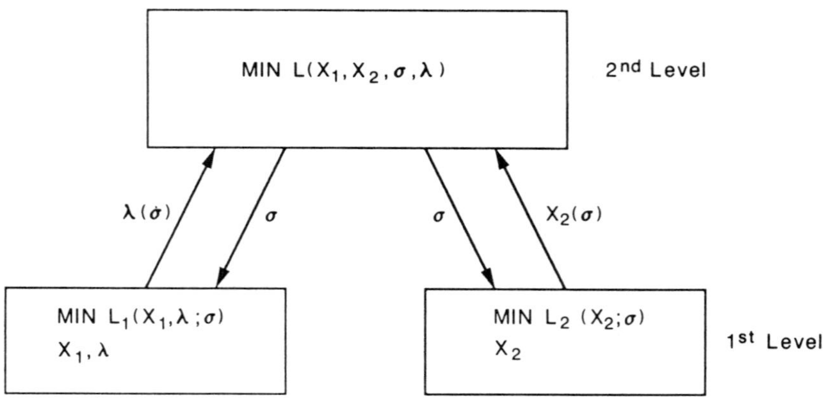

Figure 3.1 Feasible decomposition.

$$\frac{\partial L_1}{\partial \lambda} = x_1 - \sigma = 0 \tag{3.21}$$

or

$$x_1 = \sigma \tag{3.22}$$

Combining Eqs. (3.20) and (3.22) yields

$$\lambda(\sigma) = 4 - 2\sigma \tag{3.23}$$

Subsystem 2:

$$\frac{\partial L_2}{\partial x_2} = 2(x_2 - 1) + \sigma = 0 \tag{3.24}$$

or

$$x_2(\sigma) = 1 - 0.5\sigma \tag{3.25}$$

For any given value of σ determined at the second level, Eqs. (3.23) and (3.25) yield the corresponding optimal decisions at the first level in subsystems 1 and 2. Clearly, sufficiency conditions for a minimum should be checked as well.

3.7.1.2 Second-level optimization The total Lagrangian system (3.15) is optimized at the second level. Thus,

$$dL = \frac{\partial L}{\partial x_1} dx_1 + \frac{\partial L}{\partial x_2} dx_2 + \frac{\partial L}{\partial \lambda} d\lambda + \frac{\partial L}{\partial \sigma} d\sigma = 0 \tag{3.26}$$

Since x_1, x_2, and λ are chosen arbitrarily at the first level, then the following relations hold:

$$dx_1 = dx_2 = d\lambda = 0 \tag{3.27}$$

Furthermore, it is assumed that there exists a minimum for the Lagrangian L. Thus, a gradient-type algorithm may be adopted to minimize L with respect to the only decision variable at the second-level optimization, namely,

$$\sigma^{(k+1)} = \sigma^{(k)} - \Delta \frac{\partial L(\sigma^{(k)})}{\partial \sigma} \tag{3.28}$$

where $\Delta > 0$. Substituting for $\partial L/\partial \sigma$ in Eq. (3.28) yields

$$\sigma^{(k+1)} = \sigma^{(k)} - \Delta [x_2(\sigma^{(k)}) - \lambda(\sigma^{(k)})] \tag{3.29}$$

where the superscript (k) denotes the kth iteration. The results summarized by Eqs. (3.23), (3.25), and (3.29) can be used in solving the two-level optimization problem iteratively.

3.7.1.3 Iterative solution The iterative procedure may be started with an initial guessed value for σ (e.g., $\sigma^{(1)} = 1$). Equations (3.23), (3.25), and (3.29) summarize the interactions between the first and second levels.

Substituting $\sigma^{(1)} = 1$ into Eqs. (3.23) and (3.25) yields

$$\lambda^{(1)} = 2 \qquad x_2^{(1)} = 0.5 \qquad x_1^{(1)} = 1 \qquad f^{(1)}(1, 0.5) = 1.75$$

The second iteration can be initiated now with a proper choice of a step size Δ, for example, $\Delta = 0.5$. Substituting $\Delta = 0.5$ into Eq. (3.29) yields

$$\begin{aligned}\sigma^{(2)} &= 1 - 0.5(0.5 - 2)\\ &= 1.75\end{aligned}$$

Substituting $\sigma^{(2)} = 1.75$ into Eqs. (3.23) and (3.25) yields

$$\lambda^{(2)} = 0.5 \qquad x_2^{(2)} = 0.125 \qquad x_1^{(2)} = 1.75 \qquad f^{(2)}(1.75, 0.125) = 1.05$$

Finally, for the third iteration, let $\Delta = 0.4$. Then

$$\begin{aligned}\sigma^{(3)} &= 1.75 - 0.4(0.125 - 0.5)\\ &= 1.9\end{aligned}$$

Substituting $\sigma^{(3)} = 1.9$ into Eqs. (3.23) and (3.25) yields

$$\lambda^{(3)} = 0.2 \qquad x_2^{(3)} = 0.05 \quad x_1^{(3)} = 1.9 \qquad f^{(3)}(1.9, 0.05) = 1.008$$

The iterative procedure should be terminated on the basis of a prespecified criterion, for example, $|f^{(k+1)} - f^{(k)}| < 0.05\, f^{(k+1)}$. Note that the solution of the decomposed problem is rapidly converging to the optimal solution.

3.7.2 Nonfeasible Decomposition

The existence of a saddle point to the Lagrangian function is the heart of nonfeasible decomposition. The pseudo-variables are determined at the first level, where the Lagrange multipliers play the role of the second-level variables. Accordingly, the Lagrangian presented in Eq. (3.15) is decomposed into the following two sub-Lagrangians L_1 and L_2:

$$L(x_1, x_2, \sigma, \lambda) = L_1(x_1; \lambda) + L_2(x_2, \sigma; \lambda) \tag{3.30}$$

where

$$L_1(x_1; \lambda) = (x_1 - 2)^2 + \lambda x_1 \tag{3.31}$$

$$L_2(x_2, \sigma; \lambda) = (x_2 - 1)^2 + \sigma x_2 - \lambda\sigma \tag{3.32}$$

Note that the Lagrange multiplier λ, as an argument in the sub-Lagrangians, is preceded by a semicolon (;) to indicate that it should be viewed as a known parameter at the first-level optimization. Figure 3.2 depicts the transfer of information between the first and second levels.

3.7.2.1 First-level optimization Similar to the feasible decomposition, there are two subsystems at the first level, and the optimization procedure follows the same as well. The decision variable in subsystem 1 is x_1; in subsystem 2, x_2 and σ.

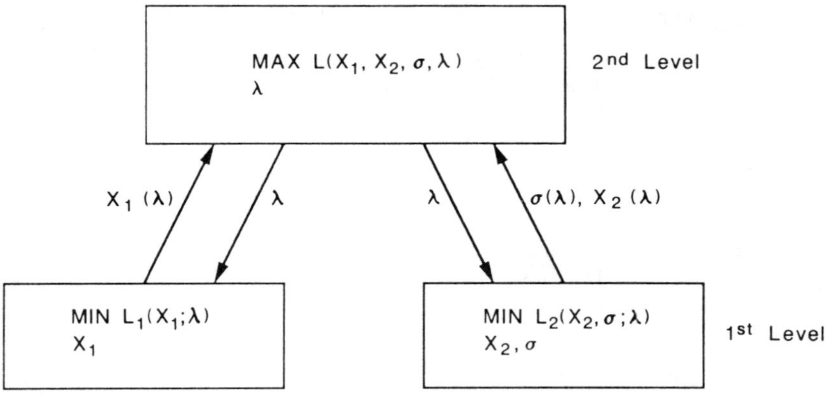

Figure 3.2 Nonfeasible decomposition.

Necessary conditions for stationarity for L_1 and L_2 are

Subsystem 1:

$$\frac{\partial L_1}{\partial x_1} = 2(x_1 - 2) + \lambda = 0 \tag{3.33}$$

$$x_1(\lambda) = 2 - 0.5\lambda \tag{3.34}$$

Subsystem 2:

$$\frac{\partial L_2}{\partial x_2} = 2(x_2 - 1) + \sigma = 0 \tag{3.35}$$

$$\sigma = 2(1 - x_2) \tag{3.36}$$

$$\frac{\partial L_2}{\partial \sigma} = x_2 - \lambda = 0 \tag{3.37}$$

$$x_2 = \lambda \tag{3.38}$$

Substituting Eq. (3.38) into (3.36) yields

$$\sigma(\lambda) = 2(1 - \lambda) \tag{3.39}$$

For any given value of λ determined at the second level, Eqs. (3.34), (3.38), and (3.39) yield the corresponding optimal decisions at the first level in subsystems 1 and 2. Here again, sufficiency conditions for a minimum should be checked. Several approaches for solving the second-level optimization are discussed in subsequent sections.

3.7.2.2 Explicit second-level coordinators When all first-level decision variables can be expressed as an explicit function of the Lagrange multipliers, as is the case in the example discussed here, the explicit second-level coordinator may be used. Since the only variable at the second level is the Lagrange multiplier λ, a necessary condition for stationarity to the Lagrangian function is given by Eq. (3.40):

$$\frac{dL}{d\lambda} = x_1 - \sigma = 0 \tag{3.40}$$

Since at the first level, both x_1 and σ are given explicitly in terms of λ by Eqs. (3.34) and (3.39), it is straightforward to determine λ directly by substituting into Eq. (3.40):

$$2 - 0.5\lambda = 2(1 - \lambda) \tag{3.41}$$

Solving Eq. (3.40) yields $\lambda = 0$. Substituting the value of $\lambda = 0$ back into the first-level variables yields

$$x_1(\lambda) = 2 \qquad x_2(\lambda) = 0 \qquad f(2, 0) = 1$$

The above solution is the same one obtained without decomposition and represents the overall optimal solution.

3.7.2.3 Dual second-level coordinator The nonfeasible decomposition provides an excellent opportunity for the use and applications of the duality theory. In particular, the first-level optimization can be viewed as solving the primal problem with respect to the original decision and the pseudo-variables. The second-level optimization can be viewed as solving the dual problem with respect to the dual variables, namely, the Lagrange multipliers. The solution of the first-level subsystems (primal problem) should converge to the solution of the second-level system (the dual problem) if the Lagrangian possesses a saddle point.

The example problem discussed in this section was carefully selected so that the conditions for a saddle point are not satisfied. Therefore, the principle of duality, where the primal problem is minimized with respect to the primal variables (x_1, x_2, and σ) and the dual problem is maximized with respect to the dual variable (λ), does not apply in this case, as is shown below.

Assuming a saddle point exists for the Lagrangian function given by Eq. (3.15), the task of the second-level controller is to maximize the Lagrangian function with respect to λ. Thus from duality theory, the second-level objective is

$$\max_{\lambda \in D} L(x_1^0, x_2^0, \sigma^0, \lambda) \tag{3.42}$$

where x_1^0, x_2^0, and σ^0 are the optimal values obtained from the subsystems' solution (primal problem solution), and D is the set of all λ's satisfying the Kuhn-Tucker conditions.

The maximization of Eq. (3.42) can be carried out with a gradient type of algorithm. The recursive equation for maximizing the Lagrangian with respect to λ can be written as

$$\lambda^{(k+1)} = \lambda^{(k)} + \Delta' \frac{\partial L}{\partial \lambda}(\lambda^{(k)}) \qquad \Delta' > 0 \tag{3.43}$$

where Δ' is a step size. Note, however, that

$$\frac{\partial L}{\partial \lambda} = x_1(\lambda) - \sigma(\lambda) = 1.5\lambda \tag{3.44}$$

Substituting Eq. (3.44) into (3.43) yields

$$\lambda^{(k+1)} = \lambda^{(k)} + \Delta\lambda \tag{3.45}$$

where $\Delta = 1.5\Delta'$.

Note that the recursive equation (3.45) will never yield to the optimal value of λ unless the initial value of λ is chosen to be zero. Any negative initial value of $\lambda^{(k)}$ will make all $\lambda^{(k+1)}$ more negative (smaller), while a positive initial value of $\lambda^{(k)}$ will make all $\lambda^{(k+1)}$ more positive (bigger). Therefore, a convergence to $\lambda = 0$ will not be achieved. The reason for the failure of the gradient type algorithm is that the necessary conditions for the existence of a saddle point are not satisfied, that is, a minimum does not exist for the Lagrangian with respect to x_1, x_2, and σ. This can be shown by examining the Hessian matrix for sufficient conditions for a minimum.

The Hessian matrix H for the Lagrangian is

$$H = \begin{bmatrix} \dfrac{\partial^2 L}{\partial x_1^2} & \dfrac{\partial^2 L}{\partial x_1\, \partial x_2} & \dfrac{\partial^2 L}{\partial x_1\, \partial \sigma} \\ \dfrac{\partial^2 L}{\partial x_2\, \partial x_1} & \dfrac{\partial^2 L}{\partial x_2^2} & \dfrac{\partial^2 L}{\partial x_2\, \partial \sigma} \\ \dfrac{\partial^2 L}{\partial \sigma\, \partial x_1} & \dfrac{\partial^2 L}{\partial \sigma\, \partial x_2} & \dfrac{\partial^2 L}{\partial \sigma^2} \end{bmatrix} \tag{3.46}$$

The Lagrangian was decomposed into two subsystems where the decision variable x_1 in subsystem 1 was uncoupled from the decisions x_2 and σ in subsystem 2. Therefore, the cross terms

$$\frac{\partial^2 L}{\partial x_1\, \partial x_2} \quad \text{and} \quad \frac{\partial^2 L}{\partial x_1\, \partial \sigma}$$

in the Hessian should vanish. The resultant Hessian is

$$H = \begin{bmatrix} 2 & 0 & 0 \\ 0 & 2 & 1 \\ 0 & 1 & 0 \end{bmatrix} \tag{3.47}$$

According to Sylvester's theorem, a sufficient condition for a minimum is that all the major minors of the Hessian be positive definite. Note, however, that the value of the determinant of the Hessian is negative (−2). Therefore, the sufficiency conditions for minimum are not satisfied, nor are the necessary conditions for a saddle point fulfilled. Note that the economic interpretations of the Lagrange multipliers make this coordinator one of the most promising in the hierarchical-multilevel approach.

3.7.2.4 Newton-Raphson second-level coordinator The Newton-Raphson second-level controller is based on the classical Newton-Raphson method. An error function $E(\lambda)$, associated with the pseudo-variable that was introduced to uncouple the subsystems interactions, is defined by Eq. (3.48). In the example problem, the reference is to Eq. (3.13), where $x_1 = x_1(\lambda)$ and $\sigma = \sigma(\lambda)$.

$$E(\lambda) = x_1(\lambda) - \sigma(\lambda) \tag{3.48}$$

Equation (3.48) can be extended to include a vector of pseudo-variables. The Newton-Raphson method seeks to find the roots to Eq. (3.49) through the recursive equation (3.50):

$$E(\lambda) = 0 \tag{3.49}$$

$$\lambda^{(k+1)} = \lambda^{(k)} - \Delta E(\lambda^{(k)}) \left[\frac{dE(\lambda^{(k)})}{d\lambda} \right]^{-1} \tag{3.50}$$

where $0 \leqslant \Delta \leqslant 1$. It is required that

$$\frac{dE(\lambda^{(k)})}{d\lambda} \neq 0 \tag{3.51}$$

Using the chain rule, it is possible to rewrite Eq. (3.50) in terms of x_1 and σ:

$$\frac{dE(\lambda)}{d\lambda} = \frac{\partial E(\lambda)}{\partial x_1} \frac{\partial x_1^{(\lambda)}}{\partial \lambda} + \frac{\partial E(\lambda)}{\partial \sigma} \frac{\partial \sigma(\lambda)}{\partial \lambda} \tag{3.52}$$

$$\frac{\partial E(\lambda)}{\partial x_1} = 1 \tag{3.53}$$

$$\frac{\partial E(\lambda)}{\partial \sigma} = -1 \tag{3.54}$$

Therefore,

$$\frac{dE(\lambda)}{d\lambda} = \frac{\partial x_1(\lambda)}{\partial \lambda} - \frac{\partial \sigma(\lambda)}{\partial \lambda} \tag{3.55}$$

Substituting Eqs. (3.52)–(3.55) into Eq. (3.50) yields

$$\lambda^{(k+1)} = \lambda^{(k)} - \Delta[x_1(\lambda^{(k)}) - \sigma(\lambda^{(k)})] \left[\frac{\partial x_1(\lambda^k)}{\partial \lambda} - \frac{\partial \sigma(\lambda^k)}{\partial \lambda} \right]^{-1} \tag{3.56}$$

Convergence is guaranteed if the minimum to the original problem exists, and if

$$\frac{\partial x_1(\lambda)}{\partial \lambda} - \frac{\partial \sigma(\lambda)}{\partial \lambda}$$

exists and is nonsingular. The value of $\lambda^{(0)}$ is initially guessed, where all other elements in the right-hand side of Eq. (3.56) are determined at the first-level optimization.

It is instructive at this stage to apply the Newton-Raphson second-level controller to the example problem. Recall Eqs. (3.34) and (3.39):

$$x_1(\lambda) = 2 - 0.5\lambda \tag{3.34}$$

$$\sigma(\lambda) = 2(1 - \lambda) \tag{3.39}$$

$$\frac{\partial x_1(\lambda)}{\partial \lambda} = -0.5 \tag{3.57}$$

$$\frac{\partial \sigma(\lambda)}{\partial \lambda} = -2 \tag{3.58}$$

$$\left[\frac{\partial x_1(\lambda)}{\partial \lambda} - \frac{\partial \sigma(\lambda)}{\partial \lambda} \right]^{-1} = 0.67 \tag{3.59}$$

$$x_1(\lambda) - \sigma(\lambda) = 1.5\lambda \tag{3.60}$$

Substituting Eqs. (3.34), (3.39), (3.59), and (3.60) into Eq. (3.56) yields

$$\lambda^{(k+1)} = (1 - \Delta)\lambda^{(k)} \qquad 0 \leqslant \Delta \leqslant 1 \tag{3.61}$$

Note that a positive initial guess of $\lambda^{(0)}$ yields a smaller value of $\lambda^{(1)}$, that is, the value of λ converges toward the optimum for $0 < \Delta < 1$, or equals the optimum value for $\Delta = 1$ (then $\lambda = 0$). Alternatively, a negative initial guess of λ yields a larger value of $\lambda^{(1)}$, that is, the value of λ converges toward the optimum for $0 < \Delta < 1$, or equals the optimum for $\Delta = 1$ (then $\lambda = 0$). It is known that if the Newton-Raphson method converges, it converges quadratically.

3.7.3 Example Problem 1

Solve the following problem using the nonfeasible decomposition:

$$\min_{x_1, x_2} [f(x_1, x_2) = x_1 + x_2 + 4x_1^2 - 4x_1x_2 + 2x_2^2 - 5]$$

subject to the constraint $x_1 + x_2 \geqslant 20$.

SOLUTION Let $x_1 = \sigma$. Then

$$\begin{aligned} L &= x_1 + x_2 + 4x_1^2 - 4\sigma x_2 + 2x_2^2 - 5 + \lambda(x_1 - \sigma) \\ &= L_1(x_1; \lambda) + L_2(x_2, \sigma; \lambda) \end{aligned}$$

where

$$\begin{aligned} L_1(x_1; \lambda) &= x_1 + 4x_1^2 + \lambda x_1 \\ L_2(x_2, \sigma; \lambda) &= x_2 + 2x_2^2 - 5 - 4\sigma x_2 - \lambda\sigma \end{aligned}$$

Subsystem 1:

$$\min\,(L_1 = x_1 + 4x_1^2 + \lambda x_1)$$

Subsystem 2:

$$\min\,(L_2 = x_2 + 2x_2^2 - 5 - 4\sigma x_2 - \lambda\sigma)$$

subject to the constraint $\sigma + x_2 \geqslant 20$.

Necessary conditions for stationarity at the first level:

$$\frac{\partial L_1}{\partial x_1} = 1 + 8x_1 + \lambda = 0 \tag{3.62}$$

or

$$x_1 = -\frac{1}{8}(\lambda + 1) \tag{3.63}$$

$$\hat{L}_2 = L_2 + \mu(20 - \sigma - x_2)$$

$$\hat{L}_2 = x_2 + 2x_2^2 - 5 - 4\sigma x_2 - \lambda\sigma + \mu(20 - \sigma - x_2) \tag{3.64}$$

$$\frac{\partial \hat{L}_2}{\partial x_2} = 1 + 4x_2 - 4\sigma - \mu = 0 \tag{3.65}$$

$$\frac{\partial \hat{L}_2}{\partial \sigma} = -4x_2 - \lambda - \mu = 0 \tag{3.66}$$

$$\frac{\partial \hat{L}_2}{\partial \mu_2} = 20 - \sigma - x_2 \leqslant 0 \tag{3.67}$$

$$\mu(20 - \sigma - x_2) = 0 \qquad \mu \geqslant 0 \tag{3.68}$$

Equating (3.65) and (3.66) yields

$$1 + 4x_2 - 4\sigma - \mu = -4x_2 - \lambda - \mu$$

$$4\sigma = 1 + \lambda + 8x_2 \tag{3.69}$$

Solving Eq. (3.68) for $\mu > 0$ (i.e., binding constraint) and substituting the result into Eq. (3.69) yields

$$4\sigma = 1 + \lambda + 8(20 - \sigma) \qquad \text{or} \qquad \sigma = \frac{1}{12}(161 + \lambda) \tag{3.70}$$

Necessary conditions for stationarity at the second level:

$$\frac{\partial L}{\partial \lambda} = x_1 - \sigma = 0$$

Substituting the values of $x_1(\lambda)$ and $\sigma(\lambda)$ from the first level [Eqs. (3.63) and (3.70)] yields

$$\frac{1}{12}(161 + \lambda) = \frac{1}{8}(\lambda + 1) \qquad \text{or} \qquad \lambda = -65 \tag{3.71}$$

Back to the first level:

Substituting the value of $\lambda = -65$ [Eq. (3.71)] into Eqs. (3.63) and (3.70) yields

$$x_1 = \frac{1}{8}(65 - 1) = 8$$

$$\sigma = \frac{1}{12}(161 - 65) = 8$$

Substituting the above values into Eq. (3.69) yields

$$x_2 = \frac{(32 - 1 + 65)}{8} = 12$$

and the value of the objective function at $x_1 = 8$, $x_2 = 12$ is $f(8, 12) = 175$.

$$x_1^* = 8 \qquad x_2^* = 12$$

and thus

$$f(x_1^*, x_2^*) = 175$$

3.7.4 Example Problem 2

Solve the following problem using (1) feasible and (2) nonfeasible decomposition:

$$\min_{\mathbf{x}} [f(\mathbf{x}) = (x_1 - 3)^2 + (x_2 - 4)^2 + (x_3 - 1)^2 - x_1x_2 - x_1x_3 + x_2x_3]$$

SOLUTION Let $x_1 = \sigma_1$, $x_2 = \sigma_2$, and $x_3 = \sigma_3$. Then the modified problem is

$$\min_{\mathbf{x}} [f(\mathbf{x}) = (x_1 - 3)^2 + (x_2 - 4)^2 + (x_3 - 1)^2 - \sigma_2x_1 - \sigma_1x_3 + \sigma_3x_2]$$

subject to the constraints $x_1 = \sigma_1$, $x_2 = \sigma_2$, and $x_3 = \sigma_3$.

Form the Lagrangian

$$L(\mathbf{x}, \boldsymbol{\sigma}, \boldsymbol{\lambda}) = (x_1 - 3)^2 + (x_2 - 4)^2 + (x_3 - 1)^2 - \sigma_2 x_1 - \sigma_1 x_3 + \sigma_3 x_2 + \lambda_1(x_1 - \sigma_1) + \lambda_2(x_2 - \sigma_2) + \lambda_3(x_3 - \sigma_3) \quad (3.72)$$

3.7.4.1 Feasible decomposition

$$L(\mathbf{x}, \boldsymbol{\lambda}; \boldsymbol{\sigma}) = L_1(x_1, \lambda_1; \sigma_1, \sigma_2) + L_2(x_2, \lambda_2; \sigma_2, \sigma_3) + L_3(x_3, \lambda_3; \sigma_1, \sigma_3)$$

where

$$L_1(x_1, \lambda_1; \sigma_1, \sigma_2) = (x_1 - 3)^2 - \sigma_2 x_1 + \lambda_1(x_1 - \sigma_1) \quad (3.73)$$

$$L_2(x_2, \lambda_2; \sigma_2, \sigma_3) = (x_2 - 4)^2 + \sigma_3 x_2 + \lambda_2(x_2 - \sigma_2) \quad (3.74)$$

$$L_3(x_3, \lambda_3; \sigma_1, \sigma_3) = (x_3 - 1)^2 - \sigma_1 x_3 + \lambda_3(x_3 - \sigma_3) \quad (3.75)$$

First level:

$$\frac{\partial L_1}{\partial x_1} = 2(x_1 - 3) - \sigma_2 + \lambda_1 = 0 \quad (3.76)$$

or

$$\lambda_1 = 2(3 - x_1) + \sigma_2$$

$$\frac{\partial L_1}{\partial \lambda_1} = x_1 - \sigma_1 = 0 \qquad \text{or} \qquad x_1 = \sigma_1 \quad (3.77)$$

From Eqs. (3.76) and (3.77):

$$\lambda_1 = 2(3 - \sigma_1) + \sigma_2 \quad (3.78)$$

$$\frac{\partial L_2}{\partial x_2} = 2(x_2 - 4) + \sigma_3 + \lambda_2 = 0 \qquad \text{or} \qquad \lambda_2 = 2(4 - x_2) - \sigma_3 \quad (3.79)$$

$$\frac{\partial L_2}{\partial \lambda_2} = x_2 - \sigma_2 = 0 \qquad \text{or} \qquad x_2 = \sigma_2 \quad (3.80)$$

Thus,

$$\lambda_2 = 2(4 - \sigma_2) - \sigma_3 \quad (3.81)$$

$$\frac{\partial L_3}{\partial x_3} = 2(x_3 - 1) - \sigma_1 + \lambda_3 = 0 \quad \text{or} \quad \lambda_3 = 2(1 - x_3) + \sigma_1 \quad (3.82)$$

$$\frac{\partial L_3}{\partial \lambda_3} = x_3 - \sigma_3 = 0 \quad \text{or} \quad x_3 = \sigma_3 \quad (3.83)$$

Thus,

$$\lambda_3 = 2(1 - \sigma_3) + \sigma_1 \quad (3.84)$$

Second level:

Using a gradient method,

$$\boldsymbol{\sigma}^{(k+1)} = \boldsymbol{\sigma}^{(k)} - \Delta \cdot \nabla_{\sigma} L(\mathbf{x}(\boldsymbol{\sigma}^{(k)}), \lambda(\boldsymbol{\sigma}^{(k)}), \boldsymbol{\sigma}^{(k)}) \tag{3.85}$$

$$\begin{bmatrix} \sigma_1^{(k+1)} \\ \sigma_2^{(k+1)} \\ \sigma_3^{(k+1)} \end{bmatrix} = \begin{bmatrix} \sigma_1^{(k)} \\ \sigma_2^{(k)} \\ \sigma_3^{(k)} \end{bmatrix} - \Delta \begin{bmatrix} 2\sigma_1^{(k)} - \sigma_2^{(k)} - \sigma_3^{(k)} - 6 \\ -\sigma_1^{(k)} + 2\sigma_2^{(k)} + \sigma_3^{(k)} - 8 \\ -\sigma_1^{(k)} + \sigma_2^{(k)} + 2\sigma_3^{(k)} - 2 \end{bmatrix}$$

where $L(\mathbf{x}(\boldsymbol{\sigma}), \boldsymbol{\lambda}(\boldsymbol{\sigma}), \boldsymbol{\sigma}) = (\sigma_1 - 3)^2 + (\sigma_2 - 4)^2 + (\sigma_3 - 1)^2 - \sigma_2\sigma_1 + \sigma_1\sigma_2 + \sigma_3\sigma_2$.

Let $\boldsymbol{\sigma}^{(1)} = (1, 1, 1)$ and $\Delta = 1$. Then

$$\begin{bmatrix} \sigma_1^{(2)} \\ \sigma_2^{(2)} \\ \sigma_3^{(2)} \end{bmatrix} = \begin{bmatrix} 1 \\ 1 \\ 1 \end{bmatrix} - \begin{bmatrix} 2 - 1 - 1 - 6 \\ -1 + 2 + 1 - 8 \\ -1 + 1 + 2 - 2 \end{bmatrix} = \begin{bmatrix} 7 \\ 7 \\ 1 \end{bmatrix}$$

Thus, $\boldsymbol{\sigma}^* = (7, 7, 1)$.

Back to first level:

$$x_1^* = \sigma_1^* = 7$$

$$x_2^* = \sigma_2^* = 7$$

$$x_3^* = \sigma_3^* = 1$$

$$f(\mathbf{x}^*) = -24$$

This is the same optimal solution obtained via calculus without decomposition.

3.7.4.2 Nonfeasible decomposition

$$L(\mathbf{x}, \boldsymbol{\lambda}, \boldsymbol{\sigma}) = L_1(x_1, \sigma_2; \lambda_1, \lambda_2) + L_2(x_2, \sigma_3; \lambda_2, \lambda_3) + L_3(x_3, \sigma_1; \lambda_1, \lambda_3)$$

where

$$\begin{aligned} L_1(x_1, \sigma_2; \lambda_1, \lambda_2) &= (x_1 - 3)^2 - x_1\sigma_2 - \lambda_2\sigma_2 + \lambda_1 x_1 \\ L_2(x_2, \sigma_3; \lambda_2, \lambda_3) &= (x_2 - 4)^2 + x_2\sigma_3 - \lambda_3\sigma_3 + \lambda_2 x_2 \\ L_3(x_3, \sigma_1; \lambda_1, \lambda_3) &= (x_3 - 1)^2 - x_3\sigma_1 - \lambda_1\sigma_1 + \lambda_3 x_3 \end{aligned}$$

First-level subsystem 1:

$$\frac{\partial L_1}{\partial x_1} = 2(x_1 - 3) - \sigma_2 + \lambda_1 = 0 \quad \text{or} \quad \sigma_2 = 2(x_1 - 3) + \lambda_1 \tag{3.86}$$

$$\frac{\partial L_1}{\partial \sigma_2} = -x_1 - \lambda_2 = 0 \quad \text{or} \quad x_1 = -\lambda_2$$

Thus, $\sigma_2 = -2(\lambda_2 + 3) + \lambda_1$.

First-level subsystem 2:

$$\frac{\partial L_2}{\partial x_2} = 2(x_2 - 4) + \sigma_3 + \lambda_2 = 0 \qquad \text{or} \qquad \sigma_3 = -2(x_2 - 4) - \lambda_2 \tag{3.87}$$

$$\frac{\partial L_2}{\partial \sigma_3} = x_2 - \lambda_3 = 0 \qquad \text{or} \qquad x_2 = \lambda_3$$

Thus, $\sigma_3 = -2(\lambda_3 - 4) - \lambda_2$.

First-level subsystem 3:

$$\frac{\partial L_3}{\partial x_3} = 2(x_3 - 1) - \sigma_1 + \lambda_3 = 0 \qquad \text{or} \qquad \sigma_1 = 2(x_3 - 1) + \lambda_3 \tag{3.88}$$

$$\frac{\partial L_3}{\partial \sigma_1} = -x_3 - \lambda_1 = 0 \qquad \text{or} \qquad x_3 = -\lambda_1$$

Thus, $\sigma_1 = -2(\lambda_1 + 1) + \lambda_3$.

Second level:

$$\frac{\partial L}{\partial \lambda_1} = x_1 - \sigma_1 = 0$$

$$\frac{\partial L}{\partial \lambda_2} = x_2 - \sigma_2 = 0$$

$$\frac{\partial L}{\partial \lambda_3} = x_3 - \sigma_3 = 0$$

Since at the first level both $\mathbf{x}$ and $\boldsymbol{\sigma}$ are given explicitly in terms of $\boldsymbol{\lambda}$, it is easy to determine $\boldsymbol{\lambda}$ directly:

$$x_1 = \sigma_1 \qquad \text{thus} \qquad -\lambda_2 = -2(\lambda_1 + 1) + \lambda_3$$

$$x_2 = \sigma_2 \qquad \text{thus} \qquad \lambda_3 = -2(\lambda_2 + 3) + \lambda_1$$

$$x_3 = \sigma_3 \qquad \text{thus} \qquad -\lambda_1 = -2(\lambda_3 - 4) - \lambda_2$$

Solving the above three equations simultaneously yields

$$\lambda_1^* = -1 \qquad \lambda_2^* = -7 \qquad \lambda_3^* = 7.$$

Returning to the first level and substituting the above values of $\boldsymbol{\lambda}^*$ into Eqs. (3.86)–(3.88) yields

$$x_1^* = 7 \qquad x_2^* = 7 \qquad x_3^* = 1 \qquad f(\mathbf{x}^*) = -24$$

This is the same result obtained via the feasible decomposition.

REFERENCES

Baumann, E. J., *Multi-Level Optimization Techniques with Application to Trajectory Decomposition,* Ph.D. Dissertation, University of California, Los Angeles, Calif., 1966.

Dantzig, G. B. and Wolfe, P., The Decomposition Algorithm for Linear Programming, *Econometrica,* vol. 9, no. 4, 1961; Also *Operations Res.,* vol. 8, Jan., Feb., Oct. 1960.

Haimes, Y. Y., Multilevel Dynamic Programming Structure for Regional Water Resource Management, in *Decomposition of Large Scale Problems,* ed. D. M. Himmelblau, pp. 269–378, North-Holland, Amsterdam, 1973.

Haimes, Y. Y., Hierarchical Modeling of Regional Total Water Resources Systems, *Automatica,* vol, 11, pp. 25–36, 1975.

Haimes, Y. Y., *Hierarchical Analyses of Water Resources Systems: Modeling and Optimization of Large Scale Systems,* McGraw-Hill, New York, 1977.

Haimes, Y. Y., and Li, D., Hierarchical Multiobjective Analysis for Large Scale Systems: Review and Current Status, *Automatica,* vol. 24, no. 1, pp. 53–69, 1988.

Haimes, Y. Y., and Macko, D., Hierarchical Structures in Water Resources Systems Management, *IEEE Syst. Man Cybern,* vol. SMC-3, pp. 396–402, 1973.

Lasdon, L., A Multilevel Technique for Optimization, Systems Research Center Report SRC-50-C-64-19, Case Western Reserve University, Cleveland, Ohio, 1964.

Lasdon, L., *Optimization Theory for Large Scale Systems,* Macmillan, London, 1970.

Lefkowitz, I., Multilevel Approach Applied to Control System Design, *Trans. ASME* 88D, 2–10, 1966.

Macko, D., General Systems Theory Approach to Multilevel Systems, Systems Research Center Report SRC 106-A-67-44, Case Western Reserve University, Cleveland, Ohio, 1967.

Mesarovic, M. D., Macko, D., and Takahara, Y., *Theory of Hierarchical Multilevel Systems,* Academic Press, New York, 1970.

Sage, A. P., *Methodology for Large Scale Systems,* McGraw-Hill, New York, 1977.

Singh, M. G., and Titli, A. (eds.), *Handbook of Large Scale Systems Engineering Applications,* North-Holland, New York, 1979.

Wismer, D. A. (ed.), *Optimization Methods for Large-Scale Systems,* McGraw-Hill, New York, 1971.

CHAPTER

FOUR

HIERARCHICAL METHODS OF GENERATING PARETO OPTIMAL ALTERNATIVES

4.1 INTRODUCTION

Multicriteria optimization methods can be put into three major classes: (1) generating methods, (2) methods based on a prior articulation of preferences, and (3) interactive methods. This chapter considers the generating methods for large-scale systems.

In the generating methods, the analyst determines the set of Pareto optimal alternatives or an approximate representation of it. After this, the decisionmaker directly selects a solution from this set based on his or her preferences. This kind of procedure is commonly used because of its several good features. First, the task of the analyst is mathematically clearly defined. Second, the decisionmaker has a view of all reasonable alternatives at hand. Third, the decisionmakers, whether one or several, are free to use any decisionmaking or voting procedure to decide on the final compromise solution. This approach has one major limitation: it becomes difficult to display the Pareto optimal solutions when the number of the alternatives is large.

This chapter begins with a review of some basics of Pareto optimality. Relationships are then derived between Pareto optimal vectors in a hierarchy of objectives. After this, we derive methods for generating Pareto optimal alternatives. We present two major formulations: one where the subproblems are multicriteria problems and another where the use of the weighting method is assumed in the subsystems.

One of the motivations for developing hierarchical methods is that such methods make it possible to decompose an overall problem into smaller subproblems which are easier to solve mathematically. The generating methods mentioned also make it possible to construct a very useful decisionmaking hierarchy to be used in real organizations. In this hierarchy, the subsystems (e.g., departments in a factory or subregions in a regional planning problem) solve Pareto optimal alternatives for a set of possible available resources. This set is given by an upper level decisionmaker, who, after the subsystem solutions have been reached, checks which of these solutions constitute Pareto optimal solutions for the overall system. This kind of procedure is simulated by a numerical example at the end of the chapter.

4.2 DISPLAYING PARETO OPTIMAL ALTERNATIVES

In this section, we will briefly review graphic ways to present the set of Pareto optimal alternatives. This set, or a finite representative subset of it, can always be presented in tabular form. However, many decisionmakers find that a graphic representation is more convenient. Figure 4.1 depicts a graphic presentation for a case with two objectives. In this case, the entire set of Pareto optimal alternatives is easily depicted graphically. The set is like a curved two-dimensional surface in a three-dimensional space. Figure 4.2 shows how the Pareto set can be represented by projecting lines in this curved surface (on which the third objective is constant) on the plane of the first two objectives.

When the number of objectives is four or more, we can use the graphic methods depicted in Figs. 4.3 and 4.4. However, in these cases only a limited number of the Pareto optimal alternatives can be shown.

4.3 REVIEW OF BASIC RESULTS IN GENERATING METHODS

There are two basic methods of generating Pareto optimal alternatives: the weighting method and the constraint method. Only the most important facts of these methods are reviewed here. For more details, see Chankong and Haimes (1983).

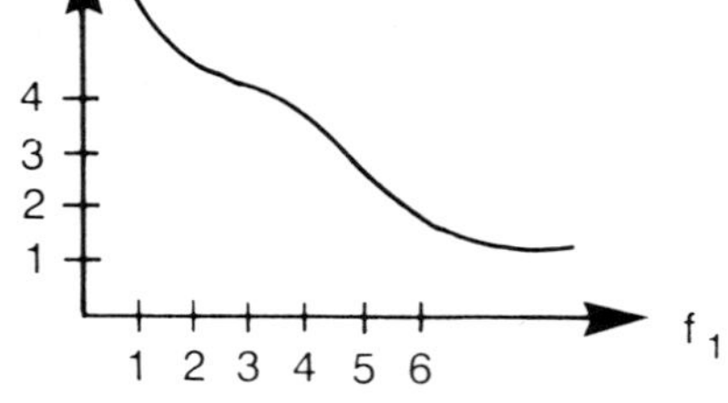

Figure 4.1 Pareto optimal set of a case with two objectives.

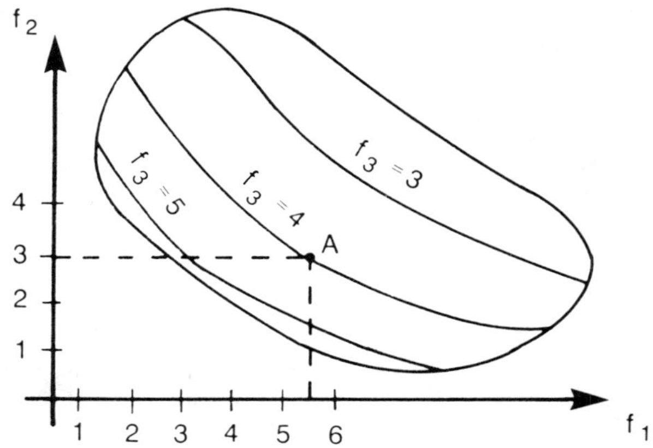

Figure 4.2 Pareto optimal set of a case with three objectives. For example, point A is a Pareto optimal point with $f_1 = 5.6$, $f_2 = 3$, and $f_3 = 4$.

4.3.1 Notion of a Proper Pareto Optimal Solution

In a proper multicriteria decisionmaking problem where the objectives are traded off, there are extreme solutions among the Pareto optimal alternatives that are never selected as preferred solutions by the decisionmaker. As an example, consider point A in Fig. 4.5. At the Pareto optimal point A, the tangent of the Pareto optimal set in this two-objective case is vertical. In other words, when we are at point A and move a little bit down the Pareto optimal curve, the improvement in f_2 can be arbitrarily many times bigger than the loss in f_1 (when the movement is small). A decisionmaker would therefore select point A only in the case where the objective f_1 totally overrides the objective f_2. But this is not a proper multicriteria optimization problem. Pareto optimal alternatives such as point A are called nonproper Pareto optimal alternatives and are excluded from most considerations in the sequel (which also makes many mathematical issues easier).

For mathematical details of the definition of a nonproper solution with an arbitrary number of objectives, see Chankong and Haimes (1983) and Sawaragi et al. (1985). In this chapter, the definition of proper Pareto optimal solution by Geoffrion (1968) is assumed.

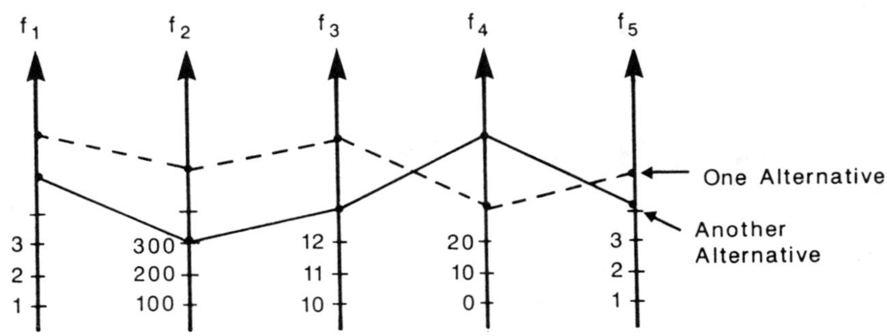

Figure 4.3 Method of presenting Pareto optimal alternatives with more than three objectives.

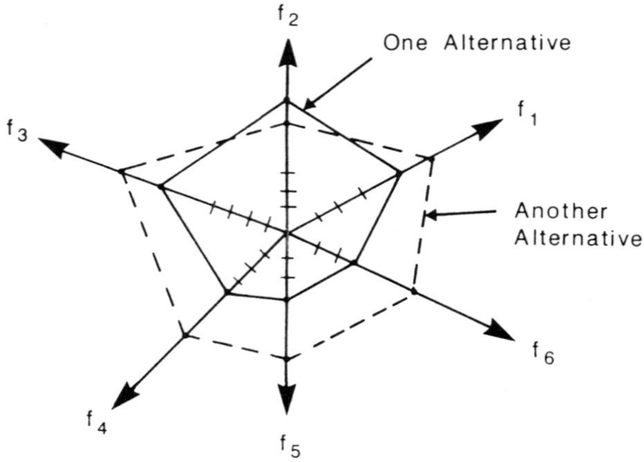

Figure 4.4 Another method of presenting Pareto optimal alternatives with more than three objectives.

4.3.2 Constraint Method for Generating Pareto Optimal Alternatives

Consider the multiobjective optimization problem given in Eq. (2.1) which is repeated here as Eq. (4.1) for convenience (with μ_i replacing g_i for further notational purposes):

$$\min \, [f_1(\mathbf{x}), f_2(\mathbf{x}), \ldots, f_n(\mathbf{x})] \tag{4.1}$$

subject to

$$\mu_i(\mathbf{x}) \leqslant 0 \qquad i = 1, 2, \ldots, m \tag{4.2}$$

where f_i and μ_i are scalar functions.

Chapter 2 showed how, by converting the above problem to an ordinary constrained optimization problem, decisionmakers can conveniently find their preferred solution.

The constraint approach can also be used to generate all Pareto optimal alternatives as expressed by Lemmas 4.1 and 4.2, which follow. For proofs, see Chankong and Haimes (1983).

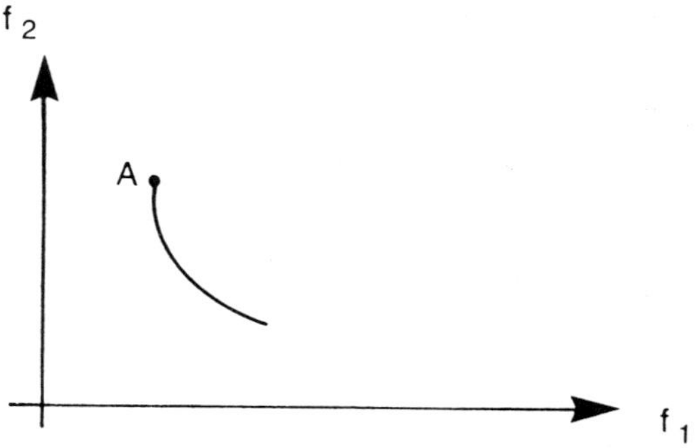

Figure 4.5 Nonproper Pareto optimal point (A).

Let the ith objective f_i be taken as a primary objective to be minimized in the constraint formulation. Then, the constrained problem associated with Eqs. (4.1) and (4.2) is the problem denoted by $p_i(\boldsymbol{\epsilon})$, where $\boldsymbol{\epsilon} = (\epsilon_1, \ldots, \epsilon_{i-1}, \epsilon_{i+1}, \ldots \epsilon_n)$:

$$p_i(\boldsymbol{\epsilon}): \min f_i(\mathbf{x}) \tag{4.3}$$

subject to

$$f_j(\mathbf{x}) \leq \epsilon_j \qquad j = 1, \ldots, n \qquad j \neq i \tag{4.4}$$

$$\mu_k(\mathbf{x}) \leq 0 \qquad k = 1, \ldots, m \tag{4.5}$$

Lemma 4.1 If $\mathbf{x}^*$ is a Pareto optimal solution of the multicriteria problem in Eqs. (4.1) and (4.2), then $\mathbf{x}^*$ solves the constraint problem $p_i(\boldsymbol{\epsilon})$ with $\epsilon_j = f_j(\mathbf{x}^*)$ for $j = 1, \ldots, n; j \neq i$.

Lemma 4.2 If $\mathbf{x}^*$ is a unique solution of $p_i(\boldsymbol{\epsilon})$, then $\mathbf{x}^*$ is a Pareto optimal solution of the problem given in Eqs. (4.1) and (4.2).

Hence, it follows directly from these lemmas that if we solve $p_i(\boldsymbol{\epsilon})$ with all possible values of $\boldsymbol{\epsilon}$, and the corresponding solutions are unique, then these solutions constitute the set of all Pareto optimal solutions of the original multicriteria problem. In the rare case where the solutions of $p_i(\boldsymbol{\epsilon})$ are not unique for some $\boldsymbol{\epsilon}$'s, the Pareto optimal alternatives must be picked from among these solutions by direct comparison, using the definition of Pareto optimality. (For more elaborate possibilities in the case of nonuniqueness, consult Sect. 4.3.2 of Chankong and Haimes, 1983.)

4.3.3 Weighting Method for Generating Pareto Optimal Alternatives

In the weighting method, ordinary optimization problems associated with the original multicriteria problem [Eqs. (4.1) and (4.2)] are denoted by $p(\mathbf{w})$ with $\mathbf{w} = w_1, \ldots, w_n$:

$$p(\mathbf{w}): \min \sum_{i=1}^{n} w_i f_i(\mathbf{x})$$

subject to

$$\mu_k(\mathbf{x}) \leq 0, \qquad k = 1, \ldots, n.$$

Note that a $p(\mathbf{w})$ problem is mathematically easier to solve than a $p_k(\boldsymbol{\epsilon})$ problem, because the number of inequalities is not increased compared to the original problem given in Eqs. (4.1) and (4.2). However, the weighting method is not a general method of generating Pareto optimal alternatives: its validity is limited to convex cases as expressed more precisely in Lemmas 4.3 and 4.4. (For proofs, see Geoffrion, 1968.)

Lemma 4.3 Assume that $f_1(\mathbf{x}), \ldots, f_n(\mathbf{x})$ and $\mu_1(\mathbf{x}), \ldots, \mu_m(\mathbf{x})$ are convex functions. If $\mathbf{x}^*$ is a proper Pareto optimal solution of the multicriteria problem in Eqs. (4.1) and (4.2), there exist $w_i^* > 0$ $(i = 1, \ldots, n)$ such that $\mathbf{x}^*$ is a solution of $p(\mathbf{w}^*)$.

Lemma 4.4 If $\mathbf{x}^*$ is a solution of $p(\mathbf{w}^*)$ where $w_i^* > 0$ $(i = 1, \ldots, n)$, then $\mathbf{x}^*$ is a proper Pareto optimal solution of the multicriteria problem in Eqs. (4.1) and (4.2).

That is, based on these two lemmas, in the case where f_i's and μ_i's are convex, the set of proper Pareto optimal alternatives can be determined by solving the ordinary optimization problem $p(\mathbf{w})$ for all strictly positive weights w_i. Since the solution of $p(\mathbf{w})$ is unaltered if the weights w_i are multiplied by a constant, it is always possible to limit things, for example, to the case where $\Sigma_{i=1}^n w_i = 1$, thus reducing the dimension of the space of weights that must be varied. Alternatively, one may, for example, normalize $w_1 = 1$.

4.3.4 Kuhn-Tucker Necessary and Sufficient Conditions for Proper Noninferiority

When Kuhn and Tucker (1951) derived the much-used conditions for ordinary optimization, they also derived corresponding conditions for Pareto optimal solutions. These conditions, which are used later in this chapter, are reviewed in Lemma 4.5.

As in the ordinary optimization, a constraint qualification or regularity condition is needed concerning the constraint functions. For practical reasons, a more easily verifiable regularity condition than that of Kuhn and Tucker is used.

An equality constraint $h(\mathbf{x}) = 0$ is equivalently represented by inequality constraints $h(\mathbf{x}) \leq 0$ and $-h(\mathbf{x}) \leq 0$, which are of the form assumed in Eqs. (4.1) and (4.2). However, for the definition of the regularity condition, we will write the equality and inequality constraints separately. Thus, instead of the original problem formulation of Eqs. (4.1) and (4.2), consider the following problem:

$$\min \, [f_1(\mathbf{x}), f_2(\mathbf{x}), \ldots, f_n(\mathbf{x})] \tag{4.6}$$

subject to

$$\mu_i(\mathbf{x}) \leq 0 \qquad i = 1, 2, \ldots, m \tag{4.7}$$

$$h_i(\mathbf{x}) = 0 \qquad i = 1, 2, \ldots, l \tag{4.8}$$

For the constraints given in Eqs. (4.7) and (4.8), a point $\bar{\mathbf{x}}$ that satisfies these equations is said to be a *regular point* if the following gradient vectors are linearly independent: $\partial h_i(\bar{\mathbf{x}})/\partial \mathbf{x}$ $(i = 1, \ldots l)$, $\partial \mu_j(\bar{\mathbf{x}})/\partial \mathbf{x}$ $(j \in J$, where J is the set of indices j for which $\mu_j(\bar{\mathbf{x}}) = 0)$. (For a discussion of alternative possibilities of defining a regular point and the corresponding constraint qualifications, see Chankong and Haimes, 1983.)

Lemma 4.5a Assume that all functions f_i, μ_i, and h_i are continuously differentiable. Then, a proper Pareto optimal point $\mathbf{x}^*$ for the multicriteria problem given in Eqs. (4.6)–(4.8), which is regular, satisfies the following relations (which are here called the Kuhn-Tucker conditions for proper noninferiority). There exist the following scalars (called Lagrange multipliers):

$$w_i > 0 \qquad i = 1, \ldots, n \tag{4.9}$$

$$\eta_i \geq 0 \qquad i = 1, \ldots, m \tag{4.10}$$

and

$$\rho_i \qquad i = 1, \ldots, l \tag{4.11}$$

such that, where $L(\mathbf{x})$ is the Lagrangian function,

$$L(\mathbf{x}) = \sum_{i=1}^{n} w_i f_i(\mathbf{x}) + \sum_{i=1}^{m} \eta_i \mu_i(\mathbf{x}) + \sum_{i=1}^{l} \rho_i h_i(\mathbf{x}) \tag{4.12a}$$

Eq. (4.12b) holds:

$$\frac{\partial L}{\partial \mathbf{x}}(\mathbf{x}^*) = 0 \tag{4.12b}$$

and, furthermore,

$$\eta_i \mu_i(\mathbf{x}^*) = 0 \qquad i = 1, \ldots, m \tag{4.13}$$

$$\mu_i(\mathbf{x}^*) \leq 0 \qquad i = 1, \ldots, m \tag{4.14}$$

$$h_i(\mathbf{x}^*) = 0 \qquad i = 1, \ldots, l \tag{4.15}$$

Lemma 4.5b On the other hand, if $\mathbf{x}^*$ (which need not be a regular point) satisfies the relations given in Eqs. (4.9)–(4.15), if all functions f_i and μ_i are continuously differentiable and convex, and if all functions h_i are linear, then $\mathbf{x}^*$ is a proper Pareto optimal point.

The following remark can be made about the proof of this lemma. Lemmas 4.3 and 4.4 establish an equivalence between proper Pareto optimality and an ordinary $p(\mathbf{w}^*)$ optimization problem in a convex case. If, in this convex case, the usual Kuhn-Tucker necessary and sufficient conditions (see, e.g., Luenberger, 1973) are applied to $p(\mathbf{w}^*)$, then this lemma is readily proved. Note, however, that in Lemma 4.5 the conditions given in Eqs. (4.9)–(4.15) hold at a proper Pareto optimal $\mathbf{x}^*$, even in a nonconvex case. For a complete proof of the lemma, see Geoffrion (1968) and Chankong and Haimes (1983). Note, by direct verification, that if the Kuhn-Tucker conditions for proper noninferiority hold for some values of the Lagrange multipliers w_i, η_i, and ρ_i, they hold also for multipliers that are obtained by dividing or multiplying the original multipliers by a strictly positive constant. In the following sections, we will sometimes normalize the multipliers such that $w_1 = 1$. That is, if $w_1 \neq 1$, we use the following multipliers in the Kuhn-Tucker conditions given in Eqs. (4.9)–(4.15):

$$w_i^s = \frac{w_i}{w_1} \qquad (i = 1, \ldots, n)$$

$$\eta_i^s = \frac{\eta_i}{w_1} \qquad (i = 1, \ldots, m)$$

$$\rho_i^s = \frac{\rho_i}{w_1} \qquad (i = 1, \ldots, l)$$

We will also use the usual Kuhn-Tucker necessary and sufficient conditions for a single-criterion optimization problem (see, e.g., Varaiya, 1972). Note that these conditions are, in fact, a special case of Lemma 4.5. Just set $n = 1$ (one objective) and replace the term "a proper Pareto optimal point" by "a minimum point." Then, Lemma 4.5 reads as a corresponding lemma for a single-criterion optimization problem. (The normalization $w_1 = 1$ is usually done in this case.)

4.4 MATHEMATICAL FORMULATION OF LARGE-SCALE MULTICRITERIA PROBLEMS WITH MULTIPLE DECISIONMAKERS

This section presents a general mathematical formulation for large-scale multicriteria problems with multiple decisionmakers. The general structure of such a complex system is depicted in Fig. 4.6. Several special cases of this formulation are considered in later sections.

This kind of mathematical problem may in practice arise, for example, in the multicriteria optimization of a company with relatively independent and accountable departments, or in a regional planning problem with a central decisionmaking body that has subregions in conflict.

In the mathematical problem formulation, a system consisting of N subsystems is considered, and $N \geqslant 2$. In each subsystem ($i = 1, \ldots, N$), let

$\mathbf{y}_i$ = the output vector of subsystem i;
$\mathbf{x}_i$ = the input vector of subsystem i from other subsystems;
$\mathbf{m}_i$ = the decision or control vector of subsystem i.

(In Fig. 4.6, the components of vector $\mathbf{x}_i$ are denoted by x_{i1}, x_{i2}, and x_{i3}.) Furthermore, let n_i = the number of objectives of the subsystem ($i = 1, \ldots, N$).

In the following discussion, it is usually tacitly assumed that $N \geqslant 2$. The case with a single objective in a subsystem is treated by making obvious modifications and by dropping vacuous expressions. In fact, one obtains most single-objective hierarchical schemes as a special case of the multicriteria procedures to be derived in later chapters.

In each subsystem i, let f_j^i = the jth objective related to subsystem i ($j = 1, \ldots, n_i$), and $\mathbf{f}^i = (f_1^i, \ldots, f_{n_i}^i)$ = vector of objectives of subsystem i. The objectives f_j^i of the subsystem i are assumed to be functions of the variables associated with the subsystems and are denoted by $f_j^i(\mathbf{x}_i, \mathbf{m}_i, \mathbf{y}_i)$. In a general case, the objective vector of the whole system does not directly consist of all f_j^i; one could call the f_j^i the subgoals or indicators. For example, when some subsystems have cost as an objective, one of the overall system's objectives may be total cost,

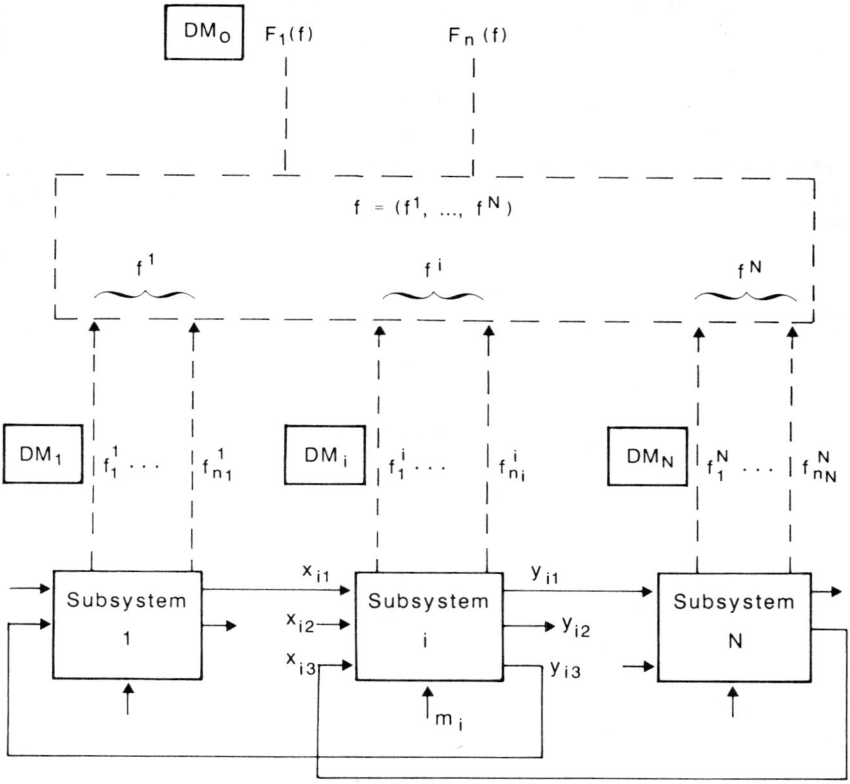

Figure 4.6 General structure of the systems (solid line), objectives (broken line) and decisionmakers DM_0, DM_1, . . ., DM_N for large-scale multicriteria problems. Three subsystems are depicted.

which is the sum of these subsystems' cost objectives. Let the following definitions hold†:

$\mathbf{f} = (\mathbf{f}^1, \ldots, \mathbf{f}^N)$ = the vector of all objectives in the subsystems,
F_k = the kth overall objective, $k = 1, \ldots, n$, where
n = number of overall objectives, and
$\mathbf{F} = (F_1, \ldots, F_n)$ = the objective vector of the overall system.

The overall objectives F_k, as functions of subsystem criteria, are denoted by $F_k(\mathbf{f})$.

In most general cases, there is for each subsystem a decisionmaker who is

†In the common mathematical notation, vectors are considered to be column vectors or $n \times 1$ matrices. Keeping to this notation, one should write, for example, $\mathbf{f} = [(f^1)^T, \ldots, (f^N)^T]^T$ where T denotes a transposition. In this text, the transposition marks are left out when we are dealing with single vectors of objectives. However, in mathematical expressions, vectors are column vectors. So, for example, if $\mathbf{a}$ and $\mathbf{b}$ are n-dimensional vectors, $\mathbf{a}^T\mathbf{b}$ is scalar, but $\mathbf{ab}^T$ is an $n \times n$ matrix.

interested in the objectives of the subsystem. Denote by DM_i the decisionmaker interested in $\mathbf{f}^i$, $i = 1, \ldots, N$. Furthermore, there may be a decisionmaker or a body of decisionmakers who are interested in the overall objectives F_k. Denote by DM_0 the decisionmaker interested in the overall objectives F_k.

In a very general case, with a decisionmaker for each subsystem and a decisionmaker for the overall system, the problem is stated mathematically by writing the objective vectors that need to be minimized one below the other, followed by the systems equations:

$$\text{DM}_0\text{: min} \begin{bmatrix} F_1(\mathbf{f}^1, \ldots, \mathbf{f}^N) \\ \vdots \\ F_n(\mathbf{f}^1, \ldots, \mathbf{f}^N) \end{bmatrix} \tag{4.16}$$

$$\text{DM}_1\text{: min} \begin{bmatrix} f_1^1(\mathbf{x}_1, \mathbf{m}_1, \mathbf{y}_1) \\ \vdots \\ f_{n_1}^1(\mathbf{x}_1, \mathbf{m}_1, \mathbf{y}_1) \end{bmatrix} \tag{4.17a}$$

$$\vdots \qquad\qquad \vdots$$

$$\text{DM}_N\text{: min} \begin{bmatrix} f_1^N(\mathbf{x}_N, \mathbf{m}_N, \mathbf{y}_N) \\ \vdots \\ f_{n_N}^N(\mathbf{x}_N, \mathbf{m}_N, \mathbf{y}_N) \end{bmatrix} \tag{4.17b}$$

subject to

$$\mathbf{y}_i = \mathbf{H}_i(\mathbf{x}_i, \mathbf{m}_i) \qquad i = 1, \ldots, N \tag{4.18}$$

$$\mathbf{x}_i = \sum_{j=1}^{N} C_{ij}\mathbf{y}_j \qquad i = 1, \ldots, N \tag{4.19}$$

$$\mathbf{g}_i(\mathbf{x}_i, \mathbf{m}_i, \mathbf{y}_i) \leq \mathbf{0}, \qquad i = 1, \ldots, N \tag{4.20}$$

$$\sum_{i=1}^{N} \mathbf{q}_i(\mathbf{x}_i, \mathbf{m}_i, \mathbf{y}_i) \leq \mathbf{0} \tag{4.21}$$

Here, Eq. (4.18) represents systems equations for the subsystems. Equation (4.19) represents present coupling among the subsystems. Usually, the C_{ij} matrices, which are here called coupling matrices, consist of zero and unity elements, where unity indicates a connection. The inputs $\mathbf{x}_i$ and outputs $\mathbf{y}_i$ are correspondingly called coupling variables. In the following, C_{ij} matrices may be any constant matrices. The equations represented by Eq. (4.20) specify constraints on the variables of each subsystem. Here $\mathbf{g}_i$ are generally vector-valued functions. The equations of Eq. (4.21), where $\mathbf{q}_i$ are generally vector-valued functions, present an overall constraint on the variables in the subsystems.

From a mathematical point of view, Eq. (4.21) is unnecessary; it can be included in the other equations. To arrive at this conversion, consider an important special case where Eq. (4.21) takes the form

$$\sum_{i=1}^{N} \bar{q}_i(\mathbf{x}_i, \mathbf{m}_i, \mathbf{y}_i) \leq b \tag{4.22}$$

Here $\bar{q}_i$ represents money required by the subsystem i, and b is the total available sum of money. Then, define a new output variable, denoted by $\bar{y}_i$, for each subsystem. This output variable actually indicates the money used by subsystems 1, 2, . . ., i. Define $\mathbf{x}_i$ as an input variable of subsystem i, indicating the money used by the previous subsystems 1, 2, . . ., $(i - 1)$. With this notation and interpretation, an equivalent form of the constraint given by Eq. (4.22) is the following set of equations:

$$\bar{y}_i = \bar{q}_i(\mathbf{x}_i, \mathbf{m}_i, \mathbf{y}_i) + \bar{x}_i \qquad i = 1, \ldots, N \tag{4.23}$$

$$\bar{x}_i = \bar{y}_{i-1} \qquad i = 2, \ldots, N \tag{4.24}$$

$$\bar{x}_1 = 0 \tag{4.25}$$

$$\bar{y}_N \leq b \tag{4.26}$$

These equations have the same form as Eqs. (4.18)–(4.20). Note that Eq. (4.25) is equivalent to the following equations of the same type as Eq. (4.20): $\bar{x}_1 \leq 0$ and $-\bar{x}_1 \leq 0$.

In a similar way, when the q_i constraints are vector-valued we can replace the constraint of Eq. (4.21) by equations of other types in the problem formulation. In the following sections and chapters, sometimes Eq. (4.21) is present in the derivations, which makes it unnecessary to do transformations such as those found in Eqs. (4.23)–(4.26).

Furthermore, note the following interesting and useful fact. Even if there are no physical couplings between the subsystems [i.e., Eq. (4.19) is absent] but there is an overall budget constraint [Eq. (4.22)], this monetary constraint can be transformed to coupling equations between the subsystems [Eq. (4.24)] and to other system and constraint equations for the subsystems.

4.5 VARYING THE NUMBER OF DECISIONMAKERS

In the previous section, a general problem formulation was given. We will also consider two special examples of this case in this chapter. Altogether we have three cases: $N + 1$, 1, and N.

Case N + 1 This is the case formulated in the previous section. A representative example is a regional planning problem where we have N lower-level decisionmakers for subregions and one higher-level decisionmaker (or one body of decisionmakers) for the whole region. In this case, we are interested in the Pareto optimality of all objectives, that is, the vector $(\mathbf{F}, \mathbf{f})$. Namely, let, for example,

$(\mathbf{F}', \mathbf{f}')$ be an inferior vector. Then, there exists a vector $(\mathbf{F}'', \mathbf{f}'')$ whose component objectives are equal to or less than those corresponding objectives of $(\mathbf{F}', \mathbf{f}')$, with at least one objective being strictly smaller. In other words, the vector $(\mathbf{F}'', \mathbf{f}'')$ is equally good as the vector $(\mathbf{F}', \mathbf{f}')$ for some decisionmakers, but it is considered to be better by at least one decisionmaker. When we assume here that the $N + 1$ decisionmakers are independent—each wanting to minimize his or her own objectives—and exclude the pathological case where a decisionmaker wants to hurt others, we see that the inferior vector $(\mathbf{F}', \mathbf{f}')$ is not a reasonable alternative for an analyst to present to the decisionmakers. Therefore, we are interested in Pareto optimal $(\mathbf{F}, \mathbf{f})$ vectors.

Case 1 We have only one decisionmaker (DM_0) for the overall system. This is uncommon in practice, where there is a large, decomposable system, but here it is assumed that the system has only one decisionmaker who is interested in the overall objectives (F_i's) that are functions of some performance criteria (f_j^i's) of the subsystems. In this case, we are interested in the Pareto optimality of $\mathbf{F}$.

Case N The higher-level decisionmaker is absent. There are N conflicting decisionmakers who are interested in their subsystems but are dependent on one another due to couplings of the subsystems. In this case, the F's can formally be identified with the f_j^i's. We are interested in the Pareto optimality of $\mathbf{f}$. Table 4.1 summarizes the three different cases considered here.

Remembering that the overall objective vector $\mathbf{F}$ is a function of the vector of all objectives in the subsystem level, we are first interested in finding out to what extent the Pareto optimality of $[\mathbf{F}(\mathbf{f}), \mathbf{f}]$ and $\mathbf{F}(\mathbf{f})$ can be reduced to the Pareto optimality of $\mathbf{f}$. The two following sections show that, indeed, in many cases, the question of Pareto optimality is reduced to generating noninferior $\mathbf{f}$ vectors. Then, in the main section of this chapter, we show how the noninferior $\mathbf{f}$ vectors can be generated in a decomposable, hierarchical manner.

4.6 CASE $N + 1$: PARETO OPTIMALITY OF ALL OBJECTIVES

Consider the general problem with $N + 1$ decisionmakers formulated in Sect. 4.4. As explained in the previous section, we are now interested in the Pareto optimality

Table 4.1 Different cases with respect to the number of decisionmakers

Case label	Decisionmakers[a]	Pareto optimality of the vector under consideration
$N + 1$	$DM_0, DM_1, \ldots, DM_N$	$(\mathbf{F}, \mathbf{f})$
1	DM_0	$\mathbf{F}$
N	$DM_1, \ldots, DM_N$	$\mathbf{f}$

[a] DM_0 = overall DM; $DM_1, \ldots, DM_N$ = subsystem DMs.

mality of the objective vector (**F**, **f**) and in the conditions under which this Pareto optimality is indicated by the Pareto optimality of **f**. The following theorem will pose a sufficient condition.

For the theorem, define the relations $\leqslant$ and $<$ in $R^n \times R^n$ ($n \geqslant 2$) as follows (iff means if and only if):

$$\mathbf{x} \leqslant \mathbf{y} \qquad \text{iff } x_i \leqslant y_i \qquad i = 1, \ldots, n, \text{ and } x_j < y_j$$

for at least one component;

$$\mathbf{x} < \mathbf{y} \qquad \text{iff } x_i < y_i \qquad i = 1, \ldots, n$$

Theorem 4.1 Assume that a monotonicity condition

$$\mathbf{f} \leqslant \mathbf{f}' \Rightarrow \mathbf{F}(\mathbf{f}) \leqslant \mathbf{F}(\mathbf{f}') \tag{4.27}$$

holds for all feasible **f** and **f**$'$. Then the set of Pareto optimal [**F**(**f**), **f**] vectors is completely determined by Pareto optimal **f** vectors, that is,

$$\{[\mathbf{F}(\mathbf{f}), \mathbf{f}] | [\mathbf{F}(\mathbf{f}), \mathbf{f}] \text{ is Pareto optimal}\} = \{[\mathbf{F}(\mathbf{f}), \mathbf{f}] | \mathbf{f} \text{ is Pareto optimal}\}$$

(A value of **f** is said to be feasible if the corresponding values of its arguments satisfy the constraint and system equations that are present.) To prove the theorem, assume $M = \{[\mathbf{F}(\mathbf{f}), \mathbf{f}] | [\mathbf{F}(\mathbf{f}), \mathbf{f}] \text{ is Pareto optimal}\}$, $N = \{[\mathbf{F}(\mathbf{f}), \mathbf{f}] | \mathbf{f} \text{ is Pareto optimal}\}$. First, take $\mathbf{x} \in M$. Then there is a vector $\mathbf{f}'$ such that $\mathbf{x} = [\mathbf{F}(\mathbf{f}'), \mathbf{f}']$. We claim that $\mathbf{x} \in N$: that is, $\mathbf{f}'$ is Pareto optimal. To get a contradiction, assume that $\mathbf{f}'' \leqslant \mathbf{f}'$. Then $[\mathbf{F}(\mathbf{f}''), \mathbf{f}''] \leqslant [\mathbf{F}(\mathbf{f}'), \mathbf{f}'] = \mathbf{x}$ by the monotonicity assumption. That is, $\mathbf{x} \notin M$, which is a contradiction. Hence, $\mathbf{x} \in N$, and $M \subset N$. Take $\mathbf{y} \in N$. Let $\mathbf{f}'$ be the corresponding objective vector: $\mathbf{y} = [\mathbf{F}(\mathbf{f}'), \mathbf{f}']$. We claim that $\mathbf{y} \in \mathbf{M}$. To get a contradiction, assume that there exists $\mathbf{f}''$ such that $[\mathbf{F}(\mathbf{f}''), \mathbf{f}''] \leqslant [\mathbf{F}(\mathbf{f}'), \mathbf{f}']$. From this equation, it directly follows that $f''_i \leqslant f'_i$. Due to the Pareto optimality of $\mathbf{f}'$, we must have $\mathbf{f}'' = \mathbf{f}'$. Hence, $[\mathbf{F}(\mathbf{f}''), \mathbf{f}''] = [\mathbf{F}(\mathbf{f}'), \mathbf{f}']$, which is a contradiction. Hence, $\mathbf{y} \in M$, and $N \subset M$. In all, $N = M$. This completes the proof.

Note that the above monotonicity condition is valid, if, for example, the F_i's are sums of f_j^i's or products of positive f_j^i's. If the assumption given in Eq. (4.27) is valid, Theorem 4.1 reduces the computational work of generating Pareto optimal points. This work is usually an exponential function of the number of objectives (see Cohon and Marks, 1975). Section 4.8 deals with the question on how the Pareto optimal **f**'s can be calculated utilizing the subsystem structure.

4.7 CASE 1: PARETO OPTIMALITY OF OVERALL OBJECTIVES

Consider the general problem formulation in Sect. 4.4 in the special case where the subsystem decisionmakers $\text{DM}_1, \ldots, \text{DM}_N$ are absent and there is a single decisionmaker DM_0 who is interested in the overall objective vector $\mathbf{F} = (F_1, \ldots, F_n)$. We are interested in conditions under which there is a correspondence between the Pareto optimality $\mathbf{F}(\mathbf{f})$ and $\mathbf{f}$. First, we have the following theorem.

Theorem 4.2 Assuming the monotonicity condition [Eq. (4.27)] holds, a necessary condition for $\mathbf{F}(\mathbf{f})$ to be Pareto optimal is the Pareto optimality of $\mathbf{f}$.

The proof follows directly from the definition of the monotonicity condition. When the monotonicity condition holds, we can thus generate a candidate set $P \triangleq \{\mathbf{F}(\mathbf{f}) | \mathbf{f} \text{ Pareto optimal}\}$ (e.g., by using the results of the next section), and then by direct elimination remove those points from P that are not Pareto optimal.

It is easy to see that, in general, the Pareto optimality of $\mathbf{f}$ does not imply the noninferiority of $\mathbf{F}(\mathbf{f})$. In many practical cases, the overall objectives F_i summarize the values of the f_j^i's. That is, the dimension of the Pareto optimal region related to $\mathbf{f}$ is higher than the dimension of the Pareto region of $\mathbf{F}$, whereby not all Pareto optimal $\mathbf{f}$'s get mapped to Pareto optimal $\mathbf{F}(\mathbf{f})$'s. As an extreme example, consider the case in which the dimension of $\mathbf{f}$ is two. Let $\mathbf{f} = (f_1^1, f_1^2)$; the dimension of $\mathbf{F}$ is one; $\mathbf{F} = f_1^1 + f_1^2$. Let the Pareto optimal set of $\mathbf{f}$ be a curved line (like that of Fig. 4.5). This set gets mapped in the one-dimensional $\mathbf{F}$ space into an interval whose minimum point is the only Pareto optimal point in the $\mathbf{F}$ space.

Next, a result will be derived concerning conditions under which the Pareto optimality of $\mathbf{f}$ implies that of $\mathbf{F}$. Let the dimensions of $\mathbf{f}$ and $\mathbf{F}$ be the same, n. In a compact form, the corresponding multicriteria problems can be stated as follows:

$$\min \mathbf{f}(\mathbf{x}) \qquad \text{subject to} \qquad \mathbf{h}(\mathbf{x}) \leq \mathbf{0} \tag{4.28}$$

$$\min \mathbf{F}(\mathbf{f}(\mathbf{x})) \qquad \text{subject to} \qquad \mathbf{h}(\mathbf{x}) \leq \mathbf{0} \tag{4.29}$$

where $\mathbf{h}$ encompasses the constraints of Eqs. (4.18)–(4.21).

Theorem 4.3 Let $\mathbf{f}(\mathbf{x}^*)$ be a proper Pareto optimal alternative for the problem stated in Eq. (4.28) and let the Kuhn-Tucker necessary and sufficient conditions (of Lemma 4.5) hold at $\mathbf{x}^*$, which is assumed to be regular. Assume that $\mathbf{F}[\mathbf{f}(\mathbf{x})]$ as a function of $\mathbf{x}$ is continuously differentiable and convex and that the dimensions of $\mathbf{F}$ and $\mathbf{f}$ are the same. Then, a sufficient condition for $\mathbf{F}[\mathbf{f}(\mathbf{x}^*)]$ to be properly Pareto optimal for the problem stated in Eq. (4.29) is the strict positivity of the elements of the matrix

$$\left\{ \frac{\partial \mathbf{F}}{\partial \mathbf{f}} [\mathbf{f}(\mathbf{x}^*)] \right\}^{-1}$$

PROOF To prove the theorem, let $\mathbf{f}(\mathbf{x}^*)$ be as assumed. Then, according to the necessary and sufficient conditions, the following equations hold at $\mathbf{x}^*$ (let n be the common dimension of $\mathbf{f}$ and $\mathbf{F}$):

$$\mathbf{w}^T \frac{\partial \mathbf{f}}{\partial \mathbf{x}}(\mathbf{x}^*) + \boldsymbol{\mu}^T \frac{\partial \mathbf{h}}{\partial \mathbf{x}}(\mathbf{x}^*) = 0 \tag{4.30}$$

$$\mathbf{h}(\mathbf{x}^*) \leq 0 \qquad \boldsymbol{\mu}^T\mathbf{h}(\mathbf{x}^*) = 0 \tag{4.31}$$

where

$$\mathbf{w} = (w_1, \ldots, w_n)^T, \text{ with } w_i > 0 \text{ and } i = 1, \ldots, n$$
$$\boldsymbol{\mu} = (\mu_1, \ldots, \mu_m)^T, \text{ with } \mu_i \geq 0 \text{ and } i = 1, \ldots, m$$

Define an n vector:

$$\mathbf{W} = \left(\left\{\frac{\partial \mathbf{F}}{\partial \mathbf{f}}[\mathbf{f}(\mathbf{x}^*)]\right\}^T\right)^{-1} \mathbf{w} \tag{4.32}$$

Note that $\mathbf{W} > \mathbf{0}$ (with all components being strictly positive) by the fact that $\mathbf{w} > \mathbf{0}$ and by the positivity assumption already made (note that, for a matrix A, $[A^T]^{-1} = [A^{-1}]^T$). By solving $\mathbf{w}$ from the definition given in Eq. (4.32) and substituting it in Eq. (4.30), one obtains

$$\mathbf{W}^T \frac{\partial \mathbf{F}}{\partial \mathbf{f}}[\mathbf{f}(\mathbf{x}^*)] \frac{\partial \mathbf{f}}{\partial \mathbf{x}}(\mathbf{x}^*) + \boldsymbol{\mu}^T \frac{\partial \mathbf{h}}{\partial \mathbf{x}}(\mathbf{x}^*) = 0$$

which is, by the chain rule of differentiation, equal to

$$\mathbf{W}^T \left.\frac{\partial \mathbf{F}[\mathbf{f}(\mathbf{x})]}{\partial \mathbf{x}}\right|_{\mathbf{x}=\mathbf{x}^*} + \boldsymbol{\mu}^T \frac{\partial \mathbf{h}}{\partial \mathbf{x}}(\mathbf{x}^*) = 0 \tag{4.33}$$

This equation, together with Eq. (4.31), plus the convexity and differentiability assumption and the fact that $\mathbf{W} > 0$, add up to sufficient conditions for $\mathbf{F}[\mathbf{f}(\mathbf{x}^*)]$ to be properly Pareto optimal. This completes the proof.

Note that Theorem 4.3 concerns the case in which dim $\mathbf{F}$ = dim $\mathbf{f}$ (so that the inverse of a matrix could exist). Hence, two cases are left out: the above-mentioned most common case in practice, where dim $\mathbf{F}$ < dim $\mathbf{f}$, and the rare case in which dim $\mathbf{F}$ > dim $\mathbf{f}$. For an example of Theorem 4.3, consider a two-dimensional case in which $F_1 = af_1^1 + bf_1^2$, and $F_2 = cf_1^1 + df_1^2$ (where a, b, c, and d are constant). Then, for all $\mathbf{x}^*$,

$$\left\{\frac{\partial \mathbf{F}}{\partial \mathbf{f}}[\mathbf{f}(\mathbf{x}^*)]\right\}^{-1} = \begin{bmatrix} a & b \\ c & d \end{bmatrix}^{-1} = \frac{1}{ad - bc}\begin{bmatrix} d & -b \\ -c & a \end{bmatrix}$$

All elements are strictly positive, if $a/D > 0$, $-b/D > 0$, $-c/D > 0$, and $d/D > 0$, where $D = ad - bc$. That is, these equalities are the key conditions of Theorem 4.3 that guarantee that the (F_1, F_2) pair—$\{F_1[f_1^1(\mathbf{x}^*), f_1^2(\mathbf{x}^*)], F_2[f_1^1(\mathbf{x}^*), f_1^2(\mathbf{x}^*)]\}$—is Pareto optimal when the pair $[f_1^1(\mathbf{x}^*), f_1^2(\mathbf{x}^*)]$ is Pareto optimal. Geometrically speaking, the positivity condition of Theorem 4.3, in the two-dimensional case, includes the condition that a Pareto optimal region in the $f_1^1 f_1^2$ space having a southeast slope gets mapped into a region with a southeast slope in the F_1F_2 space; whereas, a geometric interpretation of the monotonicity condition given in Eq. (4.27) is the situation that a vector pointing to the northeast gets mapped into a northeast vector.

Taking into account Theorem 4.2, we note that if the monotonicity condition of Eq. (4.27) and the assumptions of Theorem 4.3 hold, then the Pareto optimal $\mathbf{F}(\mathbf{f})$ vectors are exactly generated by the Pareto optimal $\mathbf{f}$ vectors.

4.8 CASE *N*: PARETO OPTIMALITY OF SUBSYSTEM OBJECTIVES

4.8.1 Introduction

In the case of N conflicting decisionmakers, we are interested in the Pareto optimality of $\mathbf{f}$, the vector of all subsystem objectives. We are particularly interested in the following question. Assume that the N local decisionmakers have solved the Pareto optimal values, denoted by $\mathbf{f}^{i*}$, for their objective vectors, given fixed values of the coupling variables; under what conditions does $\mathbf{f}^* = (\mathbf{f}^{1*}, \ldots, \mathbf{f}^{N*})$ constitute a noninferior solution of $\mathbf{f}$? If this kind of decomposed way of determining Pareto optimal $\mathbf{f}$'s is possible, considerable savings in computer time can be achieved because of the method's exponential dependency on the number of objectives. However, the calculations must be repeated for different values of the coupling variables, which should, therefore, be of low dimensions.

According to the results of the previous sections, in many other situations the question of Pareto optimality reduces to the determination of Pareto optimal $\mathbf{f}$ vectors (see Theorems 4.1–4.3). Therefore, the generating of noninferior $\mathbf{f}$ vectors is basic to many applications.

In this section, we assume the overall constraints of Eq. (4.21) are absent or, if present originally, have been replaced by constraints of the form found in Eqs. (4.18)–(4.20). (See the discussion in Sect. 4.4.)

For an insight into the main results (which are given in Sect. 4.8.3), an informal treatment of the case with two subsystems (i.e., $N = 2$) will first be made in the next subsection.

4.8.2 Case $N = 2$: An Informal Consideration

Consider a case with two subsystems and one coupling variable vector $\mathbf{p} \triangleq \mathbf{y}_2 = \mathbf{x}_1$, as depicted in Fig. 4.7. Let $\mathbf{p}$ be fixed: $\mathbf{p} = \mathbf{p}'$. Then, two separate multiobjective problems are formed. We can calculate Pareto optimal points for the following problems:

$$\min\ (f_1^1, f_2^1, \ldots, f_{n_1}^1) \tag{4.34a}$$

subject to

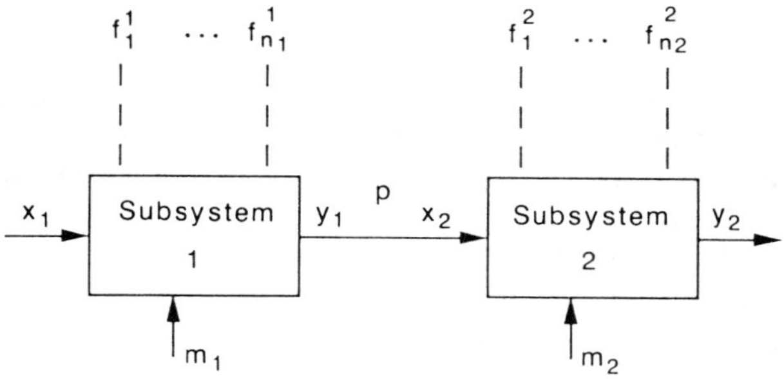

Figure 4.7 Case N with $N = 2$.

$$-\mathbf{y}_1 + H_1(\mathbf{x}_1, \mathbf{m}_1) = \mathbf{0} \tag{4.34b}$$

$$\mathbf{g}_1(\mathbf{x}_1, \mathbf{m}_1, \mathbf{y}_1) \leqslant \mathbf{0} \tag{4.34c}$$

$$-\mathbf{y}_1 + \mathbf{p}' = \mathbf{0} \tag{4.34d}$$

and

$$\min (f_1^2, f_2^2, \ldots, f_{n_2}^2) \tag{4.35a}$$

subject to

$$-\mathbf{y}_2 + H_2(\mathbf{x}_2, \mathbf{m}_2) = \mathbf{0} \tag{4.35b}$$

$$\mathbf{g}_2(\mathbf{x}_2, \mathbf{m}_2, \mathbf{y}_2) \leqslant \mathbf{0} \tag{4.35c}$$

$$-\mathbf{x}_2 + \mathbf{p}' = \mathbf{0} \tag{4.35d}$$

In the following discussion, these subproblems are assumed to be solved by a constraint method with trade-off calculations (as in the SWT method before a decisionmaker's interference). That is, corresponding to the problem given in Eq. (4.34a)–(4.34d), we solve a series of ordinary optimization problems with different upper limits ϵ_i^1 $(i = 2, \ldots, n_1)$:

$$\min f_1^1(\mathbf{x}_1, \mathbf{m}_1, \mathbf{y}_1)$$

subject to

$$f_k^1(\mathbf{x}_1, \mathbf{m}_1, \mathbf{y}_1) \leqslant \epsilon_k^1 \qquad k = 2, \ldots, n_1 \tag{4.34e}$$

and Eqs. (4.34b)–(4.34d).

Denote the Lagrange multipliers related to the parts of Eq. (4.34e) by λ_{1k}^1 $(k = 2, \ldots, n_1)$. Similarly, denote the multipliers related to the upper limits in the corresponding optimization problem of the second subsystem by $\lambda_{1k}^2 (k = 2, \ldots, n_2)$.

Assume that $\mathbf{f}^{1*}$ and $\mathbf{f}^{2*}$ are some Pareto optimal solutions obtained by this kind of optimization, corresponding to some values of the upper limits, the ϵ_k^i's. Let $\boldsymbol{\pi}^y$ and $\boldsymbol{\pi}^x$ be the corresponding multipliers related to constraints given in Eqs. (4.34d) and (4.35d), respectively. We are interested in conditions under which $\mathbf{f}^* \triangleq (\mathbf{f}^{1*}, \mathbf{f}^{2*})$ is noninferior for the whole system.

By informal, first-order considerations, we notice the following results. Consider first a case where p is a scalar. If $\boldsymbol{\pi}^x$ and $\boldsymbol{\pi}^y$ are of the same sign (e.g., positive), then, by increasing the value of p, both f_1^1 and f_1^2 would be decreased (see the usual sensitivity interpretation of Lagrange multipliers in, e.g., Luenberger, 1973). That is, if $\boldsymbol{\pi}^x$ and $\boldsymbol{\pi}^y$ are of the same sign, then $\mathbf{f}^*$ is not Pareto optimal. On the other hand, if the multipliers are of different signs, then $\mathbf{f}^*$ seems to be (locally) noninferior by these first-order considerations.

If the coupling variable $\mathbf{p}$ is a vector, then the relation corresponding to the condition of different signs is the following equation:

$$\boldsymbol{\pi}^y = -c\boldsymbol{\pi}^x \tag{4.36}$$

where c is a strictly positive scalar.

This is seen geometrically in the following way. If $\boldsymbol{\pi}^y$ and $\boldsymbol{\pi}^x$ are nonzero vectors and they are not pointing in opposite directions as in Eq. (4.36), then there exists, in the angle between $\boldsymbol{\pi}^y$ and $\boldsymbol{\pi}^x$, a change in $\mathbf{p}$, $\Delta\mathbf{p}$, such that $(\boldsymbol{\pi}^y)^T\,\Delta\mathbf{p} > 0$ and $(\boldsymbol{\pi}^x)^T\,\Delta\mathbf{p} > \mathbf{0}$. By the sensitivity interpretation, $\Delta f_1^1 = -(\boldsymbol{\pi}^y)^T\Delta\mathbf{p} < \mathbf{0}$ and $\Delta f_1^2 = -(\boldsymbol{\pi}^x)^T\Delta\mathbf{p} < \mathbf{0}$, which indicates that $\mathbf{f}^*$ is not Pareto optimal (if the other components of $\mathbf{f}^1$ and $\mathbf{f}^2$ remain the same).

In a general case with an arbitrary number of subsystems, we can fix the coupling variables and determine the corresponding Lagrange variables. Also in this general case, we then will have algebraic equations, such as Eq. (4.36), between these multipliers, where some scalars should be strictly positive. This general case will be treated in the next section. It includes rigorous mathematical assumptions that we neglected in the informal discussion in this section.

4.8.3 *N* Subsystems with a Multicriteria Problem Formulation

Consider the case with N decisionmakers without overall constraints; that is, we want to determine Pareto optimal vectors $\mathbf{f} = (f_1^1, \ldots, f_{n_1}^1; f_1^N, \ldots, f_{n_N}^N)$ subject to the constraints given in Eqs. (4.18)–(4.20).

The following hierarchical scheme for solving this problem will be considered here. Let us first fix the coupling variables between the subsystems by giving fixed values $\mathbf{p}_i'$ for the outputs of the subsystems: $\mathbf{y}_i = \mathbf{p}_i'$. Then the inputs are also fixed: $\mathbf{x}_i = \Sigma_{j=1}^N C_{ij}\mathbf{p}_j'$. And let us, then, with these constant coupling variables, determine Pareto optimal alternatives for each subsystem problem ($i = 1, \ldots, N$):

$$\min\,(f_1^i, f_2^i, \ldots, f_{n_i}^i) \tag{4.37a}$$

subject to

$$\mathbf{y}_i = \mathbf{H}_i(\mathbf{x}_i, \mathbf{m}_i) \tag{4.37b}$$

$$\mathbf{g}_i(\mathbf{x}_i, \mathbf{m}_i, \mathbf{y}_i) \leq \mathbf{0} \tag{4.37c}$$

$$-\mathbf{y}_i + \mathbf{p}_i' = \mathbf{0} \tag{4.37d}$$

$$-\mathbf{x}_i + \sum_{j=1}^{N} C_{ij}\mathbf{p}_j' = \mathbf{0} \tag{4.37e}$$

For each subproblem, the solution is generally a set of Pareto optimal values. Next—and this pertains to the main results related to this scheme—we want to determine which combinations of these solutions form Pareto optimal alternatives for the overall system. If we can do this, we then can repeat this decomposed way of determining Pareto optimal alternatives for different values of $\mathbf{p}_i$ $(i = 1, \ldots, N)$ to obtain a representative set of the solution set of the overall problem.

Theorem 4.4 presents conditions under which this kind of procedure generates the solution set of the original problem, which is referred to as the overall problem. In Theorem 4.4, a totally convex case is assumed. (The theorem is probably true with lesser convexity assumptions, but an exact formulation is not known at present.) Therefore, the subsystems may be solved by the weighting method. Corollary 4.5 in the following subsection is formulated directly for a similar hierarchical scheme, where the subsystems are solved by the weighting method. (For a scheme in which subproblems are solved by the constraint method, see Tarvainen, 1986.)

Theorem 4.4: Decomposition with Multicriteria Subproblems Consider the overall problem of determining the Pareto optimal alternatives of

$$\mathbf{f} = (\mathbf{f}^1, \ldots, \mathbf{f}^N) = (f_1^1, \ldots, f_{n_2}^1; \ldots; f_1^N, \ldots, f_{n_N}^N)$$

subject to Eqs. (4.18)–(4.20). Assume that the functions f_j^i and $\mathbf{g}_i$ are convex, the functions $\mathbf{H}_i$ are linear, and all functions are continuously differentiable. Let all proper Pareto optimal points of this problem and of every subproblem [see Eq. (4.37)] be regular. Then the set of Pareto optimal alternatives for the overall problem is exactly generated by those Pareto optimal solutions of the subproblems in Eq. (4.37) for which the following condition holds (called coordination condition).

There exist scalars $c_i > 0$, $i = 1, \ldots, N$, such that for some values of $\tilde{\boldsymbol{\pi}}_i^x$ and $\tilde{\boldsymbol{\pi}}_i^y$,

$$c_i(\tilde{\boldsymbol{\pi}}_i^y)^T + \sum_{k=1}^{N} c_k (\tilde{\boldsymbol{\pi}}_k^x)^T C_{ki} = \mathbf{0} \qquad i = 1, \ldots, N \tag{4.38}$$

where $\tilde{\boldsymbol{\pi}}_i^y$ and $\tilde{\boldsymbol{\pi}}_i^x$ are Lagrange multipliers in the Kuhn-Tucker conditions for proper noninferiority associated with the coupling equations of Eqs. (4.37d) and (4.37e), respectively. Here, the multipliers are normalized so that the multiplier associated with the first objective (f_1^i) equals one (that is, $w_1 = 1$). The C_{ki}'s are coupling matrices [see Eq. (4.19)].

Remark 1 Due to the assumptions made in the theorem, the Kuhn-Tucker conditions for proper noninferiority [see Eqs. (4.9)–(4.15)] hold at each subsystem solution. For fixed $\mathbf{p}_i'$ values, there is generally a set of Pareto optimal solutions. Therefore, for each solution, these Kuhn-Tucker conditions hold, where the multipliers associated with the coupling constraints $-\mathbf{y}_i + \mathbf{p}_i' = \mathbf{0}$ and $-\mathbf{x}_i + \Sigma_{j=1}^N C_{ij}\mathbf{p}_j' = \mathbf{0}$ are denoted by $\tilde{\boldsymbol{\pi}}_i^y$ and $\tilde{\boldsymbol{\pi}}_i^x$, respectively. To indicate the dependence on the fixed input and output variables and the particular Pareto optimal solution in question (which depends on $\mathbf{m}_i$), the notations $\boldsymbol{\pi}_i^y(\mathbf{x}_i, \mathbf{m}_i, \mathbf{y}_i)$ and $\boldsymbol{\pi}_i^x(\mathbf{x}_i, \mathbf{m}_i, \mathbf{y}_i)$ can be used.

Remark 2 One can solve the Pareto optimal points of the subproblems by any method. After each solution, the Kuhn-Tucker condition for proper noninferiority can be written down, and one can find multipliers that satisfy these conditions. The multipliers are not always unique; see the example at the end of this subsection. Then one can check whether the critical condition in Eq. (4.38) holds for some values of the $\tilde{\boldsymbol{\pi}}_i^x$ and $\tilde{\boldsymbol{\pi}}_i^y$ multipliers.

Due to the assumptions that have been made, the weighting method (which is computationally most convenient), can be used in determining the Pareto optimal points in the subsystems. Thereby the $\tilde{\boldsymbol{\pi}}_i^x$ and $\tilde{\boldsymbol{\pi}}_i^y$ multipliers can be solved as usual Lagrange multipliers in the minimization of the weighted function. Furthermore, it turns out that, in connection with the weighting method, the possibility that $\tilde{\boldsymbol{\pi}}_i^x$ and $\tilde{\boldsymbol{\pi}}_i^y$ may be nonunique does not bring any complications. The use of the weighting method is assumed in Corollary 4.5 in the next subsection. Corollary 4.5 is thus a very practical version of Theorem 4.4. The merit of Theorem 4.4 is in its theoretical elegance, which comes from the fact that the subproblems are also formulated as multicriteria problems.

Remark 3 Equation (4.38) is homogeneous with respect to the c_i's. That is, if Eq. (4.38) holds for some values of c_i, it also holds for values that are obtained by multiplying or dividing these values by a strictly positive constant. Hence, we can always normalize (e.g., c_1 to be unity).

Remark 4 If a subsystem m ($m \in \{1, \ldots, N\}$) has no output variable $\mathbf{y}_m$, then there is no $\tilde{\boldsymbol{\pi}}_m^y$ variable and Eq. (4.38) for $i = m$ (where $\tilde{\boldsymbol{\pi}}_m^y$ appears) must be left out in the formulation of the theorem. However, formally, we can apply Eq. (4.38) for all $i = 1, \ldots, N$ without paying attention to the missing equations. That is, if $\mathbf{y}_m$ is missing, then all $C_{km} = 0$, $k = 1, \ldots, N$ [see the definition of C in Eq. (4.19)], so that Eq. (4.38) is satisfied trivially for $i = m$, if we ignore the $c_m(\tilde{\boldsymbol{\pi}}_m^y)^T$ term, since $\tilde{\boldsymbol{\pi}}_m^y$ does not exist.

Remark 5 If a subsystem m ($m \in \{1, \ldots, N\}$) has an output $\mathbf{y}_m$ which is not connected to other subsystems, then $C_{km} = 0$, $k = 1, \ldots, N$. Hence, for $i = m$, Eq. (4.38) reads

$$c_m(\tilde{\boldsymbol{\pi}}_m^y)^T = \mathbf{0}$$

Since $c_m > 0$, we obtain the condition $\tilde{\boldsymbol{\pi}}_m^y = \mathbf{0}$. It is therefore quite possible to treat outputs that might be interesting variables but are not connected to other subsystems in the above formalism. However, if an output is not connected to other subsystems, it is usually better to treat it in the same manner as we treat decision variables $\mathbf{m}_i$. In this way, we avoid fixing the levels of the output and calculating the corresponding Lagrange multipliers.

Remark 6 By taking into account the fact (mentioned in Sect. 4.3.4) that the Kuhn-Tucker conditions for single-criterion problems are a direct special case of those for proper noninferiority, one can easily go through the following proof of Theorem 4.4 and see that it holds for the case where there are single-criterion problems in the subsystems.

To prove the theorem, we first show, in proof a, that every properly Pareto optimal alternative for the overall problem can be generated by the subsystem solutions, for which the condition of Eq. (4.38) holds. Second, in proof b, we show that the subsystem solutions for which Eq. (4.38) holds constitute noninferior solutions for the overall problem.

PROOF a Assume we have a properly Pareto optimal solution for the overall problem and the assumptions made in Theorem 4.4 are valid. Then, the Kuhn-Tucker conditions for proper Pareto optimality hold at this solution. These conditions [Eqs. (4.9)–(4.15)] were given for the general problem formulation given by Eqs. (4.6)–(4.8). To write down these Kuhn-Tucker conditions now, note first that $\mathbf{x}$, the vector in the general problem formulation on which the objective functions depend, is in the present case as follows:

$$\mathbf{x} = (\mathbf{x}_1, \mathbf{m}_1, \mathbf{y}_1, \ldots, \mathbf{x}_N, \mathbf{m}_N, \mathbf{y}_N)$$

Furthermore, in the present case, the Lagrangian in Eq. (4.12a) is written in a compact form using inner products of vectors, as follows:

$$\begin{aligned} L(\mathbf{x}) = {} & \sum_{i=1}^{N} (\tilde{\mathbf{w}}^i)^T \mathbf{f}^i(\mathbf{x}_i, \mathbf{m}_i, \mathbf{y}_i) \\ & + \sum_{i=1}^{N} \tilde{\boldsymbol{\delta}}_i^T[-\mathbf{y}_i + \mathbf{H}_i(\mathbf{x}_i, \mathbf{m}_i, \mathbf{y}_i)] \\ & + \sum_{i=1}^{N} \tilde{\boldsymbol{\eta}}_i^T \mathbf{g}_i(\mathbf{x}_i, \mathbf{m}_i, \mathbf{y}_i) \\ & + \sum_{i=1}^{N} \tilde{\boldsymbol{\nu}}_i^T\left(-\mathbf{x}_i + \sum_{j=1}^{N} C_{ij}\mathbf{y}_j\right), \end{aligned}$$

where $\tilde{\mathbf{w}}^i = (\tilde{w}_1^i, \ldots, \tilde{w}_{n_i}^i)$, with $\tilde{w}_j^i > 0$, $i = 1, \ldots, N$.

Here $\tilde{\boldsymbol{\delta}}_i$, $\tilde{\boldsymbol{\eta}}_i$, and $\tilde{\boldsymbol{\nu}}_i$ are vectors, because the corresponding constraint equations may have several components. The components of $\tilde{\boldsymbol{\eta}}_i$ are nonnegative [see Eq. (4.10)]. The condition $\partial L/\partial \mathbf{x} = \mathbf{0}$ will be written down componentwise. We obtain, by direct differentiation (with respect to the component vectors $\mathbf{x}_i$, $\mathbf{m}_i$, and $\mathbf{y}_i$) the following equations:

$$(\tilde{\mathbf{w}}^i)^T \frac{\partial \mathbf{f}^i}{\partial \mathbf{x}_i} + \tilde{\boldsymbol{\delta}}_i^T \frac{\partial \mathbf{H}_i}{\partial \mathbf{x}_i} + \tilde{\boldsymbol{\eta}}_i^T \frac{\partial \mathbf{g}_i}{\partial \mathbf{x}_i} - \tilde{\boldsymbol{\nu}}_i^T = \mathbf{0}$$

$$(\tilde{\mathbf{w}}^i)^T \frac{\partial \mathbf{f}^i}{\partial \mathbf{m}_i} + \tilde{\boldsymbol{\delta}}_i^T \frac{\partial \mathbf{H}_i}{\partial \mathbf{m}_i} + \tilde{\boldsymbol{\eta}}_i^T \frac{\partial \mathbf{g}_i}{\partial \mathbf{m}_i} = \mathbf{0}$$

$$(\bar{\mathbf{w}}^i)^T \frac{\partial \mathbf{f}^i}{\partial \mathbf{y}_i} - \bar{\boldsymbol{\delta}}_i^T + \bar{\boldsymbol{\delta}}_i^T \frac{\partial \mathbf{H}_i}{\partial \mathbf{y}_i} + \bar{\boldsymbol{\eta}}_i^T \frac{\partial \mathbf{g}_i}{\partial \mathbf{y}_i} + \sum_{k=1}^{N} \bar{\boldsymbol{\nu}}_k^T C_{ki} = \mathbf{0}$$

These equations are divided by $\bar{w}_1^i$, so that, for each $i = 1, \ldots, N$, the first $\bar{\mathbf{w}}^i$ multiplier equals one. Denote, for $i = 1, \ldots, N$,

$$\mathbf{w}_s^i = \frac{\bar{\mathbf{w}}^i}{\bar{w}_1^i} \tag{4.39a}$$

$$\boldsymbol{\delta}_i^s = \frac{\bar{\boldsymbol{\delta}}_i}{\bar{w}_1^i} \qquad \boldsymbol{\eta}_i^s = \frac{\bar{\boldsymbol{\eta}}_i}{\bar{w}_1^i} \qquad \boldsymbol{\nu}_i^s = \frac{\bar{\boldsymbol{\nu}}_i}{\bar{w}_1^i} \tag{4.39b}$$

Note that $\boldsymbol{\eta}_i^s \geqslant \mathbf{0}$ due to the facts that $\bar{\boldsymbol{\eta}}_i \geqslant \mathbf{0}$ and $\bar{w}_1^i > 0$. With this notation (where s stands for scaling), the above three equations take the forms of Eqs. (4.40)–(4.42) after being divided by $\bar{w}_1^i$. In the last equation, substitute $\bar{\boldsymbol{\nu}}_k = \bar{w}_1^k \bar{\nu}_k^s$. When all the Kuhn-Tucker conditions are written down, the following set of equations hold at a Pareto optimal point of the overall problem:

$$(\mathbf{w}_s^i)^T \frac{\partial \mathbf{f}^i}{\partial \mathbf{x}_i} + (\boldsymbol{\delta}_i^s)^T \frac{\partial \mathbf{H}_i}{\partial \mathbf{x}_i} + (\boldsymbol{\eta}_i^s)^T \frac{\partial \mathbf{g}_i}{\partial \mathbf{x}_i} - (\boldsymbol{\nu}_i^s)^T = \mathbf{0} \tag{4.40}$$

$$(\mathbf{w}_s^i)^T \frac{\partial \mathbf{f}^i}{\partial \mathbf{m}_i} + (\boldsymbol{\delta}_i^s)^T \frac{\partial \mathbf{H}_i}{\partial \mathbf{m}_i} + (\boldsymbol{\eta}_i^s)^T \frac{\partial \mathbf{g}_i}{\partial \mathbf{m}_i} = \mathbf{0} \tag{4.41}$$

$$(\mathbf{w}_s^i)^T \frac{\partial \mathbf{f}^i}{\partial \mathbf{y}_i} - (\boldsymbol{\delta}_i^s)^T + \boldsymbol{\delta}_i^T \frac{\partial \mathbf{H}_i}{\partial \mathbf{y}_i} + (\boldsymbol{\eta}_i^s)^T \frac{\partial \mathbf{g}_i}{\partial \mathbf{y}_i} + \sum_{k=1}^{N} \frac{\bar{w}_1^k}{\bar{w}_1^i} (\boldsymbol{\nu}_k^s)^T C_{ki} = \mathbf{0} \tag{4.42}$$

$$\mathbf{w}_s^i > \mathbf{0} \tag{4.43}$$

$$(\boldsymbol{\eta}_i^s) \geqslant \mathbf{0} \qquad (\boldsymbol{\eta}_i^s)^T \mathbf{g}_i(\mathbf{x}_i^*, \mathbf{m}_i^*, \mathbf{y}_i^*) = \mathbf{0} \qquad \mathbf{g}_i(\mathbf{x}_i^*, \mathbf{m}_i^*, \mathbf{y}_i^*) \leqslant \mathbf{0} \tag{4.44}$$

$$\mathbf{y}_i^* = \mathbf{H}_i(\mathbf{x}_i^*, \mathbf{m}_i^*) \tag{4.45}$$

$$\mathbf{x}_i^* = \sum_{j=1}^{N} C_{ij} \mathbf{y}_j^* \qquad i = 1, \ldots, N \tag{4.46}$$

In these equations, the derivatives are evaluated at $(\mathbf{x}_i^*, \mathbf{m}_i^*, \mathbf{y}_i^*)$.

Next, let us consider Pareto optimal conditions for proper noninferiority for the subproblems given in Eqs. (4.37a)–(4.37e), when the coupling variables have the same values as in the above overall solution. That is, let $\mathbf{p}_i^{\prime}$ in Eqs. (4.37d) and (4.37e) be equal to $\mathbf{y}_i^*$, $i = 1, \ldots, N$. In this case, the Lagrangian, denoted by L^i for subsystem i, $i = 1, \ldots, N$, is as follows:

$$\begin{aligned} L^i(\mathbf{x}_i, \mathbf{m}_i, \mathbf{y}_i) = (\bar{\mathbf{w}}^i)^T \mathbf{f}^i(\mathbf{x}_i, \mathbf{m}_i, \mathbf{y}_i) &+ \bar{\boldsymbol{\delta}}_i^T[-\mathbf{y}_i + \mathbf{H}_i(\mathbf{x}_i, \mathbf{m}_i, \mathbf{y}_i)] \\ &+ \bar{\boldsymbol{\eta}}_i^T \mathbf{g}_i(\mathbf{x}_i, \mathbf{m}_i, \mathbf{y}_i) + (\bar{\boldsymbol{\pi}}_i^{\mathbf{y}})^T(-\mathbf{y}_i + \mathbf{y}_i^*) \\ &+ (\bar{\boldsymbol{\pi}}_i^{\mathbf{x}})^T\left(-\mathbf{x}_i + \sum_{j=1}^{N} C_{ij} \mathbf{y}_j^*\right) \end{aligned}$$

where $\bar{\mathbf{w}}^i$, $\bar{\boldsymbol{\delta}}_i$, $\bar{\boldsymbol{\eta}}_i$, $\bar{\boldsymbol{\pi}}_i^y$, and $\bar{\boldsymbol{\pi}}_i^x$ are vectors of Lagrange multipliers. Let $\bar{w}_1^i = 1$ for $i = 1, \ldots, N$. Writing down the equations in the Kuhn-Tucker conditions for proper noninferiority, we obtain the following equations for the subsystem $i = 1, \ldots, N$:

$$(\bar{\mathbf{w}}^i)^T \frac{\partial \mathbf{f}^i}{\partial \mathbf{x}_i} + \bar{\boldsymbol{\delta}}_i^T \frac{\partial \mathbf{H}_i}{\partial \mathbf{x}_i} + \bar{\boldsymbol{\eta}}_i^T \frac{\partial \mathbf{g}_i}{\partial \mathbf{x}_i} - (\bar{\boldsymbol{\pi}}_i^x)^T = \mathbf{0} \tag{4.47}$$

$$(\bar{\mathbf{w}}^i)^T \frac{\partial \mathbf{f}^i}{\partial \mathbf{m}_i} + \bar{\boldsymbol{\delta}}_i^T \frac{\partial \mathbf{H}_i}{\partial \mathbf{m}_i} + \bar{\boldsymbol{\eta}}_i^T \frac{\partial \mathbf{g}_i}{\partial \mathbf{m}_i} = \mathbf{0} \tag{4.48}$$

$$(\bar{\mathbf{w}}^i)^T \frac{\partial \mathbf{f}^i}{\partial \mathbf{y}_i} - \bar{\boldsymbol{\delta}}_i^T \frac{\partial \mathbf{H}_i}{\partial \mathbf{y}_i} + \bar{\boldsymbol{\eta}}^T \frac{\partial \mathbf{g}_i}{\partial \mathbf{y}_i} - (\bar{\boldsymbol{\pi}}_i^y)^T = \mathbf{0} \tag{4.49}$$

$$\bar{\mathbf{w}}^i > 0 \tag{4.50}$$

$$\bar{\boldsymbol{\eta}}_i \geqslant \mathbf{0} \qquad \bar{\boldsymbol{\eta}}_i^T \mathbf{g}_i(\mathbf{x}_i, \mathbf{m}_i, \mathbf{y}_i) = \mathbf{0} \qquad \mathbf{g}_i(\mathbf{x}_i, \mathbf{m}_i, \mathbf{y}_i) \leqslant \mathbf{0} \tag{4.51}$$

$$\mathbf{y}_i = \mathbf{H}_i(\mathbf{x}_i, \mathbf{m}_i) \tag{4.52}$$

$$\mathbf{y}_i = \mathbf{y}_i^* \tag{4.53}$$

$$\mathbf{x}_i = \sum_{j=1}^{N} C_{ij} \mathbf{y}_j^* \tag{4.54}$$

We note immediately that it follows directly from Eqs. (4.40)–(4.46) that these equations hold true when we select the values of Lagrange multipliers as follows:

$$\bar{\mathbf{w}}^i = \mathbf{w}_s^i \qquad \bar{\boldsymbol{\delta}}_i = \boldsymbol{\delta}_i^s \qquad \bar{\boldsymbol{\eta}}_i = \boldsymbol{\eta}_i^s \tag{4.55}$$

$$\bar{\boldsymbol{\pi}}_i^x = \boldsymbol{\nu}_i^s \tag{4.56}$$

$$(\bar{\boldsymbol{\pi}}_i^y)^T = -\sum_{k=1}^{N} \frac{\bar{w}_1^k}{\bar{w}_1^i} (\boldsymbol{\nu}_k^s)^T C_{ki} \tag{4.57}$$

and, furthermore, select $\mathbf{x}_i = \mathbf{x}_i^*$ and $\mathbf{m}_i = \mathbf{m}_i^*$.

Note that the selection $\mathbf{x}_i = \mathbf{x}_i^*$ is trivial: when we have fixed $\mathbf{y}_i = \mathbf{y}_i^*$, then it directly follows from the system [Eq. (4.37a)] that $\mathbf{x}_i = \mathbf{x}_i^*$. However, the selection $\mathbf{m}_i = \mathbf{m}_i^*$ is not trivial: Eqs. (4.47)–(4.54) can hold for other values of $\mathbf{m}_i$ (whereby the values of Lagrange multipliers, with which the equations hold, may be different).

In any case, we have noticed that the point $(\mathbf{x}_i^*, \mathbf{m}_i^*, \mathbf{y}_i^*)$ satisfies the Kuhn-Tucker conditions for proper noninferiority in subsystem i. Due to the convexity assumptions that have been made, this implies that $(\mathbf{x}_i^*, \mathbf{m}_i^*, \mathbf{y}_i^*)$ is a proper Pareto optimal point for the ith subsystem problem. That is, if we solve all proper Pareto optimal alternatives for the subsystem problems, then each proper Pareto optimal solution for the overall problem consists of some of the subsystem solutions.

Furthermore, by substituting $\boldsymbol{\nu}_i^s = \bar{\boldsymbol{\pi}}_i^x$ [Eq. (4.56)] into Eq. (4.57) and by multiplying the resulting equation by $\bar{w}_1^i$, one obtains

$$\bar{w}_1^i(\bar{\boldsymbol{\pi}}_i^y)^T = -\sum_{k=1}^{N} \bar{w}_1^k(\bar{\boldsymbol{\pi}}_k^y)^T C_{ki} \tag{4.58a}$$

That is, by selecting

$$c_i = \bar{w}_1^i \ (>0) \qquad i = 1, \ldots, N \tag{4.58b}$$

the condition of Eq. (4.38) holds.

This completes Proof a. We have shown that an arbitrary proper Pareto optimal solution is generated by subsystem solutions, for which the condition Eq. (4.38) holds. In Proof b, we show that if we have proper Pareto optimal subsystem solutions for which Eq. (4.38) holds, these constitute a Pareto optimal solution for the overall system.

PROOF b Assume that we have a proper Pareto optimal solution ($\mathbf{x}_i^*$, $\mathbf{m}_i^*$, $\mathbf{y}_i^*$) for each subproblem $i = 1, \ldots, N$ given by Eqs. (4.37a)–(4.37e). Due to the regularity and differentiability assumptions that have been made, the Kuhn-Tucker conditions hold in each subsystem. That is, Eqs. (4.47)–(4.54) hold, with $\mathbf{x} = \mathbf{x}_i^*$ and $\mathbf{m} = \mathbf{m}_i^*$. In these equations, $\bar{\mathbf{w}}^i$, $\bar{\boldsymbol{\delta}}_i$, $\bar{\boldsymbol{\eta}}_i$, $\bar{\boldsymbol{\pi}}_i^y$ are vectors of Lagrange multipliers with some specific values (that possibly are not unique). Furthermore, let the condition of Eq. (4.38) hold.

To prove that $\mathbf{x}^* \triangleq (\mathbf{x}_1^*, \mathbf{m}_1^*, \mathbf{y}_1^*; \ldots; \mathbf{x}_N^*, \mathbf{m}_N^*, \mathbf{y}_N^*)$ is now a proper Pareto optimal point for the overall system, we must reverse the derivations of Proof a.

Therefore, define $\mathbf{w}_s^i$, $\boldsymbol{\delta}_i^s$, $\boldsymbol{\eta}_i^s$, and $\boldsymbol{\nu}_i^s$ in accordance with Eqs. (4.55) and (4.56). Then, with these selections, it directly follows that the Kuhn-Tucker condition for the overall system Eqs. (4.40)–(4.46) hold, except for Eq. (4.42).

Define the following:

$$\bar{w}_1^i = c_i \qquad i = 1, \ldots, N \tag{4.58c}$$

Then, from the condition of Eq. (4.38), it follows, with the selections already made, that

$$(\bar{\boldsymbol{\pi}}_i^y)^T = -\sum_{k=1}^{N} \frac{\bar{w}_1^k}{\bar{w}_1^i} (\boldsymbol{\nu}_k^s)^T C_{ki} \tag{4.59}$$

Substituting this expression of $\bar{\boldsymbol{\pi}}_i^y$ into Eq. (4.49), we note that Eq. (4.42) holds, too. Hence, we have found multipliers that enable all Pareto optimal conditions to hold. Due to the convexity and differentiability conditions already made, this guarantees that $\mathbf{x}^*$ is a proper Pareto optimal point for the overall system. This completes Proof b and also the proof of Theorem 4.4.

In the above proof, there is one important correspondence to be noted. Namely, in both parts of the proof, the following identification was made [see Eqs. (4.58b) and (4.58c)]:

$$\tilde{w}_1^i = c_i \qquad i = 1, \ldots, N$$

That is, if we have solved the problem by decomposition and evaluated the c_i terms satisfying the coordination condition, we know that the solution that has been obtained corresponds to an overall solution where the weight for the first objective of the ith subsystem $\tilde{w}_1^i$ equals c_i ($i = 1, \ldots, N$). In other words c_i reflects the weight given to the ith subsystem.

In ordinary single-criterion optimization practice, one becomes used to a very common uniqueness of the Lagrange multipliers. In multicriteria problems, Lagrange multipliers are not so often unique. Note that the objective vectors appear in the Kuhn-Tucker conditions for proper noninferiority [Eqs. (4.9)–(4.15)] formally in the same position as the constraint functions. Normally, the constraint functions are quite separate functions and their gradients are linearly independent. The objective functions, however, often partially measure the same things, and it is therefore not unusual for some objective functions to have linearly dependent gradients at some values of the decision vectors. Furthermore, in any case that has several objective functions, the possibility increases that the gradients of the objective and constraint functions are not linearly independent, in which the stationary equation [Eq. (4.12b)] may be satisfied with different values of Lagrange multipliers $\mathbf{w}_i$, $\boldsymbol{\eta}_i$, and $\boldsymbol{\delta}_i$. To illustrate this possibility of nonuniqueness with respect to the key multipliers $\tilde{\pi}_i^x$ and $\tilde{\pi}_i^y$ in Theorem 4.4, consider Example 4.1, depicted in Fig. 4.8.

Let us, for example, write down the Kuhn-Tucker necessary and sufficient conditions of proper noninferiority when the coupling variable equals one ($p_1' = 1$, $y_1 = 1$, $x_2 = 1$) for both subsystems.

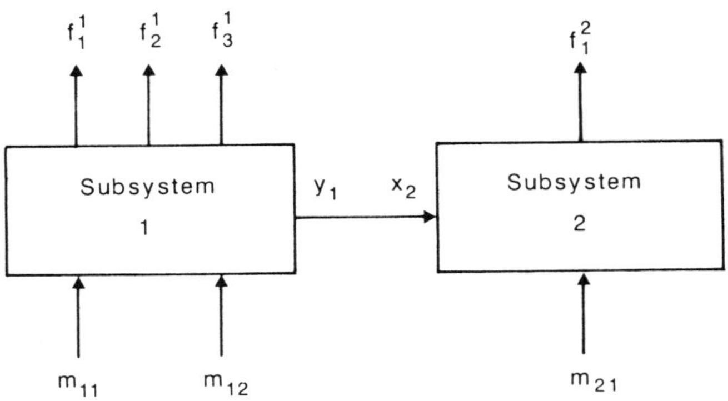

$$y_1 = m_{12} \qquad f_1^1 = 4m_{11} + y_1$$
$$x_2 = y_1 \qquad f_2^1 = m_{11}^2 + y_1^2$$
$$f_3^1 = (m_{11} - 2)^2 + 0.5(y_1 - 2)^2$$
$$f_1^2 = x_2^2 + m_{21}^2$$

Figure 4.8 Example 4.1, a case where the Lagrange multipliers are not unique.

Subsystem 1:

$$L^1 = 4m_{11} + y_1 + \bar{w}_2^1(m_{11}^2 + y_1^2) + \bar{w}_3^1[(m_{11} - 2)^2 + 0.5(y_1 - 2)^2]$$

$$+ \bar{\delta}_1(-y_1 + m_{12}) + \bar{\pi}_1^y(-y_1 + 1)$$

$$\frac{\partial L^1}{\partial m_{11}} = 4 + 2\,\bar{w}_2^1 m_{11} + 2\,\bar{w}_3^1(m_{11} - 2) = 0$$

$$\frac{\partial L^1}{\partial m_{12}} = \bar{\delta}_1 = 0$$

$$\frac{\partial L^1}{\partial y_1} = 1 + 2\,\bar{w}_2^1\, y_1 + \bar{w}_3^1\,(y_1 - 2) - \bar{\delta}_1 - \bar{\pi}_1^y = 0$$

$$y_1 = m_{12}$$

$$y_1 = 1$$

These conditions are satisfied by many solutions. For example, we have the following solution sets A and B:

A: $m_{11} = m_{12} = y = 1 \quad \bar{\delta}_1 = 0 \quad \bar{w}_2^1 = 0.5 \quad \bar{w}_3^1 = 2.5 \quad \bar{\pi}_1^y = -0.5$

B: $m_{11} = m_{12} = y = 1 \quad \bar{\delta}_1 = 0 \quad \bar{w}_2^1 = 2 \quad \bar{w}_3^1 = 4 \quad \bar{\pi}_1^y = 1$

Note that the decision variables and the values of the objectives are the same at these solutions.

Subsystem 2 (see Remark 6, above):

$$L^2 = x_2^2 + m_{21}^2 + \bar{\pi}_2^x(-x_2 + 1)$$

$$\frac{\partial L^2}{\partial m_{21}} = 2m_{21} = 0$$

$$\frac{\partial L^2}{\partial x_2} = 2x_2 - \bar{\pi}_2^x = 0$$

$$x_2 = 1$$

Subsystem 2 has a unique solution: $m_{21} = 0$, $x_2 = 1$, $\bar{\pi}_2^x = 2$.

Coordination condition Normalizing $c_1 = 1$ and noting $C_{21} = 1$ [see the definition of C in Eq. (4.19)], we have Eq. (4.38) in the following form. There exists a scalar $c_2 > 0$ such that, for some values of $\bar{\pi}_2^x$ and $\bar{\pi}_1^y$,

$$\bar{\pi}_1^y + c_2\,\bar{\pi}_2^x = 0$$

Note that we have one equation corresponding to the only output coupling (see Remark 4, above). We note that this condition holds for solution set A in subsystem 1 (whereby $c_2 = 0.25$). This indicates that the solution $m_{11} = 1$, $m_{12} = 1$, and $m_2 = 0$ is a Pareto optimal solution of $(f_1^1, f_2^1, f_3^1, f_1^2)$.

However, for solution set B, this condition cannot be satisfied due to the fact that $\bar{\pi}_1^v = 1$ and $\bar{\pi}_2^v = 2$ are of the same sign. Therefore, solution set B fails to indicate that an overall noninferior solution is at hand.

In the following section, a procedure where the weighting method is used in the subproblems is presented. It is shown that the above difficulty with nonunique multipliers is then avoided. In the light of the above example, we can informally say that the difficulty is avoided for the following reason. When we use the weighting method and let the weights systematically vary over all possible combinations, then in subsystem 1 the weights corresponding to the $\bar{w}$ values of the solution set A are also applied, which ensures that the Pareto optimal point is not missed.

4.8.4 *N* Subsystems Solved by the Weighting Method

Theorem 4.4 presents the general results of generating Pareto optimal alternatives by decomposition. In this theorem, the subproblems have been formulated as multicriteria problems for which Pareto optimal alternatives are to be determined. However, in the theorem, a totally convex case is assumed so that the weighting method can be used in the subsystems. If we explicitly assume that the weighting method is used in the subsystems, Theorem 4.4 can be presented in the form of Corollary 4.5.

Before stating and proving this corollary, which is a practical form of Theorem 4.4, let us recall that, in the overall problem in this chapter, we are minimizing $\mathbf{f}$, the vector of all subsystem objectives, subject to the subsystem and coupling constraints of Eqs. (4.18)–(4.20). In the hierarchical (feasible) approach considered in this chapter, we fix the coupling variables between the subsystems and solve the ensuing subproblems. Then, we check which of these subproblem solutions form a solution for the overall problem. By repeating this procedure for different values of the fixed coupling variables, we attempt to generate overall solutions. When the subproblems are solved by the weighting method, the following subproblems for $i = 1, \ldots, N$ are to be dealt with:

$$\min \sum_{j=1}^{n_i} w_j^i f_j^i \qquad (\text{where } w_1^i = 1) \tag{4.60a}$$

subject to

$$\mathbf{y}_i = \mathbf{H}_i(\mathbf{x}_i, \mathbf{m}_i) \tag{4.60b}$$

$$\mathbf{g}_i(\mathbf{x}_i, \mathbf{m}_i, \mathbf{y}_i) \leq \mathbf{0} \tag{4.60c}$$

$$-\mathbf{y}_i + \mathbf{p}_i' = \mathbf{0} \tag{4.60d}$$

$$-\mathbf{x}_i + \sum_{j=1}^{N} C_{ij}\mathbf{p}_j' = \mathbf{0} \tag{4.60e}$$

Here, the weights are denoted by w_j^i. Since we are interested in a proper noninferior solution, we can use strictly positive weights: $w^i_j > 0$. In Eq. (4.60d), $\mathbf{p}'_i$ is a fixed value for the output of subsystem i. The $\mathbf{p}'_i$ values, then, determine the input values of the subsystems via the couplings [see Eq. (4.60e)]. Note that the weights in each subproblem i are assumed to be normalized by keeping the first weight (w^i_1) as unity [Eq. (4.60a)].

When a completely convex and differentiable case is assumed, the Kuhn-Tucker conditions for single-criterion optimization hold for every regular subsystem solution of Eq. (4.60). In this single-criterion case a solution is defined to be regular in the same manner as in the multiobjective case: the gradients (derivatives) of the binding constraints are linearly independent. In these Kuhn-Tucker conditions, the Lagrange multipliers related to Eqs. (4.60d) and (4.60e) are denoted by $\boldsymbol{\pi}^y_i$ and $\boldsymbol{\pi}^x_i$, respectively. In this case, as will be explained later, $\boldsymbol{\pi}^y_i$ and $\boldsymbol{\pi}^x_i$ are unique for each particular Pareto optimal solution (but will naturally depend on what values of $\mathbf{p}'_i$ are used and on the particular solution indicated by the value of the decision variable $\mathbf{m}_i$).

Corollary 4.5: Decomposition Where Subsystems Are Solved by the Weighting Method Consider the overall problem of determining proper Pareto optimal alternatives of $\mathbf{f} = (\mathbf{f}^1, \ldots, \mathbf{f}^N) = (f^1_1, \ldots, f^1_{n_1}; \ldots; f^N_1, \ldots, f^N_{n_N})$, subject to Eqs. (4.18)–(4.20). Assume that the functions f^i_j and g_i are convex, the functions H_i are linear, and all functions are continuously differentiable. Let all properly Pareto optimal points of this problem and every subproblem [Eqs. (4.60)] be regular. Then the set of proper Pareto optimal alternatives for the overall problem is precisely generated by those proper Pareto optimal solutions of the Eq. (4.60) subproblems for which the following condition holds.

There exist scalars $c_i > 0$, $i = 1, \ldots, N$, such that

$$c_i(\boldsymbol{\pi}^y_i)^T + \sum_{k=1}^{N} c_k(\boldsymbol{\pi}^x_k)^T C_{ki} = \mathbf{0} \qquad i = 1, \ldots, N \tag{4.61}$$

where π^y_i and π^x_i are Lagrange multipliers, in the Kuhn-Tucker conditions for optimality, associated with Eqs. (4.60d) and (4.60e), respectively. The C_{ki}'s are coupling matrices [see Eq. (4.19)].

The proof of this corollary proceeds exactly in the same manner as the proof for Theorem 4.4, with only the following changes to it in the present case, denoted by Proofs α, β, and γ.

PROOF α The beginning of the proof exactly follows Proof a to the point where the Kuhn-Tucker conditions for the subproblems are written down [Eqs. (4.47)–(4.54)]. The Kuhn-Tucker conditions for the subproblems [Eq. (4.60)] are exactly the same as those in Eqs. (4.47)–(4.54) with the following modifications:

α1. $\mathbf{w}^i \triangleq (w^i_1, w^i_2, \ldots w^i_{n_N})$ replaces $\bar{w}^i$.
α2. $\boldsymbol{\pi}^x_i$ and $\boldsymbol{\pi}^y_i$ replace $\bar{\boldsymbol{\pi}}^x_i$ and $\bar{\boldsymbol{\pi}}^y_i$, respectively.
α3. $\boldsymbol{\delta}_i$ and $\boldsymbol{\eta}_i$ are used instead of $\bar{\boldsymbol{\delta}}$ and $\bar{\boldsymbol{\eta}}_i$, respectively.
α4. Eq. (4.50), that is, $\bar{\mathbf{w}}^i > \mathbf{0}$, is now omitted.

In the present case, we note that the subproblem conditions are satisfied by the same selections of multipliers—Eqs. (4.55)–(4.57)—and there are no changes in the rest of Proof a.

PROOF β In Proof b, we use the Kuhn-Tucker conditions for single-criterion optimization. Proof b proceeds unchanged except for the modifications just mentioned as α1–α4.

PROOF γ In the present case, when we solve a particular subproblem [Eq. (4.60)], the π_i^x and π_i^y multipliers are unique. To see this, consider, for example, the Kuhn-Tucker conditions with respect to $\mathbf{x}_i$ [corresponding to Eq. (4.47)]:

$$(\mathbf{w}^i)^T \frac{\partial \mathbf{f}^i}{\partial \mathbf{x}_i} + \boldsymbol{\delta}_i^T \frac{\partial \mathbf{H}_i}{\partial \mathbf{x}_i} + \boldsymbol{\eta}_i^T \frac{\partial \mathbf{g}_i}{\partial \mathbf{x}_i} - (\boldsymbol{\pi}_i^x)^T = \mathbf{0} \tag{4.62}$$

Now, the first term $(\mathbf{w}^i)^T(\partial \mathbf{f}^i/\partial \mathbf{x}_i)$ is fixed. If Eq. (4.62) should hold for two different values of $\boldsymbol{\pi}_i^x$, whereby the values of $\boldsymbol{\delta}_i$ and $\boldsymbol{\eta}_i$ possibly change, then writing down the corresponding equations and subtracting the one from the other would give us an equation that directly contradicts the regularity assumption.

Figure 4.9 summarizes in a schematic way a hierarchical scheme corresponding to Corollary 4.5. In the figure, the case is indicated where the problem is solved for certain representative values of the coupling variables and weights in the subproblems.

4.8.5 Numerical Considerations and Examples

The mathematical theory for generating Pareto optimal solutions by decomposition has been developed in the two previous subsections: in Sect. 4.8.3 for the case where the subsystem problems are formulated as multicriteria problems, and in Sect. 4.8.4 for the case where the subsystem problems are explicitly assumed to be solved by the weighting method. In this subsection, we consider numerical aspects of these theoretical results and give two numerical examples.

Consider how to verify the existence of strictly positive scalars c_i such that Eq. (4.38) or (4.61) holds. Note first that we may have here exactly as many equations as there are unknown scalars, or there may be more equations than unknowns. This can be seen by the following reasoning. Consider first the case where we have N subsystems which are coupled by scalar variables. Then we have equations for c_i, $i = 1, \ldots, N$. Normalizing, for example, $c_1 = 1$, we then have $N - 1$ unknown scalars $c_2, \ldots, c_N$. In the case including the least amount of couplings we have, for example, the subsystem i connected to the subsystem $i + 1$ ($i = 1, \ldots, N - 1$); that is, there are $N - 1$ scalar output variables y_i and correspondingly $N - 1$ scalar equations in Eq. (4.38) or (4.61). In other words, in this kind of scalar serial coupling, we have exactly as many unknowns as equations. But, if the Nth subsystem has been coupled to another subsystem, or the coupling variables are vectors, we have more unknowns than equations.

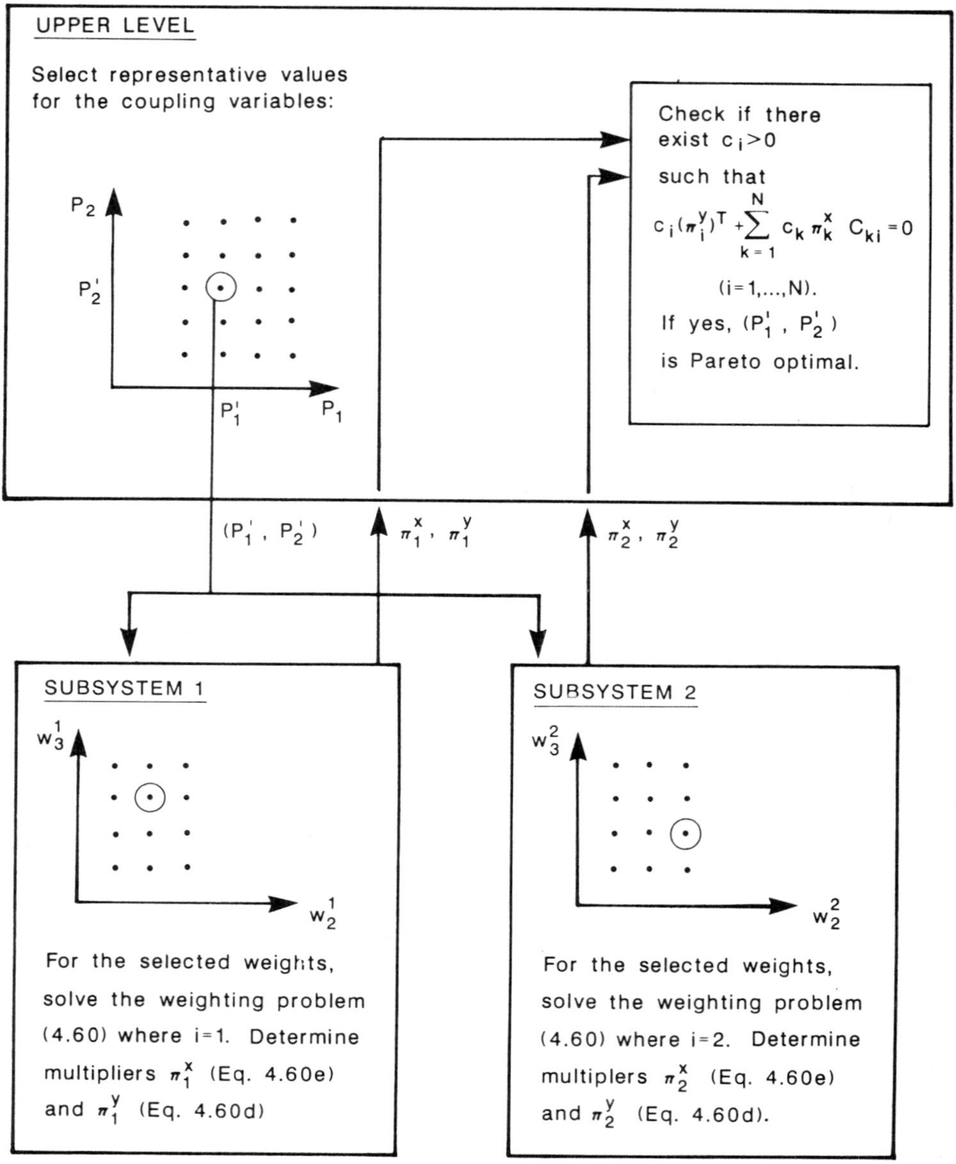

Figure 4.9 Solutions of the problems by weighting method. Discretization of variables is assumed. A case is assumed that has two subsystems ($N = 2$), two coupling variables, and three objectives in each subsystem.

To verify the existence of strictly positive scalars c_i such that Eq. (4.38) or (4.61) holds, we can use the following approaches in these two different cases:

Case a: As many unknowns as equations. If we pay no attention to the signs of the unknowns, we may have no solution, one solution, or an infinite number of solutions. We assume that there exists one solution, which can be determined by ordinary methods of linear algebra. After getting the solution, we just check if its components are strictly positive.

Case b: More unknowns than equations. In this case, we can take from the set of equations as many as there are unknowns. Assuming that this set of equations has a unique solution (whereby the signs of the unknowns are not considered), we can then check whether the solved c_i's are strictly positive. And, finally, we can check whether the remaining equations are satisfied by these c_i's.

For most practical applications, the above approaches are sufficient. Namely, if Eq. (4.38) or (4.61) has strictly positive scalars c_i as solutions, they are likely to be unique (with a normalized $c_i = 1$), because the c_i's correspond to the weights of the first objectives of the subsystems in an overall weighting problem (see the proof of Theorem 4.4). And, it is unlikely that with different values for these weights, we would have the same values for the coupling variables. Now, if the c_i's are unique, the two above approaches will work. For the unlikely case where the solutions are not unique when using these approaches, one should refer to the general literature on solving linear equations.

Example 4.2 corresponds to Case a above; Example 4.3 corresponds to Case b. The first example is solved analytically. The second example is solved numerically, which shows how the above approach is easily modified for the case where the subproblems are solved for some representative values of the coupling variables.

Example 4.2 Consider a system (depicted in Fig. 4.10) with two subsystems coupled by a scalar variable. Let the objective functions have the following expressions, where all variables are scalars:

$$f_1^1(x_1, m_{11}) = (x_1 - 2)^2 + 0.5\, m_{11}^2 \tag{4.63}$$

$$f_2^1(x_1, m_{11}) = 10 + x_1 + m_{11} \tag{4.64}$$

$$f_1^2(y_2, m_{21}, m_{22}) = (y_2 - 3)^2 + 0.5\, m_{21}^2 + 0.5\, m_{22}^2 \tag{4.65}$$

$$f_2^2(y_2, m_{21}, m_{22}) = 10 + m_{21} \tag{4.66}$$

Let

$$y_2 = m_{21} + m_{22} \tag{4.67}$$

For each subsystem, the weighting problem [see Eq. (4.60)] and its solution for a fixed coupling variable ($y_2 = p_2^i$) are as follows.

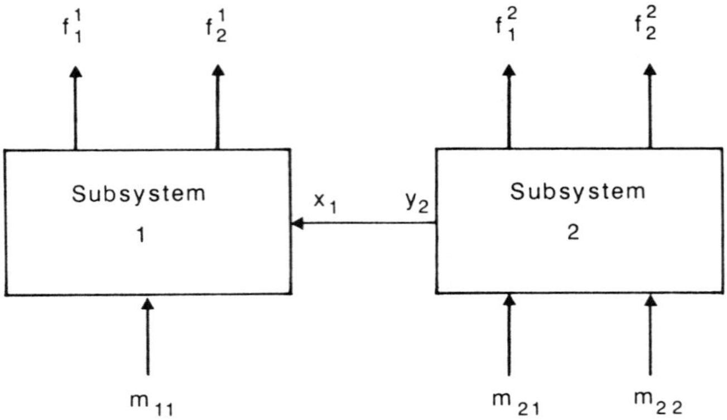

Figure 4.10 Structure of Example 4.2.

Subsystem 1:

$$\min\ (x_1 - 2)^2 + 0.5\, m_{11}^2 + w_2^1(10 + x_1 + m_{11})$$

subject to $-x_1 + p_2' = \mathbf{0}$.

The Lagrangian associated with this problem and the corresponding stationary conditions are as follows:

$$L = (x_1 - 2)^2 + 0.5\, m_{11}^2 + w_2^1(10 + x_1 + m_{11}) + \pi_1^x\,(-x_1 + p_2')$$

$$\frac{\partial L}{\partial x_1} = 2(x_1 - 2) + w_2^1 - \pi_1^x = 0 \tag{4.68}$$

$$\frac{\partial L}{\partial m_{11}} = m_{11} + w_2^1 = 0 \tag{4.69}$$

$$\frac{\partial L}{\partial \pi_1^x} = -x_1 + p_2' = 0 \tag{4.70}$$

Here we have three equations for three unknowns: x_1, m_{11}, and π_1^x. We obtain directly from Eqs. (4.70) and (4.69) and from Eq. (4.68) with $x_1 = p_2'$ from Eq. (4.70) the following unique solutions:

$$x_1 = p_2', \tag{4.71}$$

$$m_{11} = -w_2^1 \tag{4.72}$$

$$\pi_1^x = 2p_2' - 4 + w_2^1 \tag{4.73}$$

Subsystem 2:

$$\min\ (y_2 - 3)^2 + 0.5\, m_{21}^2 + 0.5\, m_{22}^2 + w_2^2(10 + m_{21})$$

subject to

$$-y_2 + p_2' = 0$$

$$-y_2 + m_{21} + m_{22} = 0$$

The Lagrangian and the stationary conditions are now:

$$L = (y_2 - 3)^2 + 0.5\, m_{21}^2 + 0.5\, m_{22}^2 + w_2^2(10 + m_{21}) + \pi_2^y(-y_2 + p_2')$$

$$+ \delta_2(-y_2 + m_{21} + m_{22})$$

$$\frac{\partial L}{\partial y_2} = 2(y_2 - 3) - \pi_2^y - \delta_2 = 0 \qquad (4.74)$$

$$\frac{\partial L}{\partial m_{21}} = m_{21} + w_2^2 + \delta_2 = 0 \qquad (4.75)$$

$$\frac{\partial L}{\partial m_{22}} = m_{22} + \delta_2 = 0 \qquad (4.76)$$

$$\frac{\partial L}{\partial \pi_2^y} = -y_2 + p_2' = 0 \qquad (4.77)$$

$$\frac{\partial L}{\partial \delta_2} = -y_2 + m_{21} + m_{22} = 0 \qquad (4.78)$$

These equations, which are easily solved, have the following unique solutions:

$$y_2 = p_2' \qquad (4.79)$$

$$m_{21} = 0.5p_2' - 0.5w_2^2 \qquad (4.80)$$

$$m_{22} = 0.5p_2' + 0.5w_2^2 \qquad (4.81)$$

$$\delta_2 = -0.5p_2' - 0.5w_2^2 \qquad (4.82)$$

$$\pi_2^y = 2.5p_2' - 6 + 0.5w_2^2 \qquad (4.83)$$

Coordination The process of determining coupling variables for the subsystem and checking which one of the subsystem solutions constitutes a proper Pareto optimal solution for the whole system is called coordination. In this example, it is easy to solve the subproblems analytically for each coupling variable. Checking which of the subsystem solutions form an overall solution is also done analytically for this case.

In this example, only one output (y_2) is connected to other subsystems. Therefore, Eq. (4.61) consists of a single equation (see Remark 4 following Theorem 4.4). When we normalize $c_1 = 1$ (note that, here, $C_{21} = 1$), the following condition will guarantee an overall solution:

There exists $c_2 > 0$ such that

$$\pi_2^y + c_2\, \pi_1^x = 0 \qquad (4.84)$$

where [see Eqs. (4.83) and (4.73)]

$$\pi_2^y = 2.5p_2' - 6 + 0.5w_2^2 \tag{4.85}$$

$$\pi_1^x = 2p_2' - 4 + w_2^1 \tag{4.86}$$

In other words, Eq. (4.84) means that π_2^y and π_1^x must be of different signs.

To take an inductive point of view, set, for example, $p_2' = 1$. Then, $\pi_2^y = -3.5 + 0.5w_2^2$ and $\pi_1^x = -2 + w_2^1$. That is, $\pi_2^y > 0$ when $w_2^2 > 7$; and $\pi_1^x > 0$ when $w_2^1 > 2$. Figure 4.11 depicts different sign combinations for the strictly positive weights w_2^2 and w_2^1.

A figure such as Fig. 4.11 often gives useful information about Pareto optimal alternatives with regard to the coupling variables. For instance, it is possible to see from Fig. 4.11, that, when $p_2' = 1$, we will have a proper Pareto optimal solution for the whole system in two cases: (a) when the second objective in the first subsystem is weighted relatively lightly ($w_2^1 < 2$) but the second objective in the second subsystem is weighted heavily ($w_2^2 > 7$); and (b) when $w_2^1 > 2$ and $w_2^2 < 7$, which is the reverse of the first situation.

The figures for values of p_2' other than 1 will resemble Fig. 4.11, where the point marked by P separates the two shaded areas. When we change the value $p_2' = 1$, point P moves. Note that the coordinates of P are those values of w_2^1 and w_2^2, for which $\pi_2^y = 0$ and $\pi_1^x = 0$: that is, Eqs. (4.85) and (4.86). Solving w_2^1 and w_2^2 from these equations, one obtains, for the w_2^1 and w_2^2 coordinates of P, as a function of p_2',

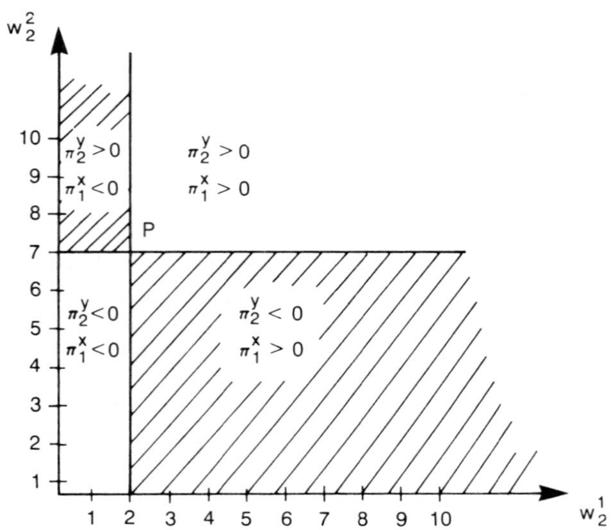

Figure 4.11 Signs of π_2^y and π_1^x, when $p_2' = 1$. Four different areas are shown. In the shaded areas, the signs are different, and they correspond to the proper Pareto optimal values.

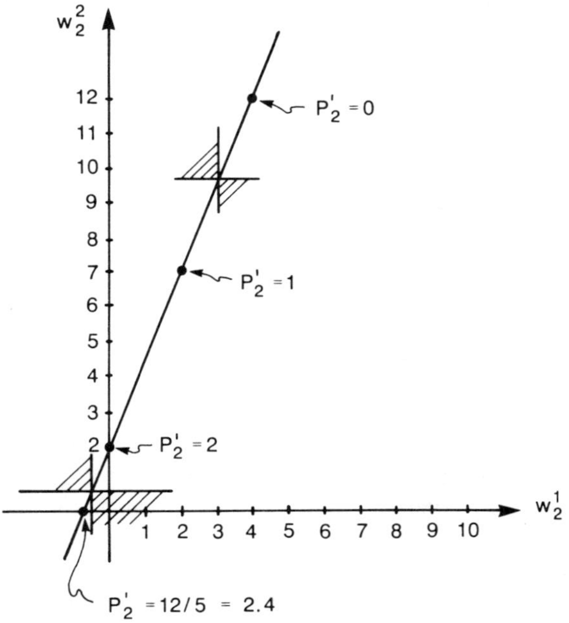

Figure 4.12 Movement of point P as a function of p_2' and the corresponding shaded areas where the signs of π_2^x and π_1^x are different.

$$P_{w_2^1}(p_2') = 4 - 2p_2' \tag{4.87}$$

$$P_{w_2^2}(p_2') = 12 - 5p_2' \tag{4.88}$$

Figure 4.12 shows the corresponding trajectory of P as a function of p_2'. From Fig. 4.12, one sees that the situation is similar to that of Fig. 4.11 (where $p_2' = 1$) for all $p_2' < 2$: there are two different shaded proper Pareto optimal areas.

However, if $p_2' \in [2, 2.4]$, there will be only one shaded area where both w_2^1 and w_2^2 are strictly positive as assumed by the weighting method. That is, when $p_2' \in [2, 2.4]$, we have a proper Pareto optimal solution for relatively small values of w_2^2 ($w_2^2 < 12 - 5p_2'$).

Furthermore, when $p_2' \geqslant 2.4$, then there are no shaded areas that have strictly positive w_2^1 and w_2^2. That is, for $p_2' \geqslant 2.4$, there is no proper Pareto optimal solution for the overall system.

The proper Pareto optimal solutions shown in Table 4.2 summarize the above considerations. Here, the values of objective functions are calculated by their expressions [Eqs. (4.63)–(4.66)] using the subsystem solutions [Eqs. (4.71)–(4.72) and (4.79)–(4.81)].

In some cases, the values of c_2 may be of interest. In the proof of Theorem 4.4, we observed that c_i is the weight given for the first objective of subsystem i ($i = 1, \ldots, N$) in the overall optimization. In this example, we have normalized $c_1 = 1$, so that c_2 reflects directly the weight of f_1^2 relative to that of f_1^1. If there were, in fact, a decisionmaker for each subsystem, then the value of c_2 would

Table 4.2 Proper Pareto optimal solutions for Example 4.2

Value of the coupling variable p_2	Values of the weights for second objectives in the first subsystem (w_2^1) and in the second subsystem (w_2^2)	Values of the objective functions
$p_2 < 2$	$0 < w_2^1 < 4 - 2p_2$ and $w_2^2 > 12 - 5p_2$ or $w_2^1 > 4 - 2p_2$ and $0 < w_2^2 < 12 - 5p_2$	$f_1^1 = (p - 2)^2 + 0.5(w_2^1)^2$ $f_2^1 = 10 + p_2 - w_2^1$ $f_1^2 = 1.25(p_2)^2 - 6p_1 + 0.25\,(w_2^2)^2 + 9$ $f_2^2 = 10 + 0.5p_2 - 0.5w_2^2$
$2 \leqslant p_2 < 2.4$	$w_2^1 > 0$ and $0 < w_2^2 < 12 - 5p_2$	
$p_2 \geqslant 2.4$	No proper Pareto optimal solutions	

indicate how much the second subsystem has been weighted. That is, the higher the value of c_2, the more importance must be given to subsystem 2.

In Fig. 4.11, the shaded areas correspond to strictly positive c_2 values for the case when the coupling variable $p_2' = 1$. If the actual values of c_2 are of interest, we can indicate these in the same figure in the following way. Substituting the expressions of π_2^y [Eq. (4.83)] and π_1^x [Eq. (4.73)] into Eq. (4.84), defining c_2, and solving for w_2^2, one obtains

$$w_2^2 = (-2c_2)\, w_2^1 + (8c_2 - 4c_2p_2' - 5p_2' + 12) \tag{4.89}$$

That is, for constant values of p_2' and c_2, w_2^2 is a linear function of w_2^1. Figure 4.13 depicts corresponding lines for a couple of c_2 values for the case $p_2' = 1$, whereby $w_2^2 = -2c_2w_2^1 + 4c_2 + 7$ by Eq. (4.89).

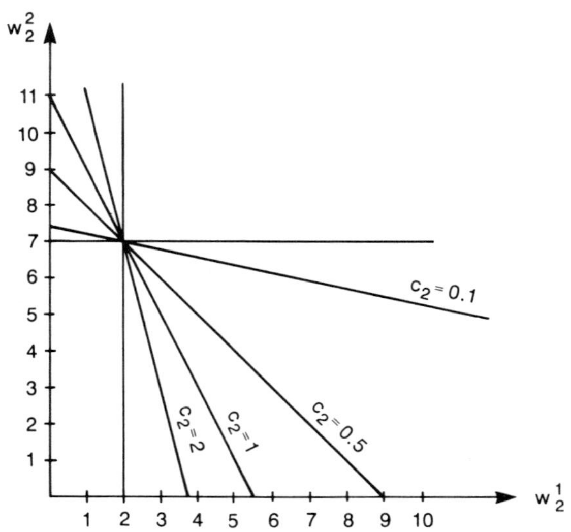

Figure 4.13 Values of c_2, reflecting the relative weight of subsystem 2, for the case when the coupling variable $p_2 = 1$.

Example 4.3 The previous example was solved analytically. This example demonstrates how the coordination condition can be checked without analytic solutions for the subsystems as a function of the coupling variables. This situation corresponds to a likely case in practice: the subsystems are solved for a finite set of representative values of the coupling variables. The subsystems in this example are solved analytically for each particular value of the coupling variables, but these solutions could easily be replaced by numerical solutions. In particular, note that the Lagrangian multipliers $\boldsymbol{\pi}_i^x$ and $\boldsymbol{\pi}_i^y$ can be estimated numerically by solving the subsystems for slightly perturbed values of the coupling variables (the sensitivity interpretation of the Lagrange multipliers; see, e.g., Luenberger, 1973).

This example also shows how to proceed when there are more equations than unknown c_i scalars in the coordination condition (see Case b in the beginning of this subsection). Moreover, this example demonstrates how the theoretical results directly apply for the case where there is a single objective in a subsystem.

Consider a system, depicted in Fig. 4.14, where there are two subsystems and two coupling variables between the subsystems. Let the objective functions and the system equations have the following expressions:

$$f_1^1(x_1, y_1, m_{11}, m_{12}) = (m_{11} + 2)^2 + (m_{12} + 1)^2 + (0.1x_1 - 10)^2 \tag{4.90a}$$

$$f_2^1(x_1, y_1, m_{11}, m_{12}) = (m_{11} + 4)^2 \tag{4.90b}$$

$$f_1^2(x_2, y_2, m_{21}, m_{22}) = (m_{21} - 5)^2 + (m_{22} - 1)^2 + 0.005(0.1x_2 - 6)^2 \tag{4.90c}$$

$$y_1 = -10m_{11} - 20m_{12} \tag{4.90d}$$

$$y_2 = 10m_{21} + 40m_{22} \tag{4.90e}$$

The subsystem solutions follow. They are begun by directly formulating the Lagrangian.

Subsystem 1:

$$L^1 = (m_{11} + 2)^2 + (m_{12} + 1)^2 + (0.1x_1 - 10)^2 + w_2^1 (m_{11} + 4)^2$$
$$+ \pi_1^x(-x_1 + p_2') + \pi_1^y(-y_1 + p_1') + \delta_1(-y_1 - 10m_{11} - 20m_{12})$$

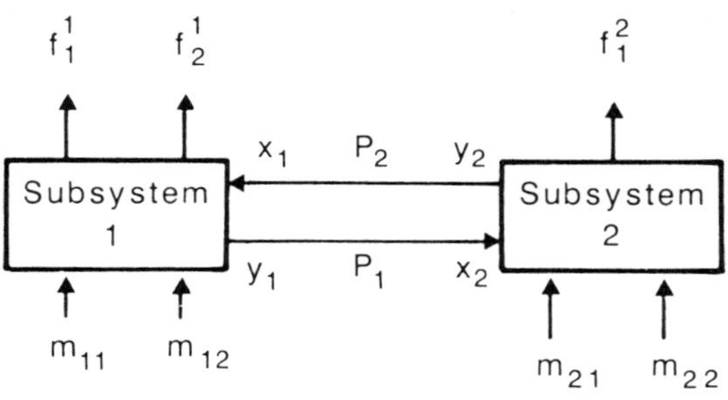

Figure 4.14 Structure of Example 4.3.

$$\frac{\partial L^1}{\partial x_1} = 0.2(0.1x_1 - 10) - \pi_1^x = 0$$

$$\frac{\partial L^1}{\partial y_1} = -\pi_1^y - \delta_1 = 0$$

$$\frac{\partial L^1}{\partial m_{11}} = 2(m_{11} + 2) + 2w_2^1(m_{11} + 4) - 10\delta_1 = 0$$

$$\frac{\partial L^1}{\partial m_{12}} = 2(m_{12} + 1) - 20\delta_1 = 0$$

$$\frac{\partial L^1}{\partial \pi_1^x} = -x_1 + p_2' = 0$$

$$\frac{\partial L^1}{\partial \pi_1^y} = -y_1 + p_1' = 0$$

$$\frac{\partial L^1}{\partial \delta_1} = -y_1 - 10m_{11} - 20m_{12} = 0$$

These equations are easily solved by the following (along with $y_1 = p_1'$, $x_1 = p_2'$, $\delta_1 = -\pi_1^y$):

$$m_{11} = \frac{-6 - 16w_2^1 - 0.1p_1'}{5 + 4w_2^1} \tag{4.91}$$

$$m_{12} = \frac{-0.2p_1' - 0.2p_1'w_2^1 + 3 + 8w_2^1}{5 + 4w_2^1} \tag{4.92}$$

$$\pi_1^x = 0.02p_2' - 2 \tag{4.93}$$

$$\pi_1^y = \frac{0.02p_1' + 0.02p_1'w_2^1 - 0.8 - 1.2w_2^1}{5 + 4w_2^1} \tag{4.94}$$

Subsystem 2:

$$L^2 = (m_{21} - 5)^2 + (m_{22} - 1)^2 + 0.005(0.1x_2 - 6)^2 + \pi_2^x(-x_2 + p_1')$$
$$+ \pi_2^y(-y_2 + p_2') + \delta_2(-y_2 + 10m_{21} + 40m_{22})$$

$$\frac{\partial L^2}{\partial x_2} = 0.001(0.1x_2 - 6) - \pi_2^x = 0$$

$$\frac{\partial L^2}{\partial y_2} = -\pi_2^y - \delta_2 = 0$$

$$\frac{\partial L^2}{\partial m_{21}} = 2(m_{21} - 5) + 10\delta_2 = 0$$

$$\frac{\partial L^2}{\partial m_{22}} = 2(m_{22} - 1) + 40\delta_2 = 0$$

$$\frac{\partial L^2}{\partial \pi_2^x} = -x_2 + p_1' = 0$$

$$\frac{\partial L^2}{\partial \pi_2^y} = -y_2 + p_2' = 0$$

$$\frac{\partial L^2}{\partial \delta_2} = -y_2 + 10m_{21} + 40m_{22} = 0$$

These equations have the following unique solutions (along with $x_2 = p_1'$, $y_2 = p_2'$, $\delta_2 = -\pi_2^y$:

$$m_{21} = \frac{0.1p_2' + 76}{17} \tag{4.95}$$

$$m_{22} = \frac{0.4p_2' - 19}{17} \tag{4.96}$$

$$\pi_2^x = 0.0001p_1' - 0.006 \tag{4.97}$$

$$\pi_2^y = \frac{0.02p_2' - 1.8}{17} \tag{4.98}$$

Coordination If $c_1 = 1$ is normalized, then the coordination conditions [Eq. (4.61)] take the following form. There exists a scalar $c_2 > 0$ such that

$$\pi_1^y + c_2\pi_2^x = 0 \tag{4.99}$$

$$c_2\pi_2^y + \pi_1^x = 0 \tag{4.100}$$

where π_1^y, π_2^x, π_2^y and π_1^x are given by Eqs. (4.94), (4.97), (4.98), and (4.93), respectively.

In other words, the subsystem solutions for given p_1', p_2', and w_2^1 form a proper Pareto optimal solution for the overall system if the corresponding π values satisfy Eqs. (4.99) and (4.100) for some strictly positive scalar c_2.

In this case, we have more equations than unknowns. Let us proceed as explained in the beginning of this subsection. That is, let us solve c_2 from Eq. (4.99). In this example, it is easily seen that a possible solution for c_2 in Eq. (4.99) is unique. A nonunique solution would give that $\pi_2^x = 0$ and $\pi_1^y = 0$. From Eq. (4.97) it follows that $\pi_2^x = 0$, provided that $p_1' = 60$. Substituting $p_1' = 60$ into

Eq. (4.94), one obtains $\pi_1^y = 0.4/(5 + 4w_2^1)$. Thus, π_2^x and π_1^y cannot be zero at the same time. Hence, a solution of c_2 for Eq. (4.99) is given by

$$c_2 = \frac{-\pi_1^y}{\pi_2^x} \tag{4.101}$$

One must then check, first, whether $c_2 > 0$ and, second, whether Eq. (4.100) is satisfied. That is, it must be determined whether

$$-\frac{\pi_1^y}{\pi_2^x}\pi_2^y + \pi_1^x = 0 \tag{4.102}$$

We will first follow this procedure, analytically. Then we will simulate a practical case, where the subproblems have been solved for a representative set of coupling variables and the coordination will be carried out numerically.

Analytical solution of coordination Substituting the expressions of π_1^y [Eq. (4.94)] and π_2^x [Eq. (4.97)] into the expression of c_2 [Eq. (4.101)], one obtains

$$c_2 = \frac{-0.02p_1' - 0.02p_1'w_2^1 + 0.8 + 1.2w_2^1}{(5 + 4w_2^1)(0.0001p_1' - 0.006)} \tag{4.103}$$

To obtain the region where $c_2 > 0$, let us first solve the points where the numerator and the denominator are zero:

Numerator: $-0.02p_1' - 0.02p_1'w_2^1 + 0.8 + 1.2\,w_2^1 = 0$

$$\Rightarrow w_2^1 = \frac{0.02p_1' - 0.8}{-0.02p_1' + 1.2} \tag{4.104}$$

Denominator: $(5 + 4w_2^1)(0.0001p_1' - 0.006) = 0$

$$\Rightarrow p_1' = 60 \qquad (\text{Note: } w_2^1 > 0)$$

When these solutions are drawn in a (p_1', w_2^1) plane, direct inspection of the values of the signs reveals that the numerator and the denominator are of the same sign (positive)—and thus $c_2 > 0$—in the shaded area of Fig. 4.15.

Next, by substituting the expression of π_2^y [Eq. (4.98)] and π_1^x [Eq. (4.93)] into Eq. (4.100), one obtains

$$p_2' = \frac{34 + 1.8c_2}{0.34 + 0.02c_2} \tag{4.105}$$

That is, given a (p_1', w_2^1) point in the shaded area of Fig. 4.15 with a strictly positive c_2, one obtains from Eq. (4.105) a p_2' value for this c_2 value such that the triple (p_1', p_2', w_2^1) satisfies the coordination equation.

Let us take a closer look at the possible values of the triple (p_1', p_2', w_2^1)—and especially the pair (p_1', p_2'). First, note that the curved left boundary of the shaded area in Fig. 4.15 corresponds to $c_2 = 0$. Thus, according to Eq. (4.105),

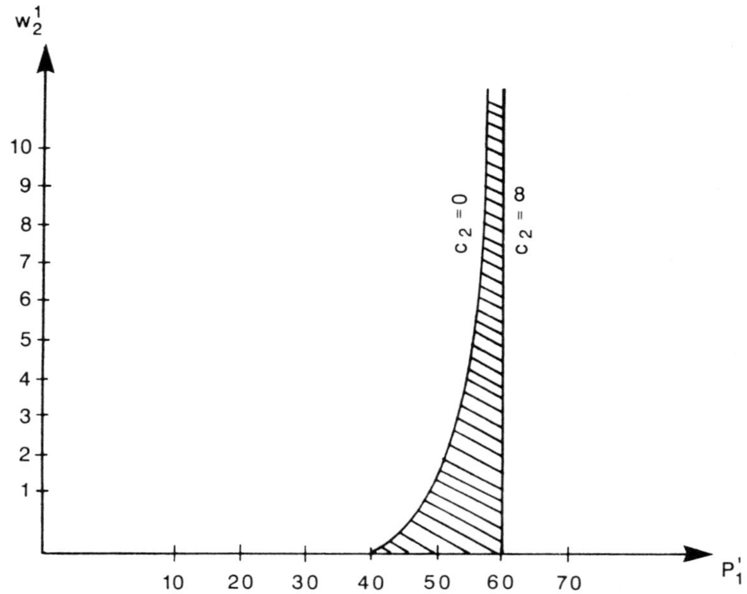

Figure 4.15 Region for positive c_2 values.

on this whole curve, $p_2' = (34/0.34) = 100$. On the curve in question, p_1' varies from 40 to 60. So, in a (p_1', p_2') plane, the horizontal line in Fig. 4.16 arises.

The right boundary corresponds to $c_2 = \infty$. Thus, according to Eq. (4.105), on this whole line, $p_2' = (1.8/0.02) = 90$. The whole right boundary gets mapped to point $p_1' = 60$ and $p_2' = 90$ in Fig. 4.16.

On the base line (interval $40 \leq p_1' \leq 60$), the value of w_2^1 is zero. Substituting $w_2^1 = 0$ into the expression of c_2 [Eq. (4.103)], gives the following values of c_2 on this baseline:

$$c_2 = \frac{-0.02p_1' + 0.8}{0.00005p_1' - 0.03}$$

Substituting this expression into Eq. (4.105) gives the following values of p_2' on this baseline:

$$p_2' = \frac{-0.019p_1' + 0.42}{-0.0023p_1' + 0.0058} \tag{4.106}$$

This equation gives the left boundary curve shown in Fig. 4.16.

When the boundaries of the shaded area of Figure 4.15 are taken into consideration, it is easy to see that the inner points of the area will be mapped between the lines in Fig. 4.16. The area surrounded by the solid line in Fig. 4.16 marks the coupling variables p_1' and p_2' for which there are proper Pareto optimal solutions for the overall system. Note that the boundaries of this area are not proper

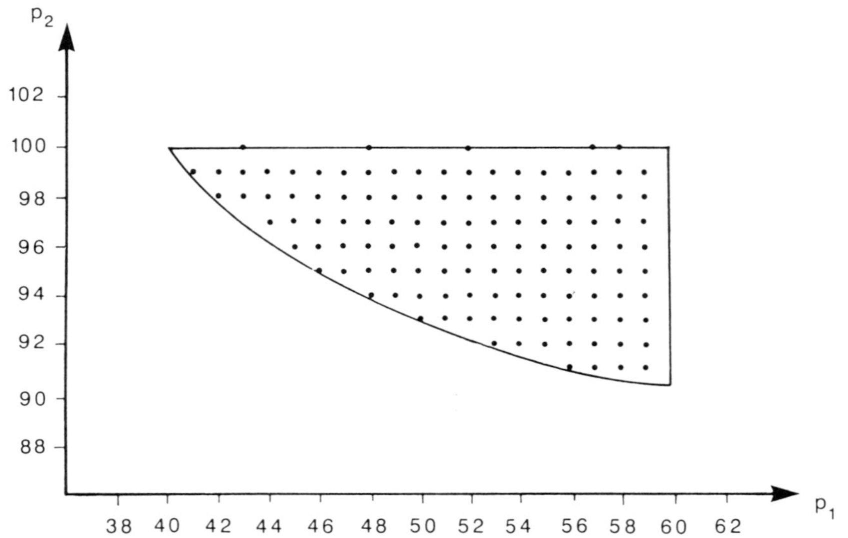

Figure 4.16 Proper Pareto optimal coupling values p'_1 and p'_2 in Example 4.3. The area surrounded by a solid line represents the analytical solution. The boundaries of this region do not belong to proper Pareto optimal solutions. The dots show discrete solutions obtained by the numerical solution procedure.

Pareto optimal values, since they correspond to values $c_2 = 0$ or ∞, and $w_2^1 = 0$ or ∞.

In rare practical cases, analytic solutions may be derived. It was possible to do so in this example, and this provides a reference point for the following numerical solution, where the values of the objective functions and c_2 and w_2^1 will be explored in more detail.

Numerical solution of coordination Let us imitate a possible practical case where, first, a central authority gives the subsystems a discrete set of possible values for the coupling variables. After this, the subsystems solve their proper Pareto optimal objectives for each value of the coupling variable vector for a set of weights. Finally, based on the coordination conditions, the central authority determines which of the subsystem solutions form a proper Pareto optimal for the overall system. This set, then, can be a starting point for further negotiations between the central authority and the decisionmakers in the subsystems on a final preferred compromise.

In principle, the numerical checking of coordination conditions [Eqs. (4.99) and (4.100)] proceeds as in the analytic solution procedure: solve c_2 from Eq. (4.99); check its positiveness and see whether Eq. (4.100) is satisfied. However, because we are dealing here with a finite set of values for coupling variables and subsystem weights, Eq. (4.100) will be satisfied only by chance. That is, we must check whether Eq. (4.100) has been satisfied for intermediate values. To satisfy this requirement, the following procedure, using simple interpolation, was used the solve this example:

1. Select a (p_1', p_2') pair.
2. Select a sequence of w_2^1 values indexed by integers $k = 0, \ldots, W$, and identified by $w_2^1(k) = k\Delta w_2^1$. That is, we consider weights $0, \Delta w_2^1, 2\Delta w_2^1, \ldots, W\Delta w_2^1$.
3. For each $w_2^1(k)$, $k = 0, \ldots, W$, calculate c_2 from Eq. (4.99) denoted by $c_2(k)$ [see Eq. (4.103)]:

$$c_2(k) = \frac{-0.02p_1' - 0.02p_1'w_2^1(k) + 0.8 + 1.2w_2^1(k)}{[5 + 4w_2^1(k)](0.0001p_1' - 0.006)} \tag{4.107}$$

   Furthermore, calculate the related equation error of Eq. (4.100), denoted by $\epsilon(k)$: $\epsilon(k) = c_2(k)[(0.02p_2' - 1.8)/17] + 0.02p_2' - 2$.
4. Interpolation: If there occurs a sign change in the sequence $\epsilon(0), \epsilon(1), \ldots, \epsilon(W)$, do the following interpolation; otherwise start from the beginning with a new (p_1', p_2') pair if one still exists. Mathematically, the change of sign in the ϵ sequence can be detected at an index $K \in \{0, 1, 2, \ldots, (W - 1)\}$ for which $\epsilon(k)\ \epsilon(K + 1) \leq 0$. By linear interpolation, $\epsilon(k)$ varies in the interval $[K, K + 1]$ according to $\epsilon(k) = \epsilon(K) + (K - K)[\epsilon(K + 1) - \epsilon(K)]$, where k now stands also for real values, not only for integers. Solving the value of k, denoted by $\bar{K}$, for which $\epsilon(k) = 0$, yields

$$\bar{K} = K - \frac{\epsilon(K)}{\epsilon(K + 1) - \epsilon(K)} \tag{4.108}$$

   provided the denominator is not zero. The corresponding interpolated values of $w_2^1(k)$ and $c_2(k)$ are

$$w_2^1(\bar{K}) = \bar{K}\Delta w_2^1$$

$$c_2(\bar{K}) = c_2(K) + (\bar{K} - K)[c_2(K + 1) - c_2(K)]$$

5. Check whether $c_2(\bar{K})$ and $w_2^1(\bar{K})$ are strictly positive. If this is the case, accept the triple $p_1', p_2', w_2^1(\bar{K})$ as a proper Pareto optimal value. Repeat the procedure from the beginning point 1 if new (p_1', p_2') pairs still exist.

Figure 4.17 presents solutions obtained by this algorithm. The proper Pareto optimal values for coupling variables p_1 and p_2 are also marked in Fig. 4.16. From Fig. 4.16, we see that the numerical solution approximates very well the solution region for p_1 and p_2 obtained by the analytic solution. However, due to discretization errors, some boundary points that are Pareto optimal but not proper Pareto optimal are included in the numerical solution set.

In Fig. 4.18, the corresponding values for objective functions f_1^1, f_2^1, and f_2^2 are tabulated. In this diagram, single-figure values for f_1^2 are marked on the $f_1^1 f_2^1$ plane. This example was also solved without decomposition with the weight-

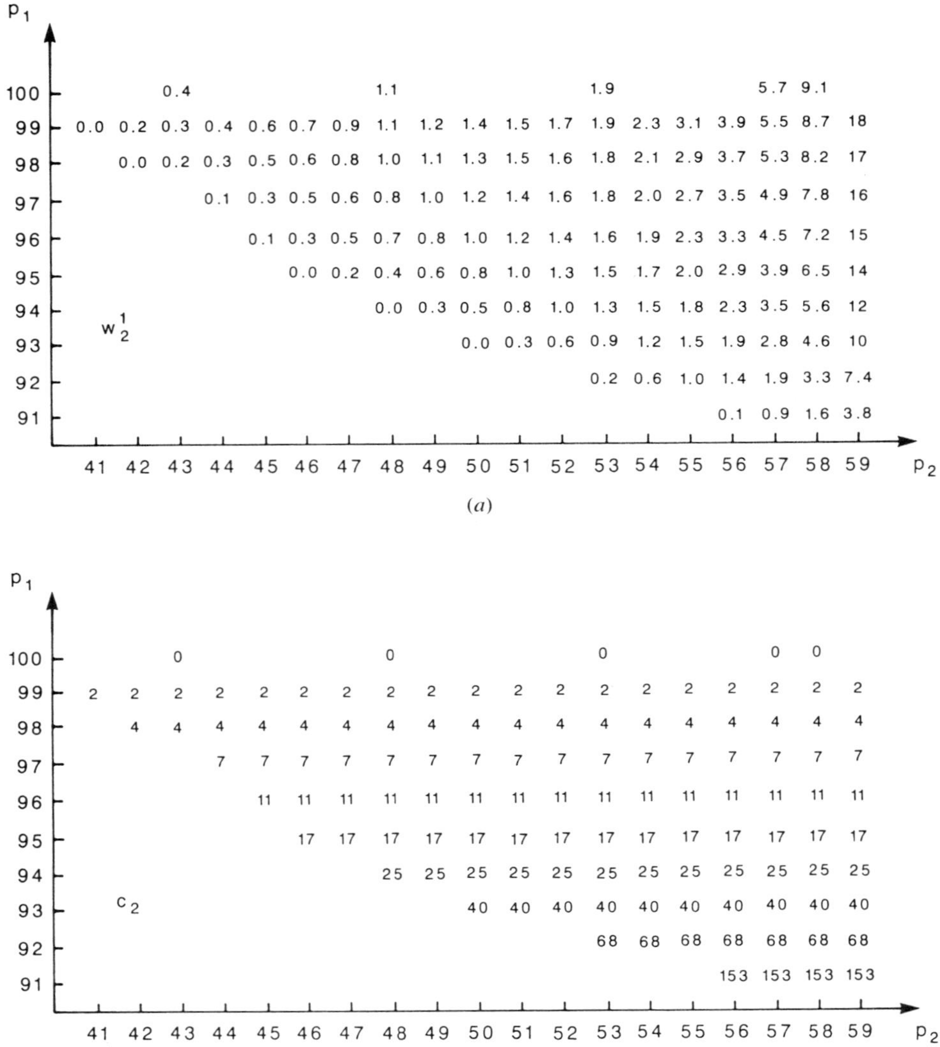

Figure 4.17 Values for proper Pareto optimal solutions reached by the numerical solution procedure in Example 4.3. (*a*) The interpolated weights w_2^1 for the first subsystem for the proper Pareto optimal variables p_1 and p_2. The discretized values for w_2^1 were 0, 0.2, 0.4, ..., 20. (*b*) Corresponding c_2 values indicating a weight for the second subsystem.

ing method. By extensively varying the weight, a similar diagram was drawn for the properly Pareto optimal values of the three objectives (lower part of Fig. 4.18). From this we see that the numerical solution algorithm has produced an approximation of the proper Pareto optimal solution set that is probably good enough for a practical application.

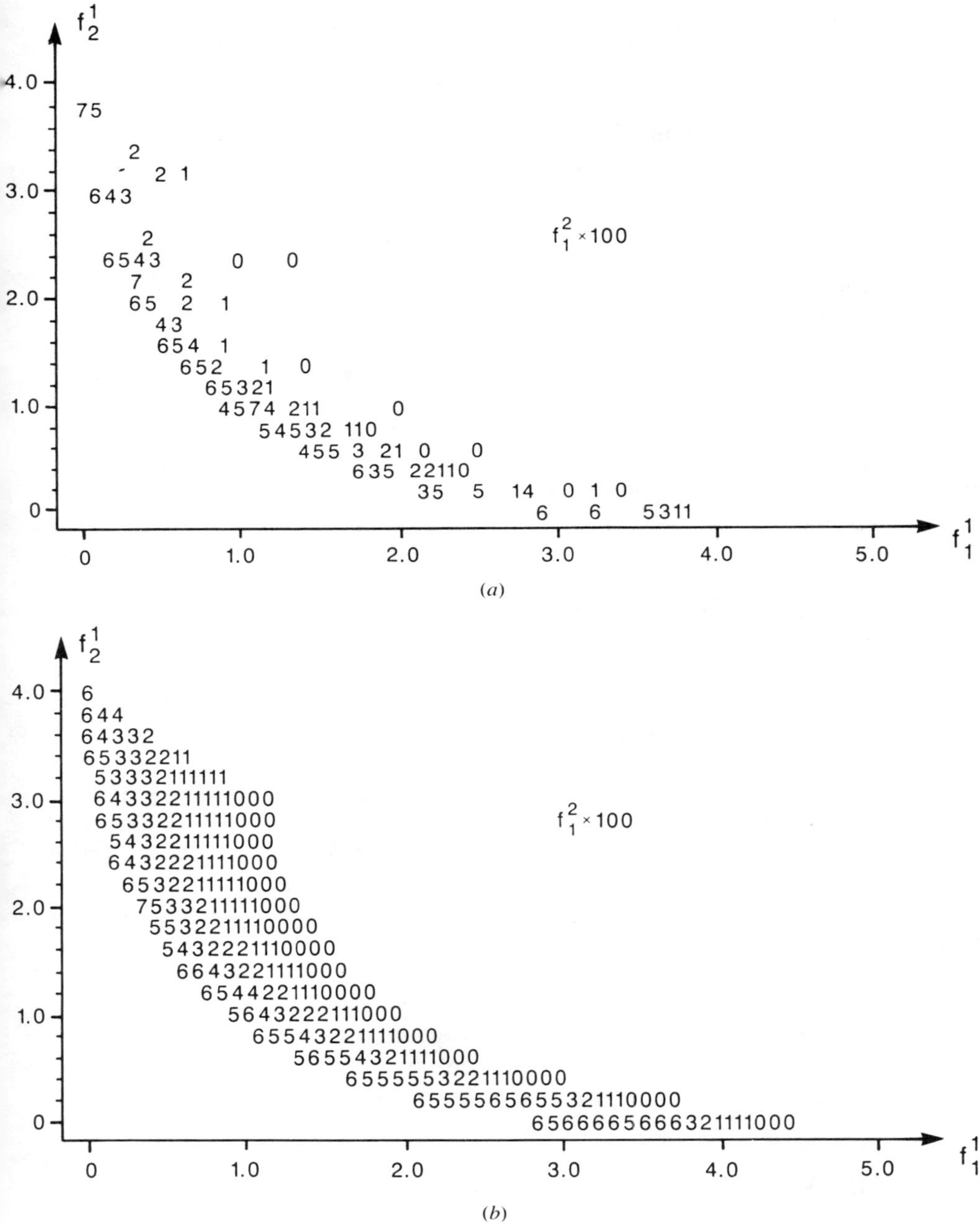

Figure 4.18 Proper Pareto optimal values of f_1^2 multiplied by 100 and given with one digit accuracy on the $f_1^1 f_2^1$ plane. (*a*) Numerical solution by decomposition (see text). (*b*) Solution by extensively varying weights in the weighting method without decomposition. Due to the rounding-off errors in placing the values of f_1^2 in the above figures, some values seem not to be locally Pareto optimal.

In this numerical solution, 11 discrete values for the weight w_2^1 were used: 0, 0.2, 0.4, ..., 2.0. In the example, the solution set was not sensitive to the discretization of w_2^1. Even by a discretization of w_2^1 with four values (0, $6\frac{2}{3}$, $13\frac{1}{3}$, 20), a good approximation of the solution set was obtained.

In connection with these simple examples demonstrating the theoretical developments, the following general remark on the use of hierarchical method is made. If one looks from a purely mathematical point of view at the subproblems of Example 4.3, they are of about the same mathematical complexity as the overall problem. This is due to the fact that the coupling variables provide additional parameters for each subsystem. Furthermore, the coordination needs computation. In all, from a purely computational point of view, decomposition brings no reduction in total computation time in a simple problem like Example 4.3. It is therefore advisable not to use decomposition whenever it is technically possible. Note, however, that decomposition makes parallel computation possible, which generally reduces the computation time.

It can easily be reckoned that the benefits from reducing the total computation time become notable when the complexity of the original problem increases, when the number of the subsystems increases, or when the number of couplings between the subsystems decreases. From a purely computational point of view, a serial system where the couplings are scalar is most suitable for decomposition. (This is actually the kind of system that is most suitable for dynamic programming, too.)

In general, the main advantages of hierarchical methods are not purely computational. Other advantages that are relevant in the hierarchical schemes considered in this chapter are the same that are listed in the attributes of decomposition in Chap. 3. In this connection, let us point out the following features. The main importance of the methods derived in this chapter lies in the possibility of using them in hierarchical organizations with multiple decisionmakers, as indicated in connection with Example 4.3. Hereby it is possible to pay attention to the values of the coupling variables and a kind of relative weights of the subsystems. The upper-level decisionmaker may consider just these factors and need not know the details of the subsystem optimizations.

Even if there is only one decisionmaker for the whole problem, the decisionmaker may get more insight into the system via the decomposition. The modular structure of the hierarchical schemes helps, for example, in keeping up with changes in the models of the subsystems. A change in a subsystem model does not affect the optimization of the other subsystems.

4.9 SUMMARY

In this chapter, methods for determining Pareto optimal solutions by decomposition have been derived along with a theoretically elegant scheme in which the subproblems are multicriteria problems. A more practical version, where the subproblems are assumed to be solved by the weighting method, was also derived and demonstrated by two numerical examples. We gave a derivation for the relations between the Pareto optimality of different possible objective vectors. In many cases the question of Pareto optimality reduces to Pareto optimality of the lower-level objectives.

The schemes derived are suitable for practical cases, where the subsystems are solved numerically for some representative values of the coupling variables. Recent results, where an analytic solution of the subsystem solutions as a function of the coupling variables is assumed, can be found in recent papers by Li and Haimes (1987a,b).

This chapter focused on decomposition methods where the subsystem structure and the multicriteria problem setting in each subsystem are preserved. The end of the previous section discussed how the derived schemes are suitable, for example, for hierarchical organizations. An alternative approach to the decomposition of multicriteria problems can be explained in one sentence: Namely, we can apply (for instance) the weighting method to the overall problem and solve the resulting single-criterion problem by the hierarchical methods explained in Chap. 3. In this approach, called the straightforward approach, the possible benefits of the decomposition are purely computational. This approach does not preserve the structure of the multicriteria subproblems and it does not lend itself to a hierarchy of decisionmakers.

REFERENCES

Chankong, V., and Haimes, Y. Y., *Multiobjective Decision Making: Theory and Methodology*, North-Holland, New York, 1983.

Cohon, J. L., and Marks, D. H., A Review and Evaluation of Multiobjective Programming Techniques, *Water Resources Res.*, vol. 2, no. 2, pp. 208–220, 1975.

Ester, J., Riedel, C., and Peschel, M., Polioptimal Approach to the Control of Hierarchical Multilevel Systems, in *Proc. 6th World Congress, Int. Federation of Automatic Control*, Boston, 1975.

Geoffrion, A. M., Proper Efficiency and Theory of Vector Maximization, *J. of Math. Anal. and Appl.*, vol. 22, pp. 618–630, 1968.

Kuhn, H. W., and Tucker, A. W., Nonlinear Programming, in *Proc. 2d Berkeley Symp. Mathematics, Statistics, and Probability*, pp. 481–492, University of California Press, Berkeley, Calif., 1951.

Li, D., and Haimes, Y. Y., The Envelope Approach for Multiobjective Optimization Problem, *IEEE Trans. Syst., Man, Cybern.*, vol. SMC-17, no. 6, 1987a.

Li, D., and Haimes, Y. Y., A Hierarchical Generating Method for Large Scale Multiobjective Systems, *J. Optimization Theory Appl.*, vol. 54, pp. 303–334, 1987b.

Luenberger, D. G., *Introduction to Linear and Nonlinear Programming*, Addison-Wesley, Reading, Mass., 1973.

Sawaragi, Y., Nakayama, H., and Tanino, T., *Theory of Multiobjective Optimization*, Academic Press, New York, 1985.

Straszak, W., Polioptimization of Large-scale Systems with Multilevel Control Structure, in *Proc. 5th World Congress, Int. Federation of Automatic Control*, 1972.

Tarvainen, K., On the Generating of Pareto Optimal Alternatives in Large Scale Systems, in *Proc. 4th IFAC Symposium on Large Scale Systems, Zurich, Switzerland*, 1986.

Varaiya, P. P., *Notes on Optimization*, Van Nostrand-Reinhold, New York, 1972.

CHAPTER

FIVE

HIERARCHICAL METHODS FOR DETERMINING PREFERRED SOLUTIONS

5.1 INTRODUCTION

Hierarchical methods that generate Pareto optimal sets were considered in the previous chapter. Although the methods have several good features, they may generate sets that are too large to handle in the communication between the analyst and the decisionmaker.

This chapter considers interactive multiobjective optimization methods. With these methods, the entire Pareto optimal set is not generated. Instead, an interactive method typically proceeds in the following manner. First, the decisionmaker is given a Pareto optimal solution and some information on the adjacent Pareto optimal solutions. Based on this information, the decisionmaker is either satisfied with the proposed solution or selects a nearby alternative. In the latter case, he or she is given local information around the new provisional solution. This process continues until the decisionmaker is satisfied.

In the interactive methods, we avoid generating the complete Pareto optimal set. However, this situation requires much more communication between the decisionmaker and the analyst. In some applications, this kind of decisionmaker participation is not possible. If this is the case, the analyst has to use, for example, the methods of Chap. 4, whereby a rough presentation of the Pareto optimal set is generated.

Even if an interactive method is used, it often must be stopped before the final preferred solution is found, in order not to bother the decisionmaker with too much time-consuming detail. In such a case, we are dealing with a suboptimal solution. In many practical applications of the interactive method, only a few iterations are actually needed. This is due to the fact that, in the multiobjective case, decisionmakers are generally faced with a set of alternatives that they are not able to put in order of preference. (The related concept of an indifference band was defined in Chap. 2.)

Since a preferred solution is selected in interactive methods, there is now a need to specify the authority relationships among several decisionmakers, if several are present. The starting point for the derivations of methods in this chapter is the authority of the decisionmaker (DM_0, using the notation of Chap. 4). That is, it is assumed that the decisionmaker who is interested in the overall objectives determines the preferences to be used in the subsystem optimizations.

Technically, this starting point means that the hierarchical schemes are derived on the basis that a single decisionmaker would select the same preferred solution when using these schemes as he or she would select when solving the problem as a whole without decomposition. In other words, when the derived schemes are directly used in an organization with several decisionmakers at different levels, the preferences of the lower-level decisionmakers will actually be determined by the upper-level decisionmaker. The preferences are transferred between the levels in the form of trade-offs.

Even though the multilevel algorithms developed in this chapter are based on the case of a single decisionmaker, the results are also useful in a real, multiple-decisionmaker case, as will be explained in the next chapter.

Like the single-objective hierarchical schemes in Chap. 3, the multiobjective methods are grouped here into two major classes: feasible and nonfeasible. In feasible methods, the intermediate results satisfy the system constraints; in nonfeasible methods, this is not the case. Note, however, that in nonfeasible methods the control or decision variables usually satisfy the constraints during the iteration and are thus feasible all the time.

In the schemes in this chapter, we assume that there is a multicriteria problem at hand. For example, the problem could be the making of a budget for the coming year. Multilevel algorithms are derived to solve this kind of a single, one-time or off-line mathematical problem. In applying the algorithms, it takes some time to transfer information between the different levels. But it should be noted that the algorithms are derived for off-line multicriteria problems.

The converse of the off-line problem is the on-line problem. In on-line problems, there is no well-defined mathematical problem at the outset. Instead, the interaction between the different levels includes system state estimation, mathematical identification of the system model, refinement of preferences, and so on. In practice, problems are often of on-line nature. Their mathematical analysis as on-line problems may incorporate many aspects and it is difficult to obtain general results. (For on-line coordination in the single-objective case, see Findeisen et al., 1980.) In many practical applications, it is best to break down an inherently on-line problem into separate problems like optimization and identification, which are carried out from time to time. The multilevel methods in this chapter are applicable for this kind of separate multiobjective problems.

In addition to the methods in this chapter, the straightforward approach for deriving multiobjective hierarchical methods, which was mentioned in Sect. 4.9, can be used in some cases. That is, one can apply an ordinary multiobjective optimization method to the whole problem. As almost every multiobjective optimization method converts the original problem into a series of single-objective problems, one can then apply the ordinary multilevel techniques discussed in Chap. 3 to the single-objective problems. In this way, we can derive a multiobjective hierarchical algorithm for almost every multiobjective optimization technique. In this straightforward approach, the decomposition usually serves for only the computational aspects; the structure of subsystems with multiple objectives is not preserved.

In the next section, we derive a general necessary condition for the preferred solution. The derivations of hierarchical schemes for preferred solutions are based on this condition.

5.2 GENERAL NECESSARY CONDITION FOR THE PREFERRED SOLUTION

Several studies deal with the necessary mathematical conditions for noninferior points (see, e.g., Kuhn and Tucker, 1951; DaCunha and Polak, 1967; Geoffrion, 1968). In the following discussion, a condition of the same form is derived. Because it incorporates the preferences of a decisionmaker, it is, moreover, a necessary condition for the preferred solution. This condition will be used in developing decomposition and coordination schemes.

To derive this necessary condition, the existence of an underlying (but not necessarily known) utility function is assumed. (For a derivation without an assumption about the utility function, see Tarvainen, 1981.) In the following discussion, the utility function is used as a theoretical vehicle; the decisionmaker is assumed to express his or her preference using trade-offs.

Ordinarily, one uses a utility function that is to be maximized, but here the corresponding loss function (obtained by multiplying the utility function by -1) is to be minimized. This is done for the mathematical reason of avoiding confusion with the signs of the Lagrangian multipliers.

Let us first rederive the mathematical relationship between utility functions and indifference trade-offs (see, e.g., Geoffrion et al., 1972). Let $U(f_1, f_2, \ldots, f_n)$ be a loss function of a decisionmaker. Let $\lambda_{ij}^*(\mathbf{f})$ be an indifference trade-off between objectives f_i and $f_j (i, j = 1, \ldots, n, i \neq j)$ at $\mathbf{f} \overset{\Delta}{=} (f_1, \ldots, f_n)^T$. The relationship between U and the λ_{ij}^*'s is derived by the following differential considerations.

Assume that the alternative $(f_1, f_2, \ldots, f_n)$ is as good as the corresponding alternative where f_i has changed by Δf_i, and f_j has changed by Δf_j, the other objectives remaining the same. That is, in terms of the loss function,

$$\Delta U = \frac{\partial U}{\partial f_i}(\mathbf{f})\,\Delta f_i + \frac{\partial U}{\partial f_j}(\mathbf{f})\,\Delta f_j = 0 \tag{5.1}$$

Let us assume that we are minimizing the objectives and Δf_i is positive. Then Δf_j is the degradation of the objective f_j. Since Δf_j compensates this degradation, Δf_i must be a negative number. Hence, the improvement (a positive number) is $-\Delta f_i$. Therefore (and similarly if Δf_j is negative), the indifference trade-off between f_i and f_j is

$$\lambda_{ij}^*(\mathbf{f}) = -\frac{\Delta f_i}{\Delta f_j} \tag{5.2}$$

From Eq. (5.1), this ratio can be solved, and one obtains

$$\lambda_{ij}^*(\mathbf{f}) = \frac{\partial U(\mathbf{f})/\partial f_j}{\partial U(\mathbf{f})/\partial f_i} \tag{5.3}$$

This is an important relation, and it will be used in converting theoretical results concerning U to more practical results that are given in terms of trade-offs.

Consider the following general multiobjective problem:

$$\min_{\mathbf{z}} \begin{bmatrix} F_1(\mathbf{z}) \\ \vdots \\ F_n(\mathbf{z}) \end{bmatrix} \tag{5.4a}$$

subject to

$$\mathbf{w}(\mathbf{z}) \leq \mathbf{0} \qquad \mathbf{r}(\mathbf{z}) = \mathbf{0} \tag{5.4b}$$

where $F_1, \ldots, F_n$, $\mathbf{w}$, and $\mathbf{r}$ are continuously differentiable functions. For brevity, define $\mathbf{F}(\mathbf{z}) \triangleq [F_1(\mathbf{z}), \ldots, F_n(\mathbf{z})]^T$.

Assume that there exists an underlying differentiable loss (negative utility) function U: $R^n \rightarrow R$. Furthermore, assume that at every value of $\mathbf{F}$, at least one of the partial derivatives $\partial U/\partial F_i$ ($i = 1, \ldots, n$) is strictly positive.

Let $\mathbf{z}^0$ be a preferred and regular solution to the problem in Eq. (5.4). At $\mathbf{z}^0$, the function $U[\mathbf{F}(\mathbf{z})]$ attains its minimum, satisfying the Kuhn-Tucker conditions for single-objective optimization:

$$\frac{\partial U[F(\mathbf{z}^0)]}{\partial \mathbf{z}} + \bar{\boldsymbol{\rho}}^T \frac{\partial \mathbf{w}}{\partial \mathbf{z}}(\mathbf{z}^0) + \bar{\boldsymbol{\xi}}^T \frac{\partial \mathbf{r}}{\partial \mathbf{z}}(\mathbf{z}^0) = \mathbf{0} \tag{5.5}$$

$$\bar{\boldsymbol{\rho}}^T \mathbf{w}(\mathbf{z}^0) = 0 \tag{5.6}$$

$$\bar{\boldsymbol{\rho}} \geqslant \mathbf{0} \tag{5.7}$$

$$\mathbf{w}(\mathbf{z}^0) \leqslant \mathbf{0} \qquad \mathbf{r}(\mathbf{z}) = \mathbf{0} \tag{5.8}$$

By the chain rule of differentiation,

$$\frac{\partial U[F(\mathbf{z}^0)]}{\partial \mathbf{z}} = \sum_{k=1}^{n} \frac{\partial U}{\partial F_k}(\mathbf{F}^0) \frac{\partial F_k(\mathbf{z}^0)}{\partial \mathbf{z}} \tag{5.9}$$

where $\mathbf{F}^0 = \mathbf{F}(\mathbf{z}^0)$.

Taking Eq. (5.9) into account, let us divide Eq. (5.5) by $\partial U/\partial F_1$, which (arbitrarily) is assumed to be a strictly positive derivative. This will transform Eq. (5.5) into the following:

$$\sum_{k=1}^{n} \left[\frac{\partial U(\mathbf{F}^0)/\partial F_k}{\partial U(\mathbf{F}^0)/\partial F_1} \right] \frac{\partial F_k}{\partial \mathbf{z}} (\mathbf{z}^0) + \boldsymbol{\rho}^T \frac{\partial \mathbf{w}}{\partial \mathbf{z}} (\mathbf{z}^0) + \boldsymbol{\xi}^T \frac{\partial \mathbf{r}}{\partial \mathbf{z}} (\mathbf{z}^0) = \mathbf{0} \tag{5.10}$$

where

$$\boldsymbol{\rho} = \frac{\bar{\boldsymbol{\rho}}}{\partial U(\mathbf{F}^0)/\partial F_1} \geqslant \mathbf{0} \qquad \boldsymbol{\xi} = \frac{\bar{\boldsymbol{\xi}}}{\partial U(\mathbf{F}^0)/\partial F_1} \tag{5.11}$$

When the relationship between trade-offs and the loss function is taken into account,

$$\lambda^*_{1k}(\mathbf{F}^0) = \frac{\partial U(\mathbf{F}^0)/\partial F_k}{\partial U(\mathbf{F}^0)/\partial F_1} \qquad k = 2, \ldots, n \tag{5.12}$$

and, by defining

$$\lambda^*_{11} = 1 \tag{5.13}$$

Eq. (5.5) takes the form of Eq. (5.14), and the other necessary conditions [Eqs. (5.6)–(5.8)] are transformed, by using Eq. (5.11), into equivalent forms in Eq. (5.15):

$$\sum_{k=1}^{n} \lambda^*_{ik}(\mathbf{F}^0) \frac{\partial F_k}{\partial \mathbf{z}} (\mathbf{z}^0) + \boldsymbol{\rho}^T \frac{\partial \mathbf{w}}{\partial \mathbf{z}} (\mathbf{z}^0) + \boldsymbol{\xi}^T \frac{\partial \mathbf{r}}{\partial \mathbf{z}} (\mathbf{z}^0) = \mathbf{0} \tag{5.14}$$

$$\boldsymbol{\rho}^T \mathbf{w}(\mathbf{z}^0) = 0 \qquad \boldsymbol{\rho} \geqslant \mathbf{0} \qquad \mathbf{w}(\mathbf{z}^0) \leqslant \mathbf{0} \qquad \mathbf{r}(\mathbf{z}^0) = \mathbf{0} \tag{5.15}$$

Equations (5.14) and (5.15) are necessary conditions for a preferred solution to the problem given in Eq. (5.4) when trade-offs are used to express the decisionmaker's preference.

5.3 MATHEMATICAL PROBLEM FORMULATION

The mathematical problems considered in this chapter are similar to those formulated in Sect. 4.4; see Fig. 4.6 and Eqs. (4.16)–(4.21). The problems here differ from those in Chap. 4 in that they now have only one decisionmaker (DM_0), and he or she is interested in a preferred solution for the overall objectives. The corresponding mathematical problem formulation is repeated here for convenience:

$$\min \begin{bmatrix} F_1(\mathbf{f}^1, \ldots, \mathbf{f}^N) \\ \vdots \\ F_n(\mathbf{f}^1, \ldots, \mathbf{f}^N) \end{bmatrix} \tag{5.16}$$

where $\mathbf{f}^i = [f^i_1(\mathbf{x}_i, \mathbf{m}_i, \mathbf{y}_i), \ldots, f^i_{n_i}(\mathbf{x}_i, \mathbf{m}_i, \mathbf{y}_i)]^T \quad (i = 1, \ldots, N)$ subject to

$$\mathbf{y}_i = \mathbf{H}_i(\mathbf{x}_i, \mathbf{m}_i) \qquad i = 1, \ldots, N \tag{5.17}$$

$$\mathbf{g}_i(\mathbf{x}_i, \mathbf{m}_i, \mathbf{y}_i) \leq \mathbf{0} \qquad i = 1, \ldots, N \tag{5.18}$$

$$\mathbf{x}_i = \sum_{j=1}^{N} C_{ij}\mathbf{y}_j \qquad i = 1, \ldots, N \tag{5.19}$$

$$\mathbf{q}_i(\mathbf{x}_i, \mathbf{m}_i, \mathbf{y}_i) = \mathbf{l}_i \qquad i = 1, \ldots, N \tag{5.20}$$

$$\sum_{i=1}^{n} \mathbf{l}_i \leq \mathbf{0} \tag{5.21}$$

Note that Eqs. (5.20) and (5.21) are an alternative form for the overall constraints in Eq. (4.21).

Here, the solution of the minimization problem given in Eq. (5.16) is a preferred solution for the decisionmaker at the upper level. The definitions of the variables are the same as in the problem formulation for generating the whole Pareto optimal set in Sect. 4.4.

5.4 NECESSARY CONDITIONS FOR THE LARGE-SCALE PROBLEM

The following sections use a technique to derive multilevel algorithms similar to that commonly used in the single-objective case. That is, first a necessary condition—in this case, Eqs. (5.14) and (5.15)—is written for the whole system. Then, some variables (coordination parameters) are fixed so that the necessary conditions decompose into separate sets of equations. Most of these sets of equations are then noticed to be necessary conditions for optimization problems (subproblems). The solutions of these subproblems thus satisfy the conditions. Overall, we achieve a scheme where the upper level fixes some variables; subproblems are solved, and the rest of the necessary conditions are checked at the upper level. If these are not satisfied, the coordination parameters are changed, and suboptimizations are repeated. This process continues until the upper-level conditions are met. Results are then ready to be implemented.

The results given by the scheme described above satisfy the necessary conditions for the whole problem. Generally, such a solution is the most satisfying one that is derived from any optimization method, without any additional assumptions that may guarantee only one solution to the necessary conditions.

For presentational convenience, we assume below that there is only one solution to the necessary conditions relating to the problem given in Eqs. (5.16)–(5.21). Hence, we want to derive schemes that produce that solution (but no other solutions).

The preferences of the decisionmaker are assumed to be articulated in the form of trade-offs. Let, for $i, j = 1, \ldots, n$ $(i \neq j)$,

$$\lambda_{ij}^*(\mathbf{F}) = \text{the indifference trade-off between } F_i \text{ and } F_j \text{ at } \mathbf{F} = (F_1, \ldots, F_n) \tag{5.22}$$

For further mathematical purposes, define an n-dimensional indifference trade-off vector $\boldsymbol{\lambda}$ that captures these trade-offs:

$$\boldsymbol{\lambda}(\mathbf{F}) = [\lambda_1(\mathbf{F}), \lambda_2(\mathbf{F}), \lambda_3(\mathbf{F}), \ldots, \lambda_n(\mathbf{F})] \tag{5.23a}$$

$$= [1, \lambda_{12}^*(\mathbf{F}), \lambda_{13}^*(\mathbf{F}), \ldots, \lambda_{1n}^*(\mathbf{F})]^T \tag{5.23b}$$

Let us now apply the necessary conditions given in Eqs. (5.14) and (5.15) to the problem given in Eqs. (5.16)–(5.21). The following notations are used:

$$\mathbf{z}_i = (\mathbf{x}_i^T, \mathbf{m}_i^T, \mathbf{y}_i^T, \mathbf{l}_i^T)^T \qquad i = 1, \ldots, N \tag{5.24}$$

$$\mathbf{z} = (\mathbf{z}_1^T, \ldots, \mathbf{z}_N^T)^T \tag{5.25}$$

Let us first develop the first term of Eq. (5.14). Suppressing arguments, we have the following (the first equality is obtained by the chain rule of differentiation):

$$\begin{aligned} \sum_{k=1}^{n} \lambda_k \frac{\partial F_k}{\partial \mathbf{z}} &= \sum_{k=1}^{n} \lambda_k \sum_{i=1}^{N} \frac{\partial F_k}{\partial \mathbf{f}^i} \frac{\partial \mathbf{f}^i}{\partial \mathbf{z}} \\ &= \sum_{i=1}^{N} \left(\sum_{k=1}^{n} \lambda_k \frac{\partial F_k}{\partial \mathbf{f}^i} \right) \frac{\partial \mathbf{f}^i}{\partial \mathbf{z}} \\ &= \sum_{i=1}^{N} (\boldsymbol{\lambda}^i)^T \frac{\partial \mathbf{f}^i}{\partial \mathbf{z}} \end{aligned} \tag{5.26}$$

where we have defined, for $i = 1, \ldots, N$, the n_i vector

$$\boldsymbol{\lambda}^i(\mathbf{f}) = \left[\sum_{k=1}^{n} \lambda_k(\mathbf{F}) \frac{\partial F_k(\mathbf{f})}{\partial \mathbf{f}^i} \right]^T \tag{5.27}$$

The vector $\boldsymbol{\lambda}^i(\mathbf{f})$ is called the internal (indifference) trade-off vector of subsystem i.

As an example, consider a system consisting of two subsystems. Let f_1^i = cost ($i = 1, 2$), f_2^i = failure probability ($i = 1, 2$), and f_3^1 = pollution index of the first subsystem.

Furthermore, let $F_1 = f_1^1 + f_1^2$ (total cost), $F_2 = f_2^1 f_2^2$ (probability that the total system is out of order), and $F_3 = f_3^1$ (pollution index of the total system).

The overall objective vector is, thus, $\mathbf{F} = (f_1^1 + f_1^2, f_2^1 f_2^2, f_3^1)$.

Usually, λ_k's evolve during the optimization in interaction with the decision-maker, but, for illustrative purposes, take, for example, $\lambda_1 = 1$, $\lambda_2 = (f_2^1 f_2^2)^\alpha$, and $\lambda_3 = \exp(\beta f_3^1)$, where α and β are some constants. Then Eq. (5.27) yields

$$\boldsymbol{\lambda}^1(\mathbf{f}) = \begin{bmatrix} 1 \\ \lambda_2 f_2^2 \\ \lambda_3 \end{bmatrix} = \begin{bmatrix} 1 \\ (f_2^1 f_2^2)^\alpha f_2^2 \\ \exp(\beta f_3^1) \end{bmatrix}$$

$$\boldsymbol{\lambda}^2(\mathbf{f}) = \begin{bmatrix} 1 \\ \lambda_2 f_2^1 \end{bmatrix} = \begin{bmatrix} 1 \\ (f_2^1 f_2^2)^\alpha f_2^1 \end{bmatrix}$$

In defining the scaled internal trade-off vectors $\boldsymbol{\lambda}_s$, we assume that, in each $\boldsymbol{\lambda}^i(\mathbf{f})$ ($i = 1, \ldots, N$), there is one strictly positive component for all values of $\mathbf{f}$. Without loss of generality, this component is assumed to be the first one, $\lambda_1^i(\mathbf{f})$ ($i = 1, \ldots, N$). Note that, if originally $\lambda_1^i(\mathbf{f}) < 0$ for every $\mathbf{f}$ and if we have a function $f(\mathbf{x}_i, \mathbf{m}_i, \mathbf{y}_i)$, then, by redefining $f_1^i = -f(\mathbf{x}_i, \mathbf{m}_i, \mathbf{y}_i)$, we will have $\lambda_1^i(\mathbf{f}) > 0$. The scaled internal trade-off vectors $\boldsymbol{\lambda}_s^i$'s are given by Eq. (5.28):

$$\boldsymbol{\lambda}_s^i(\mathbf{f}) = \frac{1}{\lambda_1^i(\mathbf{f})} \boldsymbol{\lambda}^i(\mathbf{f}) \qquad i = 1, \ldots, N \tag{5.28}$$

(Note: The subscript s denotes this scaling; it is not a component index.)

Define $\lambda^{kl}(\mathbf{f})$, the indifference trade-off between subsystems k and l:

$$\lambda^{kl}(\mathbf{f}) = \frac{\lambda_1^l(\mathbf{f})}{\lambda_1^k(\mathbf{f})} \qquad k, l = 1, \ldots, N \tag{5.29}$$

We take the rth subsystem ($r \in \{1, \ldots, N\}$) as a reference, and deal with trade-offs λ^{ri}, $i = 1, \ldots, N$. The other trade-offs are determined by these trade-offs using Eq. (5.29). Note that by Eq. (5.29), $\lambda^{rr} = 1$.

Until we get to Sect. 5.7, we assume that $\boldsymbol{\lambda}_s^i$ depends only on $\mathbf{f}^i$; we denote these vectors by $\boldsymbol{\lambda}_s^i(\mathbf{f}^i)$, $i = 1, \ldots, N$.

The cases studied first are those in which there is conflict among objectives in different subsystems through the couplings between the subsystems; objectives of different subsystems may appear in an additive (but not more complex) way in an objective of the total system. In the above example, the product of two objectives in two different subsystems causes the example not to fit into the additive class. Cases similar to this example are treated later in Sect. 5.7.

Let us now return to applying the conditions of Eqs. (5.14) and (5.15) to the problem given by Eqs. (5.16)–(5.21). The second and third terms of Eq. (5.14) will be developed. Using special symbols $\boldsymbol{\delta}_i$, $\boldsymbol{\eta}_i$, $\boldsymbol{\nu}_i$, $\boldsymbol{\mu}_i$, and $\boldsymbol{\mu}$ for the components of $\boldsymbol{\rho}$ and $\boldsymbol{\xi}$, one obtains

$$\begin{aligned}\boldsymbol{\rho}^T \frac{\partial \mathbf{w}}{\partial \mathbf{z}} + \boldsymbol{\xi}^T \frac{\partial \mathbf{r}}{\partial \mathbf{z}} = \frac{\partial}{\partial \mathbf{z}} (\boldsymbol{\rho}^T \mathbf{w} + \boldsymbol{\xi}^T \mathbf{r}) = \\ \frac{\partial}{\partial \mathbf{z}} \Bigg\{ \sum_{i=1}^{N} \Bigg[\boldsymbol{\delta}_i^T(\mathbf{H}_i(\mathbf{x}_i, \mathbf{m}_i) - \mathbf{y}_i) + \boldsymbol{\eta}_i^T \mathbf{g}_i(\mathbf{x}_i, \mathbf{m}_i, \mathbf{y}_i) \\ + \boldsymbol{\nu}_i^T\Bigg(-\mathbf{x}_i + \sum_{j=1}^{N} C_{ij}\mathbf{y}_j\Bigg) + \boldsymbol{\mu}_i^T(\mathbf{q}_i(\mathbf{x}_i, \mathbf{m}_i, \mathbf{y}_i) - \mathbf{l}_i)\Bigg] + \boldsymbol{\mu}^T \sum_{i=1}^{N} \mathbf{l}_i \Bigg\}\end{aligned} \tag{5.30}$$

Furthermore, according to Eq. (5.15),

$$\boldsymbol{\eta}_i^T \mathbf{g}_i(\mathbf{x}_i, \mathbf{m}_i, \mathbf{y}_i) = 0 \qquad i = 1, \ldots, N \tag{5.31}$$

$$\boldsymbol{\mu}^T \sum_{i=1}^{N} \mathbf{l}_i = 0 \tag{5.32}$$

$$\boldsymbol{\eta}_i \geqslant \mathbf{0} \qquad \boldsymbol{\mu} \geqslant \mathbf{0} \tag{5.33}$$

All terms in the necessary conditions given by Eqs. (5.14) and (5.15) have been derived at this stage. In the following discussion, Eq. (5.14) will be rewritten componentwise, where $\mathbf{z} = (\mathbf{x}_1, \mathbf{m}_1, \mathbf{y}_1, \mathbf{l}_1, \ldots, \mathbf{x}_N, \mathbf{m}_N, \mathbf{y}_N, \mathbf{l}_N)^T$.

All equations related to the $\mathbf{x}_i$, $\mathbf{m}_i$, and $\mathbf{y}_i$ components are divided by λ_1^i ($i = 1, \ldots, N$). The multiplier vectors are divided; and, for $i = 1, \ldots, N$, we denote

$$\boldsymbol{\delta}_i^s = \frac{\boldsymbol{\delta}_i}{\lambda_1^i} \qquad \boldsymbol{\eta}_i^s = \frac{\boldsymbol{\eta}_i}{\lambda_1^i}$$

$$\boldsymbol{\nu}_i^s = \frac{\boldsymbol{\nu}_i}{\lambda_1^i} \qquad \boldsymbol{\mu}_i^s = \frac{\boldsymbol{\mu}_i}{\lambda_1^i} \tag{5.34}$$

The equations related to the $\mathbf{l}_i$ components are divided by λ_1^r. Denote

$$\boldsymbol{\mu}^s = \frac{\boldsymbol{\mu}}{\lambda_1^r} \tag{5.35}$$

Finally, as a minor modification, all multipliers are expressed as scaled multipliers using Eqs. (5.34) and (5.35). Equation (5.29) is used to express all λ_1^i/λ_1^k trade-offs in terms of λ^{ri} ($i = 1, \ldots, N$) to reduce the number of trade-offs.

As a result, Eq. (5.14) takes the form of Eqs. (5.36)–(5.38). Condition (5.15) was developed above; see Eqs. (5.31)–(5.33). Dividing the ith equation of Eq. (5.31) by λ_1^i (>0) ($i = 1, \ldots, N$) and dividing Eq. (5.32) by λ_1^r yields Eqs. (5.40) and (5.42), respectively. The positivity condition for $\boldsymbol{\eta}_i$'s and $\boldsymbol{\mu}$ [Eq. (5.33)] takes the form of Eqs. (5.41) and (5.43) in terms of scaled multipliers. Thus, along with the systems equations, we have the following set of necessary conditions (partly suppressing arguments) for $i = 1, \ldots, N$:

$$[\boldsymbol{\lambda}_s^i(\mathbf{f}^i)]^T \frac{\partial \mathbf{f}^i}{\partial \mathbf{x}_i} + (\boldsymbol{\delta}_i^s)^T \frac{\partial \mathbf{H}_i}{\partial \mathbf{x}_i} + (\boldsymbol{\eta}_i^s)^T \frac{\partial \mathbf{g}_i}{\partial \mathbf{x}_i} - (\boldsymbol{\nu}_i^s)^T + (\boldsymbol{\mu}_i^s)^T \frac{\partial \mathbf{q}_i}{\partial \mathbf{x}_i} = \mathbf{0} \tag{5.36}$$

$$[\boldsymbol{\lambda}_s^i(\mathbf{f}^i)]^T \frac{\partial \mathbf{f}^i}{\partial \mathbf{m}_i} + (\boldsymbol{\delta}_i^s)^T \frac{\partial \mathbf{H}_i}{\partial \mathbf{m}_i} + (\boldsymbol{\eta}_i^s)^T \frac{\partial \mathbf{g}_i}{\partial \mathbf{m}_i} + (\boldsymbol{\mu}_i^s)^T \frac{\partial \mathbf{g}_i}{\partial \mathbf{m}_i} = \mathbf{0} \tag{5.37}$$

$$(\mathbf{G}_i^y)^T \triangleq [\boldsymbol{\lambda}_s^i(\mathbf{f}^i)]^T \frac{\partial \mathbf{f}^i}{\partial \mathbf{y}_i} - (\boldsymbol{\delta}_i^s)^T + (\boldsymbol{\eta}_i^s)^T \frac{\partial \mathbf{g}_i}{\partial \mathbf{y}_i} + \sum_{k=1}^{N} \frac{\lambda^{rk}}{\lambda^{ri}} (\boldsymbol{\nu}_k^s)^T C_{ki} + (\boldsymbol{\mu}_i^s)^T \frac{\partial \mathbf{q}_i}{\partial \mathbf{y}_i} = \mathbf{0} \tag{5.38}$$

$$\lambda^{ri} \boldsymbol{\mu}_i^s - \boldsymbol{\mu}^s = \mathbf{0} \tag{5.39}$$

$$(\boldsymbol{\eta}_i^s)^T \mathbf{g}_i(\mathbf{x}_i, \mathbf{m}_i, \mathbf{y}_i) = 0 \tag{5.40}$$

$$\boldsymbol{\eta}_i^s \geqslant \mathbf{0} \tag{5.41}$$

$$(\boldsymbol{\mu}^s)^T \sum_{k=1}^{N} \mathbf{l}_k = 0 \tag{5.42}$$

$$\boldsymbol{\mu}^s \geqslant \mathbf{0} \tag{5.43}$$

$$\mathbf{y}_i = \mathbf{H}_i(\mathbf{x}_i, \mathbf{m}_i) \tag{5.44}$$

$$\mathbf{g}_i(\mathbf{x}_i, \mathbf{m}_i, \mathbf{y}_i) \leq \mathbf{0} \tag{5.45}$$

$$\mathbf{x}_i = \sum_{j=1}^{N} C_{ij}\mathbf{y}_j \tag{5.46}$$

$$\mathbf{q}_i(\mathbf{x}_i, \mathbf{m}_i\mathbf{y}_i) = \mathbf{l}_i \tag{5.47}$$

$$\sum_{k=1}^{N} \mathbf{l}_k \leq \mathbf{0} \tag{5.48}$$

Equations (5.36)–(5.48) constitute the basic set of necessary conditions to derive hierarchical schemes by the method outlined at the beginning of this section.

The principal schemes derived in the following sections are the same as those in the single-objective hierarchical case—the feasible (primal, prediction) method and the nonfeasible (interaction balance, price) method. The feasible method will be considered first.

5.5 FEASIBLE METHODS

In Sect. 5.5.1 an algorithm—the feasible method with partial derivatives—is derived for the general mathematical problem formulation given in Sect. 5.3. In Sect. 5.5.2, a feasible scheme is derived for a less general case where there are no constraining boundaries for the coupling variables. In this case, the algorithm can be presented using derivatives with respect to the coupling variables. This scheme is called the feasible method with total derivatives.

5.5.1 Feasible Method with Partial Derivatives

5.5.1.1 Subsystem formulation In the feasible schemes, the coordination parameters are the interaction variables—variables that are connected to several subsystems. Here, the interaction variables consist of $\mathbf{y}_i$'s (cf. coupling equations) and $\mathbf{l}_i$'s (overall constraints).

At each iteration, specific values are determined for all $\mathbf{y}_i$'s and $\mathbf{l}_i$'s at the upper level. For the kth iteration, these values are denoted by $\mathbf{y}_i^k$, $\mathbf{l}_i^k$, $i = 1, \ldots, N$, $k \in \{1, 2, \ldots\}$.

Consider the necessary conditions given in Eqs. (5.36)–(5.48), in the way explained in Sect. 5.4, to derive a feasible scheme. Let $\mathbf{l}_i^k$'s be chosen so that Eq. (5.48) holds. If this is a strict inequality, we take $(\boldsymbol{\mu}^s)^k = \mathbf{0}$, when Eqs. (5.42) and (5.43) also hold. The values of $\mathbf{x}_i^k$, are calculated from Eq. (5.46) which is thus satisfied for $i = 1, \ldots, N$.

Next, consider Eqs. (5.37), (5.40), (5.41), (5.44), (5.45), and (5.47) for $i = 1, \ldots, N$ (where some variables have the values given above). Note that these equations are necessary conditions for the preferred solution to the following multiobjective optimization problem ($i = 1, \ldots, N$):

$$\min_{\mathbf{m}_i} \begin{bmatrix} f^i_1(\mathbf{x}^k_i, \mathbf{m}_i, \mathbf{y}^k_i) \\ \vdots \\ f^i_{n_i}(\mathbf{x}^k_i, \mathbf{m}_i, \mathbf{y}^k_i) \end{bmatrix} \tag{5.49}$$

subject to

$$\mathbf{y}^k_i = \mathbf{H}_i(\mathbf{x}^k_i, \mathbf{m}_i) \tag{5.50}$$

$$\mathbf{g}_i(\mathbf{x}^k_i, \mathbf{m}_i, \mathbf{y}^k_i) \leq \mathbf{0} \tag{5.51}$$

$$\mathbf{q}_i(\mathbf{x}^k_i, \mathbf{m}_i, \mathbf{y}^k_i) = \mathbf{l}^k_i \tag{5.52}$$

with a trade-off vector

$$\boldsymbol{\lambda}^i_s(\mathbf{f}^i(\mathbf{x}^k_i, \mathbf{m}_i, \mathbf{y}^k_i)) \tag{5.53}$$

The regularity of the above constraints in Eqs. (5.50)–(5.52) is here assumed.

Recall that the kth component of the trade-off vector $\boldsymbol{\lambda}^i_s$ is the indifference trade-off between f^i_1 and f^i_k ($k = 2, \ldots, n_i$). When carrying out the optimization of Eqs. (5.49)–(5.53) and also whenever these trade-offs are needed, $\boldsymbol{\lambda}^i_s$ is determined by Eqs. (5.28) and (5.27). For Eq. (5.27), the decisionmaker is asked for his or her indifference trade-offs. Note that the F_i's do not need to be known as explicit functions in this procedure.

Assuming that the problems given by Eqs. (5.49)–(5.53) have solutions $\mathbf{m}^k_i$, the necessary conditions of Eqs. (5.37), (5.40), (5.41), (5.44), (5.45), and (5.47) are satisfied by the $\mathbf{m}^k_i$'s and some multiplier vectors, $(\boldsymbol{\delta}^s_i)^k$, $(\boldsymbol{\eta}^s_i)^k$, $i = 1, \ldots, N$. On the basis of Eq. (5.40), a multiplier $(\eta^s_{ij})^k$ corresponding to an inactive constraint $g_{ij}(\mathbf{x}^k_i, \mathbf{m}^k_i, \mathbf{y}^k_i) < 0$ is zero. The other $\boldsymbol{\eta}$ multipliers, as well as the $\boldsymbol{\delta}$ and $\boldsymbol{\mu}_i$ multipliers, can be determined by solving Eq. (5.37), where, in the ith equation, arguments are evaluated at $(\mathbf{x}^k_i, \mathbf{m}^k_i, \mathbf{y}^k_i)$. The equations are linear with respect to the multipliers. Respective solutions are unique owing to the regularity condition for the constraints given by Eqs. (5.50)–(5.52). The nonzero $\boldsymbol{\eta}$ multipliers automatically satisfy Eq. (5.41), which is then satisfied by all $\boldsymbol{\eta}$ multipliers.

Equation (5.36) is satisfied directly by determining the $(\boldsymbol{\nu}^s_i)^k$ multiplier vector from these equations. [Transfer the $\boldsymbol{\nu}^s_i$ term into the right-hand side of Eq. (5.36).]

We are left with Eqs. (5.38), (5.39), (5.42), and (5.43). We can, for example determine $(\boldsymbol{\mu}^s)^k$ from the rth equation of Eq. (5.39), $(\boldsymbol{\mu}^s)^k = (\boldsymbol{\mu}^s_r)^k$, and then directly check whether the other $N - 1$ equations and Eqs. (5.38), (5.42), and (5.43) are satisfied. If they are not satisfied, new values $\mathbf{y}^{k+1}_i$, $\mathbf{l}^{k+1}_i$ ($i = 1, \ldots, N$) need to be determined, and the above procedure is repeated.

5.5.1.2 Coordination Next, consider the coordination process—the change of coordination parameters $\mathbf{y}_i$ and $\mathbf{l}_i$ ($i = 1, \ldots, N$). The coordination can be based on a straightforward idea of moving in a direction where an underlying loss function decreases most quickly. Again, the loss function will be a theoretical vehicle: the indifference trade-offs are actually elicited from the decisionmaker.

The optimizaton of the utility function will be presented here using the simplified Zoutendijk method (see Luenberger, 1973). This optimization method is a simple one which may be enhanced for better convergence using the modifications invented by Zoutendijk (1960) and described concisely by Luenberger (1973).

In the simplified Zoutendijk method, the problem

$$\min_{\mathbf{z}} U(\mathbf{z}) \tag{5.54}$$

subject to

$$\mathbf{r}(\mathbf{z}) = \mathbf{0} \tag{5.55}$$

$$\mathbf{g}(\mathbf{z}) \leqslant \mathbf{0} \tag{5.56}$$

is transformed into a series of problems, where the terms are evaluated at a feasible $\mathbf{z}^k$:

$$\min_{\Delta\mathbf{z}} \frac{\partial U}{\partial \mathbf{z}} \Delta\mathbf{z} \tag{5.57}$$

subject to

$$\frac{\partial \mathbf{r}}{\partial \mathbf{z}} \Delta\mathbf{z} = \mathbf{0} \tag{5.58}$$

$$\frac{\partial \mathbf{g}_i}{\partial \mathbf{z}} \Delta\mathbf{z} \leqslant \mathbf{0} \qquad \text{for} \qquad i \in I \triangleq \{i | \mathbf{g}_i(\mathbf{x}) = \mathbf{0}\} \tag{5.59}$$

$$\|\Delta\mathbf{z}\| \leqslant 1 \tag{5.60}$$

Here, we aim at solving a direction $\Delta\mathbf{z}$ such that U decreases the fastest subject to the feasibility constraints given by Eqs. (5.58) and (5.59) and the normalizing equation given by Eq. (5.60). These ensure a bounded solution to the problem given in Eq. (5.57) where the decrease of U is approximated to first-order accuracy. [Function $\mathbf{r}$ should be linear for a $\Delta\mathbf{z}$ determined by Eq. (5.58) to be feasible. If $\mathbf{r}$ is nonlinear, we may still be able to obtain a feasible solution by a projection onto the feasible region.] After having solved $\Delta\mathbf{z}$, a scalar c^k is determined by searching so that $\mathbf{z}^{k+1} = \mathbf{z}^k + c^k\Delta\mathbf{z}$ is feasible, and $U(\mathbf{z}^{k+1}) < U(\mathbf{z}^k)$. For details, see the references mentioned.

After this general discussion, let us go back to the point we were at the end of the previous section: we had selected some values for the coordination variables $\mathbf{y}_i^k$ and $\mathbf{l}_i^k$ ($i = 1, \ldots, N$), and had solved some subproblems whereby several equations of the necessary conditions given by Eqs. (5.36)–(5.48) were satisfied. Assume that all the equations are not yet satisfied.

The subproblem solutions actually minimize the underlying loss function with respect to subsystem decision variables $\mathbf{m}_i$, given the fixed coupling variables. The underlying loss function has some value at the values of all variables we have insofar specified. Let us try to find new $\mathbf{y}_i$ and $\mathbf{l}_i$ values which decrease the values of the underlying loss function, which is denoted by $U(F_1, F_2, \ldots, F_n)$. Consider the minimization of U in terms of the mathematical decision vector $\mathbf{z} = (\mathbf{z}_1^T, \ldots,$

$\mathbf{z}_N^T)^T$, where $\mathbf{z}_i = (\mathbf{x}_i^T, \mathbf{m}_i^T, \mathbf{y}_i^T, \mathbf{l}_i^T)^T$ [cf. Eqs. (5.24) and (5.25)]. Let us find a direction $\Delta\mathbf{z} = (\Delta\mathbf{z}_1^T, \ldots, \Delta\mathbf{z}_N^T)^T$ with $\Delta\mathbf{z}_i = (\Delta\mathbf{x}_i^T, \Delta\mathbf{m}_i^T, \Delta\mathbf{y}_i^T, \Delta\mathbf{l}_i^T)^T$ in the same way as in the simplified feasible direction method.

Corresponding to Eqs. (5.58) and (5.59), the following set of equations specify a feasible change $\Delta\mathbf{z}$ to first-order accuracy (cf. problem formulation in Sect. 5.3):

$$\Delta\mathbf{y}_i = \frac{\partial\mathbf{H}_i}{\partial\mathbf{x}_i}\Delta\mathbf{x}_i + \frac{\partial\mathbf{H}_i}{\partial\mathbf{m}_i}\Delta\mathbf{m}_i \tag{5.61}$$

$$\frac{\partial g_{ij}}{\partial\mathbf{z}_i}\Delta\mathbf{z}_i \leqslant 0 \qquad j \in \{j | g_{ij}(\mathbf{z}_i^k) = 0\} \tag{5.62}$$

$$\Delta\mathbf{x}_i = \sum_{j=1}^{N} C_{ij}\Delta\mathbf{y}_j \tag{5.63}$$

$$\frac{\partial\mathbf{q}_i}{\partial\mathbf{z}_i}\Delta\mathbf{z}_i = \Delta\mathbf{l}_i \tag{5.64}$$

$$\sum_{j=1}^{N}\Delta\mathbf{l}_j \leqslant \mathbf{0} \qquad \text{if} \qquad \sum_{j=1}^{N}\mathbf{l}_j = \mathbf{0} \tag{5.65}$$

for $i = 1, \ldots, N$. The arguments in Eqs. (5.61)–(5.65) are the values of variables at the kth iteration.

Next, let us evaluate the gradient of the loss function with respect to $\mathbf{z}$:

$$\begin{aligned}
\frac{\partial U[\mathbf{F}(\mathbf{f})]}{\partial\mathbf{z}} &= \sum_{k=1}^{n}\frac{\partial U}{\partial F_k}\frac{\partial F_k}{\partial\mathbf{z}} \\
&= \frac{\partial U}{\partial F_1}\sum_{k=1}^{n}\left(\frac{\partial U/\partial F_k}{\partial U/\partial F_1}\right)\sum_{i=1}^{N}\frac{\partial F_k}{\partial\mathbf{f}^i}\frac{\partial\mathbf{f}^i}{\partial\mathbf{z}} && (5.66) \\
&= \frac{\partial U}{\partial F_1}\sum_{k=1}^{n}\lambda_k\sum_{i=1}^{N}\frac{\partial F_k}{\partial\mathbf{f}^i}\frac{\partial\mathbf{f}^i}{\partial\mathbf{z}} && (5.67) \\
&= \frac{\partial U}{\partial F_1}\sum_{i=1}^{N}\sum_{k=1}^{n}\lambda_k\frac{\partial F_k}{\partial\mathbf{f}^i}\frac{\partial\mathbf{f}^i}{\partial\mathbf{z}} && (5.68) \\
&= \frac{\partial U}{\partial F_1}\sum_{i=1}^{N}(\boldsymbol{\lambda}^i)^T\frac{\partial\mathbf{f}^i}{\partial\mathbf{z}} && (5.69)
\end{aligned}$$

In Eq. (5.66), $\partial U/\partial F_1$ is again assumed to be strictly positive. In Eq. (5.67), Eq. (5.3) has been taken into account (note $\lambda_k = \lambda_{1k}^*$). In Eq. (5.69), the definition of $\boldsymbol{\lambda}^i$ has been used [see Eq. (5.27)].

Since $\partial U/\partial F_1$ was assumed to be strictly positive, it can be left out when writing the minimization problem corresponding to the problem given by Eq. (5.57):

$$\min_{\Delta z} \sum_{i=1}^{N} (\boldsymbol{\lambda}^i)^T \frac{\partial \mathbf{f}^i}{\partial \mathbf{z}} \Delta \mathbf{z} \tag{5.70}$$

This expression can be evaluated by taking into account the satisfied equations at the point in question. Multiply Eqs. (5.36)–(5.38) [in the form $[\boldsymbol{\lambda}_s^i(\mathbf{f}^i)]^T \ldots = (\mathbf{G}_i^y)^T$ by $(\lambda^{ri})^k = (\lambda_1^i)^k/(\lambda_1^r)^k$ $(i = 1, \ldots, N)$]; postmultiply the resulting equations by $\Delta \mathbf{x}_i$, $\Delta \mathbf{m}_i$, $\Delta \mathbf{y}_i$, respectively $(i = 1, \ldots, N)$. Sum up all resulting equations. Accounting for the constraints in Eqs. (5.61), (5.63), and (5.64) and recalling $\boldsymbol{\lambda}^i = \lambda_1^i \boldsymbol{\lambda}_s^i$, we will get, after a rearrangement of terms,

$$\begin{aligned}\min_{\Delta z} \Sigma(\boldsymbol{\lambda}^i)^T \frac{\partial \mathbf{f}^i}{\partial \mathbf{z}} \Delta \mathbf{z} &= \min_{\Delta z} (\lambda_1^r)^k \left\{ \sum_{i=1}^{N} (\lambda^{ri})^k \left[(\mathbf{G}_i^y)^T \Delta \mathbf{y}_i - (\boldsymbol{\eta}_i^k)^T \frac{\partial g_i}{\partial \mathbf{z}_i} \Delta \mathbf{z}_i \right.\right. \\ &\qquad \left.\left. - (\boldsymbol{\mu}_i^k)^T \Delta \mathbf{l}_i \right]\right\} \\ &= \min_{\Delta z} \sum_{i=1}^{N} (\lambda^{ri})^k \left[(\mathbf{G}_i^y)^T \Delta \mathbf{y}_i - (\boldsymbol{\eta}_i^k)^T \frac{\partial g_i}{\partial \mathbf{z}_i} \Delta \mathbf{z}_i - (\boldsymbol{\mu}_i^k)^T \Delta \mathbf{l}_i \right]\end{aligned} \tag{5.71}$$

In the last equation, the assumption $\lambda_1^r > 0$ has been used.

Finally, instead of the normalizing Eq. (5.60), we may use the following linear constraints:

$$-\mathbf{1} \leqslant \Delta z \leqslant \mathbf{1} \tag{5.72}$$

where $\mathbf{1} = (1, \ldots, 1)^T$.

Furthermore, note that if the $\Delta \mathbf{y}_i$'s are limited, so are the $\Delta \mathbf{x}_i$'s by Eq. (5.63). If we still assume the usual case in practice that the limited $\Delta \mathbf{y}_i$'s and $\Delta \mathbf{x}_i$'s will also constrain the $\Delta \mathbf{m}_i$'s by Eqs. (5.61) and (5.62), we can, instead of Eq. (5.72), take the following simpler normalizing equations:

$$-\mathbf{1} \leqslant \Delta \mathbf{y}_i \leqslant \mathbf{1} \qquad -\mathbf{1} \leqslant \Delta \mathbf{l}_i \leqslant \mathbf{1} \qquad i = 1, \ldots, N \tag{5.73}$$

In summary, a direction for searching for new values for the decision variables can be determined by solving a minimization problem with a criterion given by Eq. (5.71) and with constraints given by Eqs. (5.61)–(5.65), (5.73). Actually, we are interested in the coordination parameters $\mathbf{y}_i$, $\mathbf{l}_i$ $(i = 1, \ldots, N)$ and, so, from the result of the optimization, we take the corresponding $\Delta \mathbf{y}_i$, $\Delta \mathbf{l}_i$ direction to be used in determining new values for the next iteration. (Since only these components of $\mathbf{z}$ are used, a possible nonlinearity of the $\mathbf{H}_i$'s is not critical in many cases. That is, even if $\Delta \mathbf{z}$ might not be a feasible direction, the $\Delta \mathbf{y}_i$ and $\Delta \mathbf{l}_i$ directions are more likely to be feasible.)

Note that, in a general case, determining new coordination variables includes solving a linear programming (LP) problem that covers the entire system. In less general cases, the LP problem may split naturally into separate LP problems—depending on the connections between subsystems.

Due to the simple feasible method used, convergence of the algorithm is not always guaranteed. That is, this simple algorithm may stop before all necessary equations given by Eqs. (5.36)–(5.48) are satisfied. In any case, it is easy to verify that if we are at a point where these equations are satisfied, the above minimization problem will yield $\Delta\mathbf{z} = \mathbf{0}$ or that $\Delta\mathbf{z}$ is arbitrary (corresponding to a zero gradient of U without active constraints.)

The convergence of the algorithm can be improved by the same techniques as in ordinary feasible methods (see Luenberger, 1973).

Figure 5.1 summarizes the feasible method with partial derivatives.

The scheme discussed in this section is a multiobjective counterpart of the single-objective feasible method, reported by Brosilow et al. (1965).

The main theoretical difficulty in feasible methods is that the $\mathbf{y}_i$'s should be kept feasible all the time. In practice, in the single-objective case, this issue often seems not to be critical. Instead, more practical difficulties have been encountered with slow convergence and with the amount of computation that is required when the number of multipliers is large. The nonfeasible methods do not generally have these difficulties, but they demand strict convexity.

5.5.2 Feasible Method with Total Derivatives

Consider the problem formulated in Sect. 5.3 in the special case where the subsystem inequalities given by Eq. (5.18) and the overall inequalities (5.20) and (5.21) are absent. Furthermore, assume that all values of the $\mathbf{y}_i$'s are feasible.

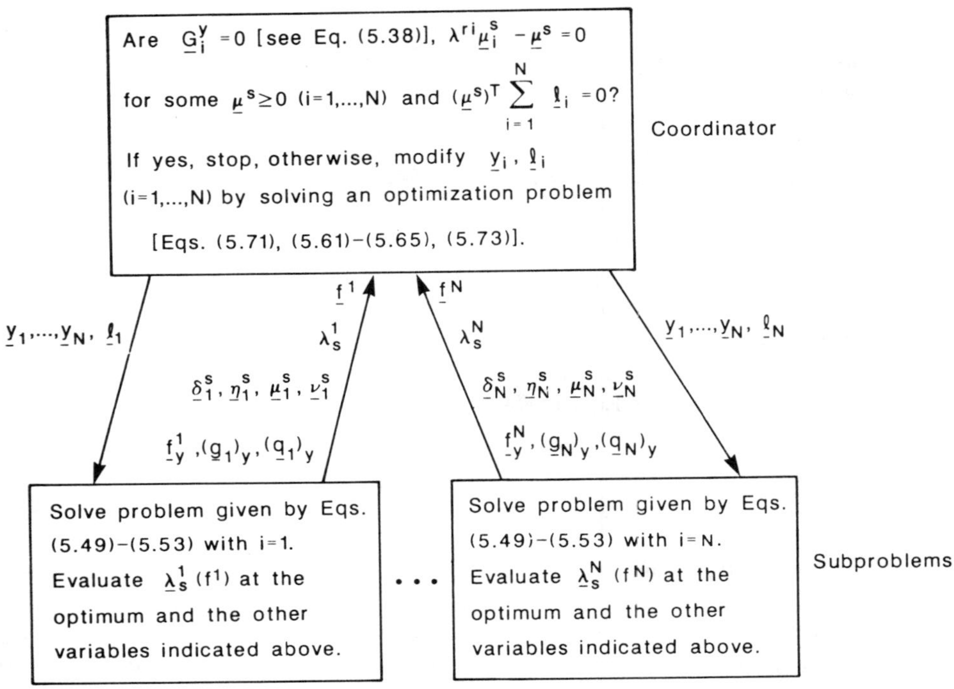

Figure 5.1 Feasible method with partial derivatives (indicated by subscripts).

In this case, the coordination variable consists of only $\mathbf{y}_i$'s. Denote $\mathbf{y} = (\mathbf{y}_1, \ldots, \mathbf{y}_N)^T$. As in Sect. 5.5.1.1, given a value $\mathbf{y}^k$ for the coordination variable $\mathbf{y}$, the subproblems given by Eqs. (5.49), (5.50), and (5.53) are solved. The solutions, $\mathbf{m}_i(\mathbf{y})$ ($i = 1, \ldots, N$) are functions of $\mathbf{y}$. The values of the $\mathbf{x}_i$'s were determined by Eq. (5.46). Denote

$$\mathbf{x}_i(\mathbf{y}) = \sum_{j=1}^{N} C_{ij}\mathbf{y}_j \qquad i = 1, \ldots, N$$

The values of the objective vectors $\mathbf{f}^i$ are functions of $\mathbf{y}$, denoted by $\tilde{\mathbf{f}}^i(\mathbf{y}) = \mathbf{f}^i(\mathbf{x}_i(\mathbf{y}), \mathbf{m}_i(\mathbf{y}), \mathbf{y}_i)$. Using this notation, a new form for the coordination equations given by Eq. (5.38) is derived.

First, note by direct differentiation that for $i, j = 1, \ldots, N$,

$$\frac{\partial \tilde{\mathbf{f}}^i(\mathbf{y})}{\partial \mathbf{y}_j} = \frac{\partial \mathbf{f}^i}{\partial \mathbf{x}_i}\frac{\partial \mathbf{x}_i}{\partial \mathbf{y}_j} + \frac{\partial \mathbf{f}^i}{\partial \mathbf{m}_i}\frac{\partial \mathbf{m}_i}{\partial \mathbf{y}_j} + \begin{cases} \dfrac{\partial \mathbf{f}^i}{\partial \mathbf{y}_j} & j = i \\ \mathbf{0} & j \neq i \end{cases} \tag{5.74}$$

Note that $\partial \mathbf{x}_i/\partial \mathbf{y}_j = C_{ij}$. Also note that, because $-\mathbf{y}_i + \mathrm{H}_i(\mathbf{x}_i(\mathbf{y}), \mathbf{m}_i(\mathbf{y})) = \mathbf{0}$ for all values of $\mathbf{y}$, the derivative of this equation with respect to $\mathbf{y}$ is zero. That is,

$$\frac{\partial \mathbf{H}_i}{\partial \mathbf{x}_i}\frac{\partial \mathbf{x}_i}{\partial \mathbf{y}_j} + \frac{\partial \mathbf{H}_i}{\partial \mathbf{m}_i}\frac{\partial \mathbf{m}_i}{\partial \mathbf{y}_j} = \begin{cases} I & \text{if } j = i \\ 0 & \text{if } j \neq i \end{cases} \tag{5.75}$$

where I and O denote the identity and null matrices.

Consider Eqs. (5.36)–(5.38), where the suppressed arguments are $\mathbf{x}_i^k = \mathbf{x}_i(\mathbf{y}^k)$, $\mathbf{m}_i^k = \mathbf{m}_i(\mathbf{y}^k)$, and $\mathbf{y}_i^k$ ($i = 1, \ldots, N$). For brevity, assume the definitions $\mathbf{f}^{ik} = \mathbf{f}^i(\mathbf{x}_i^k, \mathbf{m}_i^k, \mathbf{y}_i^k)$, $i = 1, \ldots, N$ and $\mathbf{f}^k = (\mathbf{f}^{1k}, \ldots, \mathbf{f}^{Nk})^T$. Equation (5.37) is satisfied as the necessary conditions, and Eq. (5.36) is trivially satisfied for some $\boldsymbol{\nu}_i^s$'s (which are not of interest here). It remains to be seen whether Eq. (5.38) holds. In the following step, Eq. (5.38), with $i = j$, is modified, where j is an arbitrary index belonging to $\{1, \ldots, N\}$.

Multiply Eqs. (5.36) and (5.37) by $\lambda^{ri}(\mathbf{f}^k) = \lambda_1^i(\mathbf{f}^k)/\lambda_1^r(\mathbf{f}^k)$, $i = 1, \ldots, N$, and Eq. (5.38) with $i = j$ by $\lambda^{rj}(\mathbf{f}^k)$, and return to the unscaled multipliers [Eq. (5.34)]. Postmultiply the new Eq. (5.36) by $\partial \mathbf{x}_i/\partial \mathbf{y}_j = C_{ij}$ and the new Eq. (5.37) by $\partial \mathbf{m}_i/\partial \mathbf{y}_j$, $i = 1, \ldots, N$. Add all the resulting equations to the new Eq. (5.38), with $i = j$. Taking into account Eqs. (5.74) and (5.75) and recalling $\mathbf{g}_i \equiv \mathbf{0}$, $\mathbf{q}_i \equiv \mathbf{0}$ ($i = 1, \ldots, N$) we will get

$$\sum_{i=1}^{N} \lambda^{ri}(\mathbf{f}^k)\boldsymbol{\lambda}_s^i(\mathbf{f}^{ik})^T \frac{\partial \tilde{\mathbf{f}}^i(\mathbf{y})}{\partial \mathbf{y}_j} = \mathbf{0} \tag{5.76}$$

Here, j is arbitrary, and thus Eq. (5.76) holds for $j = 1, \ldots, N$. These equations are presented compactly in a vector form in Fig. 5.2, which summarizes this scheme. Note that, as in Fig. 5.1, the subproblems may be solved by any multicriteria optimization method. Only the indifference trade-offs at the solutions are needed for the upper level.

In this scheme, the total derivatives $d\mathbf{f}^i/d\mathbf{y}$ may be calculated numerically by

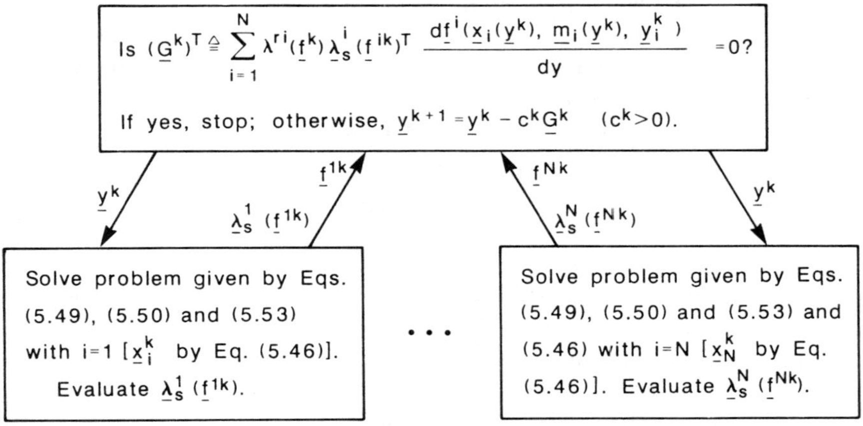

Figure 5.2 Feasible method with total derivatives with respect to the coordination variables.

repeating the suboptimizations for perturbed values of $\mathbf{y}$ and approximating the derivatives by ratios of differences. Alternatively, the $\partial\mathbf{x}_i/\partial\mathbf{y}_i$ and $\partial\mathbf{m}_i/\partial\mathbf{y}_j$ derivatives are evaluated numerically, and Eq. (5.74) is used.

The coordination scheme becomes simple now that the inequality constraints are absent [i.e., $\mathbf{g}_i \equiv \mathbf{0}$, $\mathbf{q}_i \equiv \mathbf{0}$ $(i = 1, \ldots, N)$] and the system equations do not impose any constraints on the $\Delta\mathbf{y}_i$'s (i.e., interaction variables between the subsystems can take any value). In this case, we have to minimize [cf. Eq. (5.71)] the expression $\Sigma_{i=1}^{N}(\lambda^{ri})^k(\mathbf{G}_i^y)^T\Delta\mathbf{y}_i$ subject to a normalizing equation, such as Eq. (5.60). This is a simple optimization task, with the following analytic solution (where c is a normalizing constant):

$$\Delta\mathbf{y}_i = -c(\lambda^{ri})^k(\mathbf{G}_i^y) \qquad i = 1, \ldots, N \tag{5.77}$$

Recall the derivation of Eq. (5.76): Eq. (5.76) is Eq. (5.38) (with $i = j$) multiplied by $\lambda^{rj}(\mathbf{f}^k)$ $[=(\lambda^{rj})^k]$; hence,

$$(\lambda^{rj})^k(\mathbf{G}_j^y)^T = \sum_{i=1}^{N} \lambda^{ri}(\mathbf{f}^k)\boldsymbol{\lambda}_s^i(\mathbf{f}^{ik})^T \frac{\partial\tilde{\mathbf{f}}^i(\mathbf{y})}{\partial\mathbf{y}_j} \qquad j = 1, \ldots, N \tag{5.78}$$

Substituting Eq. (5.78) into Eq. (5.77) yields, in vector form, the coordination scheme given in Fig. 5.2. In this scheme, the parameter c^k counts for the step length, whose determination was discussed above.

This includes the development of basic feasible schemes. The derived schemes generalize those presented by Geoffrion and Hogan (1972) with respect to problem formulation and assumptions.

5.6 NONFEASIBLE METHODS

To derive nonfeasible schemes, Eqs. (5.36)–(5.48) are used as in developing feasible ones. However, in this case, the $\boldsymbol{\nu}_i^s$'s, $\boldsymbol{\mu}^s$, and the λ^{ri}'s are coordination variables with assigned values denoted by $\boldsymbol{\nu}_i^k$ $(i = 1, \ldots, N)$, $\boldsymbol{\mu}^k \geqslant 0$ [so that Eq. (5.43) holds], and $(\lambda^{ri})^k$ $(i = 1, \ldots, N)$ at the beginning of the kth iteration. Recall that $\lambda^{rr} = 1$. Also, other λ^{ri}'s may be known constants, as in the example following Eq. (5.27). For presentational convenience, the same notation, $(\lambda^{ri})^k$, is also used for those λ^{ri}'s that are known constants.

First, specify [so that Eq. (5.39) is satisfied]

$$\boldsymbol{\mu}_i^k = \frac{\boldsymbol{\mu}^k}{(\lambda^{ri})^k} \qquad i = 1, \ldots, N \tag{5.79}$$

Then, consider Eqs. (5.36)–(5.38), (5.40), (5.41), (5.44), and (5.45). Recalling that the first component of $\boldsymbol{\lambda}_s^i$, $(\boldsymbol{\lambda}_s^i)_1$ is unity $(i = 1, \ldots, N)$ and rearranging terms using identities of the type $(\boldsymbol{\nu}_i^k)^T = \partial[(\boldsymbol{\nu}_i^k)^T\mathbf{x}_i]/\partial\mathbf{x}_i$, we have Eqs. (5.36)–(5.38) (where we have some values specified, as explained above) in the following equivalent compact form, where $\mathbf{z}_i = (\mathbf{x}_i, \mathbf{m}_i, \mathbf{y}_i)^T$:

$$\frac{\partial}{\partial \mathbf{z}_i}\left\{f_1^i + \sum_{l=1}^{N}\left[\frac{(\lambda^{rl})^k}{(\lambda^{ri})^k}(\boldsymbol{\nu}_l^k)^T C_{li}\right]\mathbf{y} - (\boldsymbol{\nu}_i^k)^T\mathbf{x}_i + \frac{(\boldsymbol{\mu}^k)^T}{(\lambda^{ri})^k}\mathbf{q}_i\right\}$$

$$+ \sum_{j=2}^{n_i}\left[(\boldsymbol{\lambda}_s^i)_j(\mathbf{f}^i)\frac{\partial f_j^i}{\partial \mathbf{z}_i}\right] + (\boldsymbol{\delta}_i^s)^T\frac{\partial}{\partial \mathbf{z}_i}(\mathbf{y}_i - \mathbf{H}_i) + (\boldsymbol{\eta}_i^s)^T\frac{\partial \mathbf{g}_i}{\partial \mathbf{z}_i} = \mathbf{0} \tag{5.80}$$

for $i = 1, \ldots, N$. We note that this equation, along with Eqs. (5.40), (5.41), (5.44), and (5.45), is a first-order characterization for the solution to the following multiobjective optimization problem $(i = 1, \ldots, N)$:

$$\min_{\mathbf{x}_i,\mathbf{m}_i,\mathbf{y}_i}\begin{bmatrix} f_1^i(\mathbf{x}_i, \mathbf{m}_i, \mathbf{y}_i) + \left[\displaystyle\sum_{l=1}^{N}\frac{(\lambda^{rl})^k}{(\lambda^{ri})^k}(\boldsymbol{\nu}_l^k)^T C_{li}\right]\mathbf{y}_i - (\boldsymbol{\nu}_i^k)^T\mathbf{x}_i + \dfrac{(\boldsymbol{\mu}^k)^T}{(\lambda^{ri})^k}\mathbf{q}_i \\ f_2^i(\mathbf{x}_i, \mathbf{m}_i, \mathbf{y}_i) \\ \vdots \\ f_{n_i}^i(\mathbf{x}_i, \mathbf{m}_i, \mathbf{y}_i) \end{bmatrix} \tag{5.81}$$

subject to

$$\mathbf{y}_i = \mathbf{H}_i(\mathbf{x}_i, \mathbf{m}_i) \tag{5.82}$$

$$\mathbf{g}_i(\mathbf{x}_i, \mathbf{m}_i, \mathbf{y}_i) \leqslant \mathbf{0} \tag{5.83}$$

with a trade-off vector

$$\boldsymbol{\lambda}_s^i[\mathbf{f}^i(\mathbf{x}_i, \mathbf{m}_i, \mathbf{y}_i)] \tag{5.84}$$

Note that the trade-offs in the problem given by Eqs. (5.81)–(5.84) are evaluated using the trade-off vector $\boldsymbol{\lambda}_s^i$ at $\mathbf{f}^i$, rather than at the values of the objectives in Eq. (5.81). This modification can be directly implemented in methods dealing with trade-offs.

Assume, for a moment, that the problems given by Eqs. (5.81)–(5.84) have solutions denoted by $\mathbf{x}_i^k$, $\mathbf{m}_i^k$, $\mathbf{y}_i^k$, $i = 1, \ldots, N$. The subsystem solutions satisfy the necessary conditions given by Eqs. (5.36)–(5.38). Furthermore, Eq. (5.47) is trivially satisfied by specifying $\mathbf{l}_i^k = \mathbf{q}_i(\mathbf{x}_i^k, \mathbf{m}_i^k, \mathbf{y}_i^k)$, $i = 1, \ldots, N$. Thus, Eqs.

(5.46), (5.48), and (5.42), which comprise variables from several subsystems, are checked at the upper level. In addition, the values of the guessed $(\boldsymbol{\lambda}^{ri})^k$'s are checked against the values of objectives for consistency. The satisfaction of the following equalities and inequalities is thus checked:

$$\sum_{j=1}^{N} C_{ij}\mathbf{y}_j^k - \mathbf{x}_i^k = \mathbf{0} \qquad i = 1, \ldots, N \tag{5.85}$$

$$\sum_{i=1}^{N} \mathbf{q}_i(\mathbf{x}_i^k\ \mathbf{m}_i^k, \mathbf{y}_i^k) \leqslant \mathbf{0} \tag{5.86}$$

$$(\boldsymbol{\mu}^k)^T \sum_{i=1}^{N} \mathbf{q}_i(\mathbf{x}_i^k, \mathbf{m}_i^k, \mathbf{y}_i^k) = 0 \tag{5.87}$$

$$\lambda^{ri}(\mathbf{f}^k) - (\lambda^{ri})^k = 0 \qquad i = 1, \ldots, N \qquad i \neq r \tag{5.88}$$

If these equations are satisfied, Eqs. (5.36)–(5.48) have been solved, and the original multiobjective problem has also been solved. Otherwise, new values are assigned for the coordination variables, and the subsystem optimization is repeated until Eqs. (5.85)–(5.88) hold. The scheme is summarized in Fig. 5.3 (for the case $\mathbf{q}_i \equiv \mathbf{0}$, $i = 1, \ldots, N$).

An alternative subproblem formulation is also presented in Fig. 5.3. Rearranging the terms in Eq. (5.80) yields the necessary conditions of the alternative subproblem formulation:

$$(\boldsymbol{\lambda}_s^i)^T \frac{\partial \mathbf{f}^i}{\partial \mathbf{z}_i} + \mathbf{1} \cdot \frac{\partial}{\partial \mathbf{z}_i} \left\{ \left[\sum_{l=1}^{N} \frac{(\lambda^{rl})^k}{(\lambda^{ri})^k} (\boldsymbol{\nu}_l^k)^T C_{li} \right] \mathbf{y}_i - (\boldsymbol{\nu}_i^k)^T \mathbf{x}_i + \frac{(\boldsymbol{\mu}^k)^T}{(\lambda^{ri})^k} q_i \right\}$$

$$+ (\boldsymbol{\delta}_i^s)^T \frac{\partial}{\partial \mathbf{z}_i} (\mathbf{y}_i - \mathbf{H}_i) + (\boldsymbol{\eta}_i^s)^T \frac{\partial \mathbf{g}_i}{\partial \mathbf{z}_i} = 0 \qquad i = 1, \ldots, N \tag{5.89}$$

The corresponding optimization problem (for $i = N$) is given in Fig. 5.3. The difference between this formulation and the earlier one is technical; however, when using a specific multiobjective method to solve the subproblems, either of these formulations may fit.

Assuming the existence of an underlying loss function, the coordination procedures of the single-objective case can be evoked (see Lasdon, 1970). That is, in terms of the scaled multipliers, the following updating formula for $\boldsymbol{\nu}_i$ and $\boldsymbol{\mu}$ can be used:

$$\boldsymbol{\nu}_i^{k+1} = \boldsymbol{\nu}_i^k + \frac{c^k}{\lambda^{ri}} \mathbf{e}_i(\mathbf{z}^k) \qquad i = 1, \ldots, N \tag{5.90}$$

with

$$\mathbf{e}_i(\mathbf{z}^k) = \sum_{l=1}^{N} C_{il}\mathbf{y}_l^k - \mathbf{x}_i^k$$

where $c^k > 0$. The parameter c^k is determined so that $\|\mathbf{e}(\mathbf{z}^{k+1})\| < \|\mathbf{e}(\mathbf{z}^k)\|$.

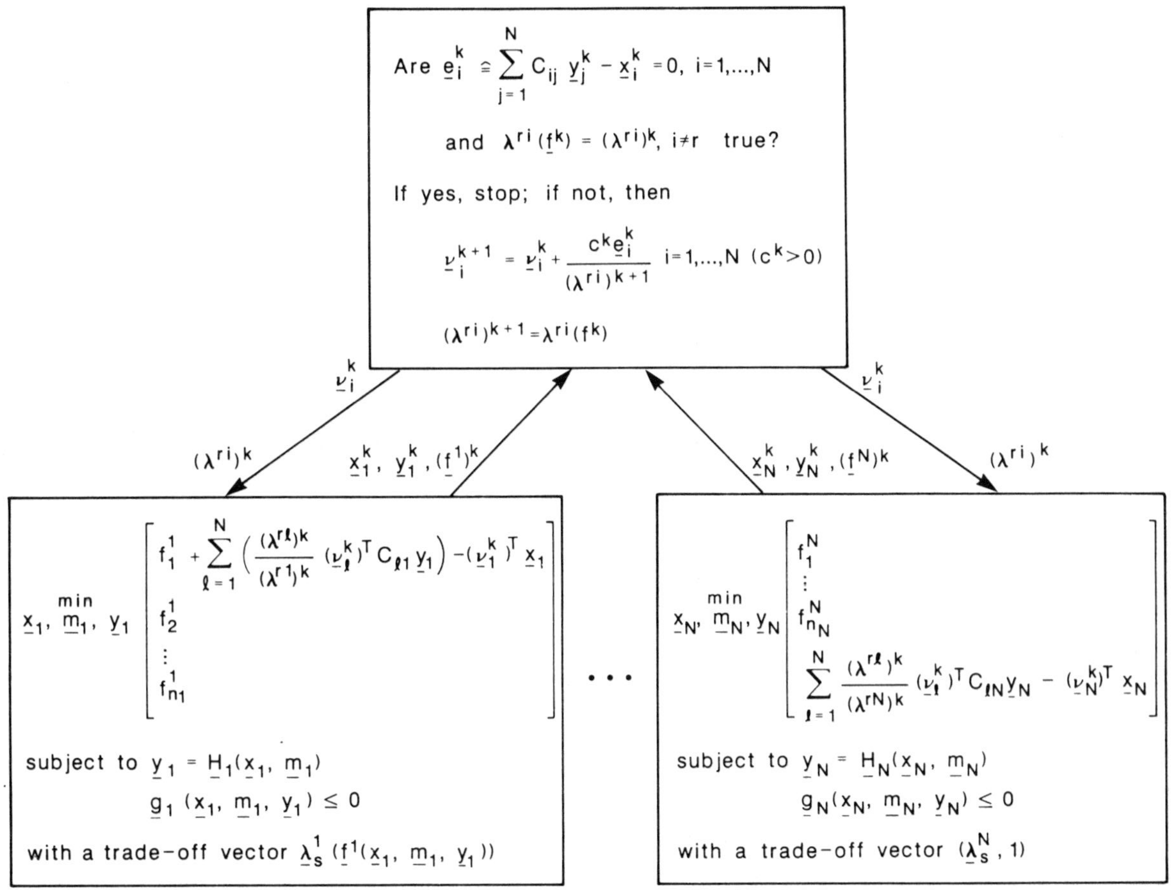

Figure 5.3 Nonfeasible method with two different subproblem formulations, $q_i \equiv 0$ $(i = 1, \dots, N)$.

$$\mu_j^{k+1} = \mu_j^k + \begin{cases} c_q^k \sum_{i=1}^{N} q_{ij}(\mathbf{z}^k) & \text{if } \mu_j^k > \mathbf{0} \\ c_q^k \max\left\{0, \sum_{i=1}^{N} q_{ij}(\mathbf{z}^k)\right\} & \text{if } \mu_j^k = \mathbf{0} \end{cases} \tag{5.91}$$

where $c_q^k > 0$.

The most convenient way to update λ^{ri} is to use a technique often called the relaxation method, which is common in engineering and physics, and specify that

$$(\lambda^{ri})^{k+1} = \lambda^{ri}[f(\mathbf{z}^k)] \qquad i \neq r \tag{5.92}$$

If this technique causes instabilities due to major changes in λ^{ri}, use Eq. (5.93):

$$(\lambda^{ri})^{k+1} = c_\lambda(\lambda^{ri})^k + (1 - c_\lambda)\lambda^{ri}[f(\mathbf{z}^k)] \qquad i \neq r \qquad 0 < c_\lambda < 1 \tag{5.93}$$

When λ^{ri} are changed, the new values are used in the updating formula for $\boldsymbol{\nu}$ [Eq. (5.90)]. However, there is no guarantee that, with a parallel change of λ^{ri}, the strategy of using Eq. (5.90) would decrease the norm of the unbalance $\mathbf{e} \triangleq (\mathbf{e}_1, \ldots, \mathbf{e}_N)^T$. Mathematically, one can determine an optimal change $\boldsymbol{\nu}^{k+1} - \boldsymbol{\nu}^k$ for the case with a change of λ^{ri}. Usually, if $\|\mathbf{e}\|$ increases after a change of λ^{ri}, it is easier to keep λ^{ri} constant until $\|\mathbf{e}\|$ decreases. The safest policy is to change λ^{ri} only when $\mathbf{e} = \mathbf{0}$ (and $\mathbf{q} \leq \mathbf{0}$); however, it is useful to try to change λ^{ri} concurrently when driving the unbalance to zero.

Let us return to the assumption that the subproblems given by Eqs. (5.81)–(5.84) have solutions. A similar existence assumption was made in connection with the derivation of feasible methods. However, in the feasible methods, the subproblems can be considered as constrained versions of the original problem where the coupling variables are fixed. The subproblems given by Eqs. (5.81)–(5.84) are not now in such a direct relationship to the original problem.

As in the single-objective case, convexity assumptions must be made. Otherwise, the linear $\mathbf{y}_i$ and $\mathbf{x}_i$ terms in the first objective of Eq. (5.81) may either drive the solutions to infinity or there may be an infinite number of solutions (among which the correct overall solution resides). If the existence of an underlying loss function is assumed, sufficient conditions for the existence of subsystem solutions and the convergence of the scheme can be derived directly based on the results in the single-objective optimization (see Lasdon, 1970).

Linearity of the system equations ($\mathbf{H}_i$), convexity of the inequality constraints ($\mathbf{g}_i$, $\mathbf{q}_i$), and strict convexity of the underlying loss function are generally needed. Because the loss function is not assumed to be known beforehand, the corresponding assumption cannot be verified. In general, a trial has to be given to the method.

5.7 EXTENSIONS TO BASIC SCHEMES

5.7.1 Other Coordination Schemes

From the above treatment, it is evident that much of the research conducted on hierarchical single-objective optimization can be extended to the multiobjective case. For example, the mixed method (see e.g., Smith and Sage, 1973; Singh and Titli, 1978; Singh, 1980) can be directly generalized to the multiobjective case. It is a nonfeasible method where the outputs, the $\mathbf{y}_i$ (or the inputs), are given by the upper level and changed in the same manner, based on unbalances, as in the nonfeasible method.

The feasible and nonfeasible methods can also be combined so that some interconnections between subsystems are handled by the feasible method and others by the nonfeasible method, as suggested in Sect. 1 in Wismer (1971).

5.7.2 Trade-Offs That Depend on Objectives in Other Subsystems

So far we have been concerned with the case where the internal trade-offs in each subsystem do not depend on the objectives in other subsystems. When this is not the case, one may proceed exactly in the same manner as for λ^{ri} above. That is, for a trade-off that depends on outside objectives, one must guess a value at the beginning. When all subsystem optimizations are carried out, check whether the guessed value corresponds to the value of objectives. If not, change the guessed value, as λ^{ri} was changed in Eq. (5.92) or (5.93). This procedure may be applied directly to the objectives that are outside a given subsystem and affect trade-offs.

Alternatively, treat the objectives from other subsystems that affect a trade-off as if they were additional coordination parameters. If, for example, an objective $f_j^i(\mathbf{z}_i)$ affects a trade-off in a subsystem $k \neq i$, take a coordination parameter α. For each value of α as a constraint level, let $f_j^i(\mathbf{z}_i) = \alpha$ in the ith subsystem. When evaluating the trade-off in question in the ith subsystem, the value of α is used for the objective $f_j^i(\mathbf{z}_i)$. In the feasible method, this modification is directly implemented. In the nonfeasible method, change the value of α, as in the feasible method; in effect, the method will be mixed. Such a way of handling this kind of trade-off is the same as handling so called nonseparable problems in the single-objective case (see, e.g., Singh and Titli, 1978).

5.8 NUMERICAL EXAMPLES

Let us consider a static multiobjective problem. It will be solved as a whole (without decomposition), then by a feasible method, and finally by a nonfeasible method in Sects. 5.8.1–5.8.3, respectively. In Sect. 5.8.4, the problem will be modified to a general one, where a subsystem's trade-offs depend on objectives of other subsystems. This modified problem will then be solved using a nonfeasible scheme. We will use a mathematical model as the decisionmaker—skipping the proper, often iterative, decisionmaking process by analytic solutions.

5.8.1 The Problem and Its Integrated Solution

Consider the following multiobjective problem:

$$\min \begin{bmatrix} F_1 \\ F_2 \\ F_3 \end{bmatrix} \quad \text{with} \quad \begin{aligned} F_1 &= f_1^1 + f_1^2 \\ F_2 &= f_2^1 \\ F_3 &= f_2^2 \end{aligned} \tag{5.94}$$

where $\mathbf{m}_1 = (m_{11}, m_{12}, m_{13})^T$, $\mathbf{m}_2 = (m_{21}, m_{22}, m_{23})^T$,

$$f_1^1 = f_1^1(\mathbf{m}_1, x_1) = (m_{11} - 1)^2 - 2m_{11}m_{12} + 2m_{12}^2 + m_{13}^2 + 10x_1^2 \qquad (5.95a)$$

$$f_2^1 = f_2^1(\mathbf{m}_1, x_1) = m_{11} + m_{12} + 5 + x_1 \qquad (5.95b)$$

$$f_1^2 = f_1^2(\mathbf{m}_2, x_2) = m_{21}^2 + 2(m_{22} - 5)^2 + (x_2 - 10)^2 + m_{23}^2 \qquad (5.95c)$$

$$f_2^2 = f_2^2(\mathbf{m}_2, x_2) = m_{21} + m_{22} \qquad (5.95d)$$

subject to

$$y_1 = 2m_{11} - m_{12} + m_{13} + x_1 \qquad (5.95e)$$

$$y_2 = m_{21} + m_{22} - m_{23} + x_2 + 2 \qquad (5.95f)$$

$$x_1 = y_2 \qquad (5.95g)$$

$$x_2 = y_1 \qquad (5.95h)$$

The system is depicted in Fig. 5.4. Let the acceptable trade-off vector of the decisionmaker be

$$\boldsymbol{\lambda}(\mathbf{f}) = [\lambda_1(\mathbf{f}), \lambda_2(\mathbf{f}), \lambda_3(\mathbf{f})]^T = [1, \lambda_{12}^*(\mathbf{f}), \lambda_{13}^*(\mathbf{f})]^T \qquad (5.96a)$$

$$\lambda_{12}^*(\mathbf{f}) = 0.1f_2^1 + 0.1 \qquad (5.96b)$$

$$\lambda_{13}^*(\mathbf{f}) = 0.1f_2^2 + 1 \qquad (5.96c)$$

For illustrative purposes, we have assumed the uniqueness of acceptable trade-offs—without taking into account that, in practice, there is usually an indifference band, as explained in the SWT method in Chap. 2.

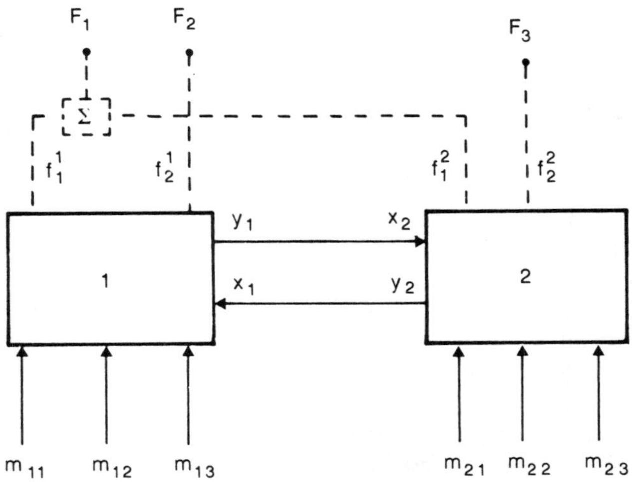

Figure 5.4 Example system.

We can immediately simplify the problem, eliminating the system equations (5.95e) and (5.95f) by substituting m_{13} given by Eq. (5.95e) and m_{23} given by Eq. (5.95f) into other equations. We have, therefore, as new expressions for objectives, the following functions, whose symbols differ through arguments from the corresponding previous ones:

$$f_1^1(m_{11}, m_{12}, x_1, y_1) = (m_{11} - 1)^2 - 2m_{11}m_{12} + 2m_{12}^2 + (y_1 - x_1 - 2m_{11} + m_{12})^2 + 10x_1^2 \tag{5.97a}$$

$$f_2^1(m_{11}, m_{12}, x_1, y_1) = f_2^1(\mathbf{m}_1, x_1) = m_{11} + m_{12} + 5 + x_1 \tag{5.97b}$$

$$f_1^2(m_{21}, m_{22}, x_2, y_2) = m_{21}^2 + 2(m_{22} - 5)^2 + (x_2 - 10)^2 + (m_{21} + m_{22} + x_2 + 2 - y_2)^2 \tag{5.97c}$$

$$f_2^2(m_{21}, m_{22}, x_2, y_2) = f_2^2(\mathbf{m}_2, x_2) = m_{21} + m_{22} \tag{5.97d}$$

Thus, in the following examples, we treat the problem given by Eq. (5.94) with expressions for objectives given by Eqs. (5.97a)–(5.97d), subject to the coupling constraints of Eqs. (5.95g) and (5.95h), and with a mathematical model for the decisionmaker's preferences in the parts of Eq. (5.96).

Using this mathematical model in place of the decisionmaker, we can solve the problem analytically as a whole and in multilevel techniques. In practice, we usually have a process of interaction with a decisionmaker. During this process, a trade-off function $\boldsymbol{\lambda}(\mathbf{f})$, such as Eq. (5.96), is evaluated by the decisionmaker—only at some values of objectives, partly based on the information the decisionmaker gains during the interaction. (See Sect. 5.9.)

The problem will be solved, first, as a whole. For this solution, the problem can be further simplified by eliminating the coupling equations (5.95g) and (5.95h) by substituting x_1, given by Eq. (5.95g), and x_2, given by Eq. (5.95h), into other expressions in Eq. (5.97).

The problem now assumes the following unconstrained form:

$$\min_{\mathbf{z}} \begin{bmatrix} G(\mathbf{z}) \\ F_1(\mathbf{z}) \\ \vdots \\ F_r(\mathbf{z}) \end{bmatrix} \quad \text{with} \quad \begin{aligned} G(\mathbf{z}) &= \mathbf{z}^T A\mathbf{z} + \mathbf{b}^T\mathbf{z} \\ F_1(\mathbf{z}) &= \mathbf{c}_1^T\mathbf{z} + d_1 \\ &\vdots \\ F_r(\mathbf{z}) &= \mathbf{c}_r^T\mathbf{z} + d_r \end{aligned} \tag{5.98a}$$

with a trade-off vector

$$(1, \boldsymbol{\alpha}_1^T\mathbf{F} + \beta_1, \ldots, \boldsymbol{\alpha}_r^T\mathbf{F} + \beta_r)^T \tag{5.98b}$$

where $\mathbf{F} = [F_1(\mathbf{z}), \ldots, F_r(\mathbf{z})]^T$; $\mathbf{z}$, $\mathbf{b}$, $\mathbf{c}_i$'s, and $\boldsymbol{\alpha}_i$'s are vectors; A is a positive definite matrix; and the d_i's and β's are scalars.

We solve this general problem. Later on, in the decomposition methods, the subproblems will be of the same form.

The problem given by Eqs. (5.98a) and (5.98b) can be solved, for example, by using the necessary conditions given in Eqs. (5.14) and (5.15). Because there are no constraints, we have an equality to be solved:

$$\frac{\partial}{\partial \mathbf{z}}(\mathbf{z}^T A\mathbf{z} + \mathbf{b}^T\mathbf{z}) + \sum_{i=1}^{r} [\boldsymbol{\alpha}_i^T\mathbf{F}(\mathbf{z}) + \beta_i]\frac{\partial}{\partial \mathbf{z}}(\mathbf{c}_i^T\mathbf{z} + d_i) = \mathbf{0} \tag{5.99}$$

Taking the derivatives, taking into account the $\mathbf{F}(\mathbf{z})$'s expression, and denoting

$$\mathbf{C}^T = \begin{bmatrix} \mathbf{c}_1^T \\ \vdots \\ \mathbf{c}_r^T \end{bmatrix} \quad \mathbf{d} = \begin{bmatrix} d_1 \\ \vdots \\ d_r \end{bmatrix} \quad \boldsymbol{\alpha} = \begin{bmatrix} \boldsymbol{\alpha}_1^T \\ \vdots \\ \boldsymbol{\alpha}_r^T \end{bmatrix} \quad \boldsymbol{\beta} = \begin{bmatrix} \beta_1 \\ \vdots \\ \beta_r \end{bmatrix} \tag{5.100}$$

we have

$$2\mathbf{z}^T A + \mathbf{b}^T + \mathbf{z}^T\mathbf{C}\boldsymbol{\alpha}^T\mathbf{C}^T + \mathbf{d}^T\boldsymbol{\alpha}^T\mathbf{C}^T + \boldsymbol{\beta}^T\mathbf{C}^T = \mathbf{0} \tag{5.101}$$

From this linear equation in $\mathbf{z}$, we achieve, as a unique solution to the problem given by Eq. (5.98),

$$\mathbf{z}^0 = -(2A + \mathbf{C}\boldsymbol{\alpha}\mathbf{C}^T)^{-1}(\mathbf{b} + \mathbf{C}\boldsymbol{\alpha}\mathbf{d} + \mathbf{C}\boldsymbol{\beta}) \tag{5.102}$$

if the inverse exists.

In our specific problem (after the substitutions mentioned) and in terms of the general formulation, we have

$$\mathbf{z} = \begin{bmatrix} m_{11} \\ m_{12} \\ m_{21} \\ m_{22} \\ y_2 \\ y_1 \end{bmatrix} \qquad A = \begin{bmatrix} 5 & -3 & 0 & 0 & 2 & -2 \\ -3 & 3 & 0 & 0 & -1 & 1 \\ 0 & 0 & 2 & 1 & -1 & 1 \\ 0 & 0 & 1 & 3 & -1 & 1 \\ 2 & -1 & -1 & -1 & 12 & -2 \\ -2 & 1 & 1 & 1 & -2 & 3 \end{bmatrix}$$

$$\mathbf{b} = \begin{bmatrix} -2 \\ 0 \\ 4 \\ -16 \\ -4 \\ -16 \end{bmatrix} \qquad \mathbf{C}^T = \begin{bmatrix} 1 & 1 & 0 & 0 & 1 & 0 \\ 0 & 0 & 1 & 1 & 0 & 0 \end{bmatrix}$$

$$\mathbf{d} = \begin{bmatrix} 5 \\ 0 \end{bmatrix} \qquad \boldsymbol{\alpha} = \begin{bmatrix} 0.1 & 0 \\ 0 & 0.1 \end{bmatrix} \qquad \boldsymbol{\beta} = \begin{bmatrix} 0.1 \\ 1 \end{bmatrix}$$

Applying Eq. (5.102), we establish

$$\mathbf{z}^0 = \begin{bmatrix} m_{11} \\ m_{12} \\ m_{21} \\ m_{22} \\ y_2 \\ y_1 \end{bmatrix} = \begin{bmatrix} 2.318 \\ 0.6359 \\ -4.787 \\ 2.606 \\ 0.4502 \\ 5.027 \end{bmatrix} \tag{5.103}$$

and

$$m_{13} = y_1 - y_2 - 2m_{11} + m_{12} = 0.5763 \tag{5.104a}$$

$$m_{23} = m_{21} + m_{22} + y_1 - y_2 + 2 = 4.396 \tag{5.104b}$$

$$f_1^1 = 1.958 \qquad f_1^2 = 78.431 \qquad f_1^1 + f_1^2 = 80.389$$

$$f_2^1 = 8.404 \qquad f_2^2 = -2.181$$

$$\lambda_{12}^* = 0.9404 \tag{5.105a}$$

$$\lambda_{13}^* = 0.7819 \tag{5.105b}$$

5.8.2 Solution Using the Feasible Method

Let us first determine the internal trade-off vectors $\boldsymbol{\lambda}^i$ ($i = 1, 2$) and the trade-off between these two subsystems, λ^{ri}, to be used in the multilevel methods. From Eq. (5.27) we have

$$\boldsymbol{\lambda}^1(\mathbf{f}) = \lambda_1 \left[\frac{\partial F_1}{\partial f_1^1}, \frac{\partial F_1}{\partial f_2^1} \right]^T + \lambda_2 \left[\frac{\partial F_2}{\partial f_1^1}, \frac{\partial F_2}{\partial f_2^1} \right]^T + \lambda_3 \left[\frac{\partial F_3}{\partial f_1^1}, \frac{\partial F_3}{\partial f_2^1} \right]^T$$

$$= \begin{bmatrix} \lambda_1 \\ \lambda_2 \end{bmatrix} = \begin{bmatrix} 1 \\ 0.1 f_2^1 + 0.1 \end{bmatrix} \tag{5.106}$$

and, correspondingly,

$$\boldsymbol{\lambda}^2(\mathbf{f}) = \begin{bmatrix} \lambda_1 \\ \lambda_3 \end{bmatrix} = \begin{bmatrix} 1 \\ 0.1 f_2^2 + 1 \end{bmatrix} \tag{5.107}$$

Now we take $r = 1$; and we have [cf. Eqs. (5.29) and (5.28)]

$$\lambda^{12} = 1 \qquad \boldsymbol{\lambda}_s^i = \boldsymbol{\lambda}^i \qquad i = 1, 2 \tag{5.108}$$

Here, we have again used an analytic method. In practice, the $\boldsymbol{\lambda}^i$'s are evaluated according to Eq. (5.27) pointwise when they are needed in the multicriteria method. The F_i's thus do not need to be known analytically.

Let us now apply the feasible method with partial derivatives to the problem given by Eqs. (5.94)–(5.96). The coordination variable in this case is (y_1, y_2), which is given by the upper level to the subproblems formulated in the following. (Compare this to Eqs. (5.49)–(5.53) and note that, in this scheme, the $\mathbf{x}_i$'s are specified by the coupling equations; in this example, $x_1^k = y_2^k$ and $x_2^k = y_1^k$.)

Subproblem 1:

$$\min \begin{bmatrix} f_1^1(m_{11}, m_{12}, y_2^k, y_1^k) \\ f_2^1(m_{11}, m_{12}, y_2^k, y_1^k) \end{bmatrix}$$

with a trade-off vector $\boldsymbol{\lambda}_s^1(\mathbf{f}) = (1, 0.1 f_2^1 + 0.1)^T$ [cf. Eqs. (5.106) and (5.108)]. Here the objectives are given by the expressions in Eq. (5.97). This problem is of the general form of Eq. (5.98), with

$$\mathbf{z} = \begin{bmatrix} m_{11} \\ m_{12} \end{bmatrix} \qquad A = \begin{bmatrix} 5 & -3 \\ -3 & 3 \end{bmatrix} \qquad \mathbf{b} = \begin{bmatrix} -2 - 4(y_1^k - y_2^k) \\ 2(y_1^k - y_2^k) \end{bmatrix}$$

$$\mathbf{C}^T = [1 \quad 1] \qquad d = 5 + y_2 \qquad \alpha = 0.1 \qquad \beta = 0.1$$

Applying Eq. (5.102), we achieve

$$\mathbf{z}^0 = \begin{bmatrix} m_{11}^k \\ m_{12}^k \end{bmatrix} = \begin{bmatrix} 0.1866 - 0.5149y_2^k + 0.4701y_1^k \\ 0.08209 - 0.1866y_2^k + 0.1269y_1^k \end{bmatrix} \tag{5.109}$$

and

$$\boldsymbol{\lambda}_s^1 = (1,\ 0.62687 + 0.029851y_2^k + 0.059701y_1^k)^T \tag{5.110}$$

According to the general scheme, we would now solve the $\boldsymbol{\delta}_1^s$, $\boldsymbol{\eta}_1^s$, and $\boldsymbol{\mu}_1^s$ multipliers in Eq. (5.37) with $i = 1$. In this case, the respective constraints are not present, and we take $\boldsymbol{\delta}_1^s = \boldsymbol{\eta}_1^s = \boldsymbol{\mu}_1^s = \mathbf{0}$.

Next we solve $\boldsymbol{\nu}_1^s$ from Eq. (5.36) with $i = 1$; that is,

$$\begin{aligned} \nu_1^s &= (\boldsymbol{\lambda}_s^1)^T \frac{\partial \mathbf{f}^1}{\partial x_1} \\ &= (\boldsymbol{\lambda}_s^1)^T \begin{bmatrix} \dfrac{\partial f_1^1(m_{11}, m_{12}, x_1, y_1)}{\partial x_1} \\[2ex] \dfrac{\partial f_2^1(m_{11}, m_{12}, x_1, y_1)}{\partial x_1} \end{bmatrix} (m_{11}^k, m_{12}^k, y_2^k, y_1^k) \\ &= 1.209897 + 20.343\, y_2^k - 0.3134\, y_1^k \end{aligned} \tag{5.111}$$

Here we used Eq. (5.110) and performed differentiation analytically [Eqs. (5.97a) and (5.97b)]. In practice, differentiation is usually performed numerically.

For the upper level, we need to evaluate the following term [cf. Eq. (5.38)]:

$$(\boldsymbol{\lambda}_s^1)^T \frac{\partial \mathbf{f}^1}{\partial y_1} = (\boldsymbol{\lambda}_s^1)^T \begin{bmatrix} \dfrac{\partial f_1^1(m_{11}, m_{12}, x_1, y_1)}{\partial y_1} \\[2ex] \dfrac{\partial f_2^1(m_{11}, m_{12}, x_1, y_2)}{\partial y_1} \end{bmatrix} (m_{11}^k, m_{12}^k, y_2^k, y_1^k)$$

to get

$$(\boldsymbol{\lambda}_s^1)^T \frac{\partial \mathbf{f}^1}{\partial y_1} = -0.5821 - 0.3134\, y_2^k + 0.3731\, y_1^k \tag{5.112}$$

Subproblem 2:

This subsystem is quite similar to the first one; in the following, the results of this subproblem are given:

$$m_{21}^k = -3.0097 + 0.38835\, y_2^k - 0.38835\, y_1^k \tag{5.113}$$

$$m_{22}^k = 3.49515 - 0.194175\, y_2^k + 0.194175\, y_1^k \tag{5.114}$$

$$\boldsymbol{\lambda}_s^2 = (1,\ 1.0485 + 0.05825\, y_2^k - 0.05825\, y_1^k)^T \tag{5.115}$$

$$\nu_2^s = -15.029 - 0.8350\, y_2^k + 2.835\, y_1^k \tag{5.116}$$

$$(\boldsymbol{\lambda}_s^2)^T \frac{\partial \mathbf{f}^2}{\partial y_2} = -4.971 + 0.835\, y_2^k - 0.835\, y_1^k \tag{5.117}$$

Coordination The upper level checks whether Eq. (5.38) holds. Note that, in this example, $C_{11} = C_{22} = 0$ and $C_{12} = C_{21} = 1$. The parts of Eq. (5.38) for $i = 1, 2$ are now, in this case,

$$(\boldsymbol{\lambda}_s^1)^T \frac{\partial \mathbf{f}^1}{\partial y_1} + \nu_2^s = 0 \tag{5.118a}$$

$$(\boldsymbol{\lambda}_s^2)^T \frac{\partial \mathbf{f}^2}{\partial y_2} + \nu_1^s = 0 \tag{5.118b}$$

or, taking into account Eqs. (5.112), (5.116), (5.117), and (5.111),

$$(G_1^y)^k \triangleq -15.611 - 1.148 y_2^k + 3.208 y_1^k = 0 \tag{5.119a}$$

$$(G_2^y)^k \triangleq -3.762 + 21.178 y_2^k - 1.148 y_1^k = 0 \tag{5.119b}$$

We can solve this set of linear equations analytically; denoting the optimal solution by $\mathbf{y}^0 = (y_1^0, y_2^0)^T$, we have

$$y_1^0 = 5.027 \tag{5.120a}$$

$$y_2^0 = 0.4502 \tag{5.120b}$$

These values are the same as those given by the integrated solution [Eq. (5.103)]. Other values determined by the subsystems at $\mathbf{y}^0$ agree also with the corresponding values of the integrated solution.

In practice, $\mathbf{y}^0$ is found by an iterative procedure. According to Eq. (5.77), the following updating scheme can be used:

$$\mathbf{y}^{k+1} = \mathbf{y}^k - c^k(\mathbf{G}^y)^k \tag{5.121}$$

where $(\mathbf{G}^y)^k = [(G_1^y)^k, (G_2^y)^k]^T$.

We can now study the behavior of this scheme. Let $c^k = c$ (a constant for all k). Denote the error in the coordination variable at the kth iteration by

$$\Delta \mathbf{y}^k = \mathbf{y}^k - \mathbf{y}^0 \tag{5.122}$$

We get a difference equation for $\Delta \mathbf{y}^k$:

$$\Delta \mathbf{y}^{k+1} = \mathbf{y}^{k+1} - \mathbf{y}^0 = \mathbf{y}^k - c(\mathbf{G}^y)^k - \mathbf{y}^0 = \mathbf{y}^k - \mathbf{y}^0 - c(\mathbf{G}^y)^k = \Delta \mathbf{y}^k - c(\mathbf{G}^y)^k$$

where, as verified by Eq. (5.119),

$$(\mathbf{G}^y)^k = \begin{bmatrix} -1.148 & 3.208 \\ 21.178 & -1.148 \end{bmatrix} \Delta \mathbf{y}^k$$

Taking this into account, we obtain

$$\Delta \mathbf{y}^{k+1} = L \Delta \mathbf{y}^k \tag{5.123}$$

where

$$L = \begin{bmatrix} 1.148c & 1 - 3.208c \\ 1 - 21.178c & 1.148c \end{bmatrix}$$

The eigenvalues of L are $1 - 3.135c$ and $1 - 21.251c$. These are <1 in modulus, when $0 < c < 0.094$. When eigenvalues of L are <1 in modulus, $\Delta\mathbf{y}^k \to 0$, or $\mathbf{y}^k \to \mathbf{y}^0$, as $k \to \infty$. Therefore, the algorithm given by Eq. (5.121) converges from all initial errors, when a constant step $c < 0.094$. If $c > 0.094$, convergence does not occur from all initial errors.

5.8.3. Solution Using the Nonfeasible Method

In the following discussion, the subproblem formulation given by Eqs. (5.81)–(5.84) is used. The coordination parameter $\boldsymbol{\nu}^k = (\nu_1^k, \nu_2^k)^T$ is given by the upper level to the subproblems.

Subproblem 1:

$$\min \begin{bmatrix} f_1^1(m_{11}, m_{12}, x_1, y_1) + \nu_2^k y_1 - \nu_1^k x_1 \\ f_2^1(m_{11}, m_{12}, x_1, y_1) \end{bmatrix}$$

with a trade-off vector $\boldsymbol{\lambda}_s^1(\mathbf{f}) = (1, 0.1f_2^1 + 0.1)^T$. The objectives are given by Eq. (5.97). This problem is of the general form of Eq. (5.98), with

$$\mathbf{z} = \begin{bmatrix} m_{11} \\ m_{12} \\ x_1 \\ y_1 \end{bmatrix} \qquad A = \begin{bmatrix} 5 & -3 & 2 & -2 \\ -3 & 3 & -1 & 2 \\ 2 & -1 & 11 & -1 \\ -2 & 1 & -1 & 1 \end{bmatrix} \qquad \mathbf{b} = \begin{bmatrix} -2 \\ 0 \\ -\nu_1^k \\ \nu_2^k \end{bmatrix}$$

$$\mathbf{C}^T = [1 \quad 1 \quad 1 \quad 0] \qquad d = 5 \qquad \alpha = 0.1 \qquad \beta = 0.1$$

Applying Eq. (5.102), we get

$$\mathbf{z}^0 = \begin{bmatrix} m_{11}^k \\ m_{12}^k \\ x_1^k \\ y_1^k \end{bmatrix} = \begin{bmatrix} 0.9243 - 0.005976\,\nu_1^k - 1.255\,\nu_2^k \\ 0.2829 - 0.003984\,\nu_1^k - 0.3367\,\nu_2^k \\ -0.03586 + 0.04980\,\nu_1^k - 0.04183\,\nu_2^k \\ 1.530 + 0.0418\,\nu_1^k - 2.715\,\nu_2^k \end{bmatrix} \begin{matrix} (5.124a) \\ (5.124b) \\ (5.124c) \\ (5.124d) \end{matrix}$$

Subproblem 2:

This is quite similar to subproblem 1; the results are given directly in the following:

$$m_{21}^k = 0.6977 - 0.4651\,\nu_1^k \tag{5.125a}$$

$$m_{22}^k = 4.651 - 0.2326\,\nu_1^k \tag{5.125b}$$

$$x_2^k = 10 - 0.5\nu_1^k + 0.5\,\nu_2^k \tag{5.125c}$$

$$y_2^k = 15.953 - 1.698\,\nu_1^k + 0.5\,\nu_2^k \tag{5.125d}$$

Coordination The task of the upper level here is to determine a coordination parameter such that the unbalances [cf. Eqs. (5.85)] are zero. In this case

$$e_1^k = y_2^k - x_1^k = 0 \tag{5.126a}$$

$$e_2^k = y_1^k - x_2^k = 0 \tag{5.126b}$$

or, using Eqs. (5.125d), (5.124c), (5.124d), and (5.125c),

$$15.989 - 1.7474\,\nu_1^k + 0.5418\nu_2^k = 0 \tag{5.127a}$$

$$-8.470 + 5.418\,\nu_1^k - 3.215\nu_2^k = 0 \tag{5.127b}$$

Denoting the solution of these equations, the optimal values of the coordination parameters, by $\boldsymbol{\nu}^0 = (\nu_1^0, \nu_2^0)$, we have $\nu_1^0 = 8.7925$ and $\nu_2^0 = -1.1527$. Now we can calculate corresponding values for the other variables using Eqs. (5.124) and (5.125). They agree with the values given by the integrated solution and the feasible method.

Next, consider the coordination algorithm given by Eq. (5.90), which in this case is

$$\boldsymbol{\nu}^{k+1} = \boldsymbol{\nu}^k + c^k \begin{bmatrix} 15.989 - 1.747\,\nu_1^k + 0.5418\,\nu_2^k \\ -8.470 + 0.5418\,\nu_1^k - 3.215\,\nu_2^k \end{bmatrix} \tag{5.128}$$

In the same manner as in the feasible method, we can verify that this algorithm works when the c^k's are sufficiently small. Here $0 < c^k < 1.04$.

5.8.4. Modified Problem

Consider the problem of Sect. 5.8.1 with the change that, instead of Eqs. (5.96b) and (5.96c), we have

$$\lambda_{12}^*(\mathbf{f}) = 0.1f_2^1 + 0.1 + 0.2f_2^2 \qquad \lambda_{13}^*(\mathbf{f}) = 0.1f_2^2 + 1 + 0.2f_2^1$$

The scaled internal trade-off vectors will now be the following [cf. Eqs. (5.106)–(5.108)]:

$$\boldsymbol{\lambda}_s^1(\mathbf{f}) = \begin{bmatrix} 1 \\ 0.1f_2^1 + 0.1 + 0.2f_2^2 \end{bmatrix}$$

$$\boldsymbol{\lambda}_s^2(\mathbf{f}) = \begin{bmatrix} 1 \\ 0.1f_2^2 + 1 + 0.2f_2^1 \end{bmatrix}$$

$$\lambda^{12} = 1$$

That is, the trade-offs in subsystem 1 depend on f_2^2, which is an objective in subsystem 2; and in subsystem 2, the trade-offs depend on f_2^1 of subsystem 1.

In the following discussion, the problem is solved by the nonfeasible method of the previous section with an additional relaxation approach that was mentioned in Sect. 5.7.2. That is, every time subproblem 1 (2) is solved, a constant value equal to that of the latest completed iteration is used for $f_2^2(f_2^1)$. Hence, the only difference in the subproblems of Sect. 5.8.3 is the constants in the trade-offs. For example, we now have the following subproblem at the kth iteration.

Subproblem 1:

$$\min_{\substack{m_{11}, m_{12} \\ x_1, y_1}} \begin{bmatrix} f_1^1(m_{11}, m_{12}, x_1, y_1) + \nu_2^k y_1 - \nu_1^k x_1 \\ f_2^1(m_{11}, m_{12}, x_1, y_1) \end{bmatrix}$$

with a trade-off vector $\boldsymbol{\lambda}_s^1 = (1,\ 0.1\, f_2^1 + 0.1 + 0.2(f_2^2)^{k-1})^T$.

This problem can be solved analytically by Eq. (5.102), yielding

$$\mathbf{z}^0 = \begin{bmatrix} m_{11}^k \\ m_{12}^k \\ x_1^k \\ y_1^k \end{bmatrix} = \begin{bmatrix} 0.9243 - 0.005976\, \nu_1^k - 1.255\, \nu_2^k - 0.239(f_2^2)^{k-1} \\ 0.2829 - 0.003984\, \nu_1^k - 0.3367\, \nu_2^k - 0.1593(f_2^2)^{k-1} \\ -0.03586 + 0.0498\, \nu_1^k - 0.04183\, \nu_2^k - 0.00797(f_2^2)^{k-1} \\ 1.530 + 0.0418\, \nu_1^k - 2.715\, \nu_2^k - 0.3267(f_2^2)^{k-1} \end{bmatrix} \tag{5.129}$$

A new value for f_2^1 is calculated from Eqs. (5.97b) and (5.129), which yields

$$(f_2^1)^k = 6.171 + 0.0398\, \nu_1^k - 1.633\, \nu_2^k - 0.4064\, (f_2^2)^{k-1} \tag{5.130}$$

Subproblem 2:

Proceeding as in subsystem 1, we have, at the kth iteration:

$$m_{21}^k = -0.6977 - 0.4651\, \nu_1^k - 0.093(f_2^1)^{k-1} \tag{5.131a}$$

$$m_{22}^k = 4.651 - 0.2326\, \nu_1^k - 0.04651(f_2^1)^{k-1} \tag{5.131b}$$

$$x_2^k = 10 - 0.5\, \nu_1^k + 0.5\, \nu_2^k \tag{5.131c}$$

$$y_2^k = 15.953 - 1.698\, \nu_1^k + 0.5\, \nu_2^k - 0.1395(f_2^1)^{k-1} \tag{5.131d}$$

From Eqs. (5.97d) and (5.131), a new value for f_2^2 is obtained:

$$(f_2^2)^k = 3.953 - 0.6977\nu_1^k - 0.1395(f_2^1)^{k-1} \tag{5.132}$$

Coordination The unbalances e_1^k and e_2^k [cf. Eq. (5.126)] are now as follows:

$$e_1^k = 15.989 - 1.747\, \nu_1^k + 0.5418\, \nu_2^k - 0.1395\, (f_2^1)^{k-1} + 0.007968\, (f_2^2)^{k-1} \tag{5.133}$$

$$e_2^k = -8.470 + 0.5418\, \nu_1^k - 3.215\, \nu_2^k - 0.3267\, (f_2^2)^{k-1} \tag{5.134}$$

Substituting $(f_2^1)^*$ for $(f_2^1)^k$ and $(f_2^1)^{k-1}$, $(f_2^2)^*$ for $(f_2^2)^{k-1}$ and $(f_2^2)^k$, ν_1^* for ν_1^k, and ν_2^* for ν_2^k in Eqs. (5.130) and (5.132)–(5.134) and solving this set of equations with $e_1^k = 0$ and $e_2^k = 0$ yields, as an equilibrium point,

$$\nu_1^* = 8.0951 \qquad \nu_2^* = -0.96639$$

$$(f_2^1)^* = 27.62 \qquad (f_2^2)^* = -16.29$$

Consider the same coordination algorithm as in Sect. 5.8.3:

$$\begin{bmatrix} \nu_1^{k+1} \\ \nu_2^{k+1} \end{bmatrix} = \begin{bmatrix} \nu_1^k \\ \nu_2^k \end{bmatrix} + c^k \begin{bmatrix} e_1^k \\ e_2^k \end{bmatrix} \tag{5.135}$$

The convergence of the algorithm toward the solution point can be studied as above. Denote, for $k = 1, 2, \ldots,$

$$\Delta\mathbf{E}^k \triangleq \begin{bmatrix} \Delta\nu_1^k \\ \Delta\nu_2^k \\ \Delta f_2^{1k} \\ \Delta f_2^{2k} \end{bmatrix} \triangleq \begin{bmatrix} \nu_1^k - \nu_1^* \\ \nu_2^k - \nu_2^* \\ (f_2^1)^{k-1} - (f_2^1)^* \\ (f_2^2)^{k-1} - (f_2^2)^* \end{bmatrix}$$

For this error vector $\Delta\mathbf{E}$, we have from Eqs. (5.130) and (5.132)–(5.135) the following difference equation for $k = 1, 2, \ldots$:

$$\Delta\mathbf{E}^{k+1} = A^k \Delta\mathbf{E}^k$$

with

$$A^k = \begin{bmatrix} 1 - 1.747c^k & 0.5418c^k & -0.1395c^k & 0.007968c^k \\ 0.5418c^k & 1 - 3.215c^k & 0 & -0.3267c^k \\ 0.0398 & -1.633 & 0 & -0.4064 \\ -0.6977 & 0 & -0.1395 & 0 \end{bmatrix}$$

By numerical eigenvalue evaluations, it can be determined that the eigenvalues are all <1 in modulus when $0 < c^k < 0.59$. Hence, when c^k is kept in this interval, the algorithm converges to the solution of the problem.

5.8.5 Summary

The above examples illustrate the structure and mathematical aspects of hierarchical multiobjective optimization techniques. Despite the relatively small size of these examples, the numerical benefits of the hierarchical approach were evident in the analytic method used. Only some 2×2 matrices in the feasible method (Sect. 5.8.2) and two 4×4 matrices in the nonfeasible method (Sect. 5.8.3) had to be inverted, compared to a 6×6 matrix to be inverted in the solution without decomposition (Sect. 5.8.1).

5.9 INTERACTION BETWEEN ANALYST AND DECISIONMAKER

In the following discussion, we give our attention to the interaction between the analyst and the decisionmaker. Section 5.9.1 deals with a case that directly corresponds to the mathematical development in the previous sections, and Sect. 5.9.2 considers the possibility of limited interaction between analyst and decisionmaker.

5.9.1 Unlimited Interaction

The feasible and nonfeasible schemes derived above use the notion of indifference trade-offs. Hence, the method of Geoffrion et al. (1972) can be directly applied in subproblems. That is, whenever indifference trade-offs are needed in subsystem solutions, they are determined by Eqs. (5.27) and (5.28). If subsystem trade-offs depend on the variables of other subsystems, a simple relaxation method—as in the previous example—can possibly be used.

Next, consider the use of the SWT method (see Chap. 2) for solving the subproblems. The main way this differs from the above method is that the decisionmaker does not express indifference trade-offs. Instead, actual trade-offs are determined as Lagrange multipliers and the decisionmaker reacts to them through varying degrees of indifference, acceptance, or rejection.

As an example, consider the ith subproblem at the kth iteration in the feasible scheme outlined in Sect. 5.5.1. The subproblem [Eqs. (5.49)–(5.52)] is first transformed into the following single-objective problem (see Chap. 2):

$$\min_{\mathbf{m}_i} f_1^i(\mathbf{x}_i^k, \mathbf{y}_i^k)$$

subject to

$$f_j^i(\mathbf{x}_i^k, \mathbf{m}_i, \mathbf{y}_i^k) \leq \epsilon_j \qquad j = 2, \ldots, n_i$$

$$\mathbf{y}_i^k = \mathbf{H}_i(\mathbf{x}_i^k, \mathbf{m}_i) \qquad \mathbf{g}_i(\mathbf{x}_i^k, \mathbf{m}_i, \mathbf{y}_i^k) \leq \mathbf{0} \qquad \mathbf{q}_i(\mathbf{x}_i^k, \mathbf{m}_i, \mathbf{y}_i^k) = \mathbf{l}_i^k$$

Let $\lambda_{i2}(\epsilon), \ldots, \lambda_{in}(\epsilon)$ be the resulting Lagrange multipliers where $\boldsymbol{\epsilon} = (\epsilon_2, \ldots, \epsilon_{n_i})^T$. When the constraints are not active, an $\boldsymbol{\epsilon}$, denoted by $\boldsymbol{\epsilon}^0$, is sought, so that

$$\lambda_{ij}(\boldsymbol{\epsilon}^0) = (\boldsymbol{\lambda}_s^i)_j \qquad j = 2, \ldots, n_i \tag{5.136}$$

where $(\boldsymbol{\lambda}_s^i)_j$ is the indifference trade-off between f_1^i and f_j^i [see Eq. (5.53)].

Consider, first, a case where each f_j^i $(j = 2, \ldots, n_i)$ is an argument in only one overall objective. For example, let f_2^i be an argument in only F_3. Then, for $j = 2$ and taking into account Eqs. (5.27) and (5.28), Eq. (5.136) becomes

$$\lambda_{i2}(\boldsymbol{\epsilon}^0) = \frac{\lambda_3}{\lambda_1^i} \frac{\partial F_3}{\partial f_2} \tag{5.137}$$

That is, one can divide this equation by $(\partial F_3/\partial f_2)/\lambda_1^i$ (assuming this is nonzero) and search for an $\boldsymbol{\epsilon}^0$ in order that

$$\frac{\lambda_1^i}{\partial F_3/\partial f_2} \lambda_{i2}(\boldsymbol{\epsilon}^0) = \lambda_3 \tag{5.138}$$

In other words, we can calculate the expression on the left-hand side for any $\boldsymbol{\epsilon}$ and show it to the decisionmaker—asking whether he or she accepts that value as an indifference trade-off between F_1 and F_3. When $\boldsymbol{\epsilon} = \boldsymbol{\epsilon}^0$, the decisionmaker regards the calculated value as indifferent, according to Eq. (5.138). Note that if dealing with overall objectives, the decisionmaker cannot be asked to react to the trade-offs in Eq. (5.136).

The interactive SWT method can be applied to find $\boldsymbol{\epsilon}^0$, giving the preferred solution. When, at an arbitrary $\boldsymbol{\epsilon}$, the expression on the left-hand side is evaluated, it is shown to the decisionmaker, who is then asked, "Would you like to decrease F_1 by b units $[b = \lambda_1^i(\boldsymbol{\epsilon})\lambda_{i2}(\boldsymbol{\epsilon})/(\partial F_3/\partial f_2)]$ per one-unit increase in F_3?" The decisionmaker gives a number, denoted here by W_{i2}^{ϵ}, between -10 and $+10$, for example. A positive number means that the decisionmaker likes the trade (see Chap. 2).

Assume that the same question is made with respect to all f_j^i ($n = 2, \ldots, n_i$). The decisionmaker's response will yield a surrogate worth valued W_{ij}^{ϵ} ($j = 2, \ldots, n_i$). The signs of these values determine the desired directions in the overall objective space. For example, if W_{i2}^{ϵ} is positive, the decisionmaker wants to increase F_3. The relative magnitudes of W_{ij}^{ϵ} are used to determine the relative changes of the overall objectives. Corresponding changes in the $\mathbf{f}^i$ space must be calculated. The interactive method may end up at a boundary where the wanted directions point outward from the feasible set. In this case, Eq. (5.136) does not hold at the preferred solution.

Next consider the case where lower-level objectives appear as arguments in several overall objectives. Because the overall objectives usually aggregate lower-level objectives, the previous case is probably more common. Let f_2 be an argument in F_3 and F_4 (a general formulation can be seen from the following discussion). Then, instead of Eq. (5.137) we have

$$\lambda_{i2}(\boldsymbol{\epsilon}^0) = \frac{1}{\lambda_1^i}\left(\lambda_3 \frac{\partial F_3}{\partial f_2} + \lambda_4 \frac{\partial F_4}{\partial f_2}\right)$$

or

$$[1, \lambda_3, \lambda_4]\left[\lambda_{i2}(\boldsymbol{\epsilon}^0), -\frac{1}{\lambda_1^i}\frac{\partial F_3}{\partial f_2}, -\frac{1}{\lambda_1^i}\frac{\partial F_4}{\partial f_2}\right]^T = 0$$

That is, a change $\Delta\mathbf{F}$, with $\Delta F_1 = \lambda_{12}(\boldsymbol{\epsilon}^0)$, $\Delta F_3 = -(\partial F_3/\partial f_2)/\lambda_1^i$, and $\Delta F_4 = -(\partial F_4/\partial f_2)/\lambda_1^i$, is indifferent to the decisionmaker at $\boldsymbol{\epsilon}^0$ [cf. the necessary condition given by Eq. (5.14) without constraints].

Hence, in this case, questions of the following type are asked: "Would you like to decrease F_1 by $\lambda_{i2}(\boldsymbol{\epsilon})$ units while increasing F_3 by $(\partial F_3/\partial f_2)/\lambda_1^i$ units and F_4 by $(\partial F_4/\partial f_2)\lambda_1^i$ units?" From this point on, the treatment of this case is the same as that of the first case.

5.9.2 Limited Interaction

A way to try to reduce the interaction between the analyst and the decisionmaker is to use constant trade-offs in the subproblems in each iteration. That is, at the beginning of an iteration, the $\boldsymbol{\lambda}_s^i$ are evaluated but are then kept constant. The subproblems become single-objective problems. The subproblem solutions are suboptimal, and convergence difficulties may arise. Heal (1973) presented a feasible algorithm of this type in iterative planning.

REFERENCES

Brosilow, C. G., Lasdon, L., and Pearson, J. D., Feasible Optimization Methods for Interconnected Systems, in *Proc. 1965 JACC Conf.*, Troy, New York, June 22–25, 1965.

DaCunha, N. O., and Polak, E., Constrained Minimization under Vector-valued Criteria in Finite Dimensional Spaces, *J. Math. Anal. Appl.*, vol. 19, pp. 103–124, 1967.

Findeisen, W., Brdys, M., Malinowski, K., Tatjevski, P., and Wozniak, A., *Control and Coordination in Hierarchical Systems*, Wiley, New York, 1980.

Geoffrion, A. M., Proper Efficiency and Theory of Vector Maximization, *J. Math. Anal. Appl.*, vol. 22, pp. 618–630, 1968.

Geoffrion, A. M., Dyer, J., and Feinberg, A., An Interactive Approach for Multicriterion Optimization with an Application to the Operation of an Academic Department, *Management Sci.*, vol. 19, pp. 357–368, 1972.

Geoffrion, A. M., and Hogan, W. W., Coordination of Two-level Organizations with Multiple Objectives, in *Techniques of Optimization*, ed. A. V. Balakrishnan, Academic Press, New York, 1972.

Heal, G. M., *The Theory of Economic Planning*, North-Holland and American Elsevier, New York, 1973.

Kuhn, H. W., and Tucker, A. W., Nonlinear Programming, in *Proc. 2d Berkeley Symp. Mathematics, Statistics, and Probability*, University of California Press, Berkeley, Calif., pp. 481–492, 1951.

Lasdon, L., *Optimization Theory for Large Systems*, Macmillan, London, 1970.

Luenberger, D. G., *Introduction to Linear and Nonlinear Programming*, Addison-Wesley, Reading, Mass., 1973.

Singh, M. G., *Dynamic Hierarchical Control*, Elsevier, Amsterdam, 1980.

Singh, M. G., and Titli, A., *System-Decomposition, Optimization, and Control*, Pergamon, Oxford, 1978.

Smith, N. J., and Sage, A. P., An Introduction to Hierarchical Systems Theory, *Comput. Elect. Eng.*, vol. 1, pp. 55–71, 1973.

Tarvainen, K., *Hierarchical Multiobjective Optimization*, Ph.D. dissertation, Systems Engineering Department, Case Western Reserve University, Cleveland, Ohio (available from University Microfilms International, Ann Arbor, Mich.), 1981.

Wismer, D. A. (ed.), *Optimization Methods for Large Scale Systems with Applications*, McGraw-Hill, New York, 1971.

Zoutendijk, G., *Methods of Feasible Directions*, Elsevier, Amsterdam, 1960.

CHAPTER

SIX

MULTIPLE DECISIONMAKERS

6.1 INTRODUCTION

Iterative negotiation procedures involving multiple decisionmakers are considered in this chapter. The procedures we consider here are generalizations of the single-decisionmaker cases in the previous chapter. Other types of procedures with multiple decisionmakers are dealt with in other chapters.

In Sect. 6.2, Stackelberg schemes are derived by fixing the order in which decisions are made. Section 6.3 considers totally direct generalizations of the schemes given in the previous chapter. We consider, in Sect. 6.4, a more elaborate algorithm in which trade-offs are used instead of indifference trade-offs.

6.2 STACKELBERG SCHEME

When there are several decisionmakers, they may agree on the order in which decisions are made. For example, consider the system and objective structures that were outlined in Fig. 4.6. Let each subsystem i have its own decisionmaker DM_i; let there be a higher-level decisionmaker DM_0, who is interested in the overall objectives. In a Stackelberg scheme (see Stackelberg, 1952; Cruz, 1978), it would now be agreed that DM_0 makes the decisions concerning the interaction variables, after which the DM_i's decide $\mathbf{m}_i$ ($i = 1, \ldots, N$). After the DM_i's ($i = 1, \ldots, N$) have made their decisions, the DM_0 may change the values of the interaction variables, and so on, until the DM_0 no longer wants to make changes.

For the lower-level decisionmakers, the subproblems are ordinary multiobjective problems for given values of the interaction variables. The trade-offs are now chosen completely by the local decisionmakers. For the DM_0, the problem of determining the interaction variables is the same as that cited for the single decisionmaker in the previous chapter, and the same algorithms can be applied. The algorithm with total derivatives (Fig. 5.2) is the simplest with respect to information transfer between the levels. In using it, DM_0 evaluates the λ^{ri}'s and $\boldsymbol{\lambda}_s^i$'s, just as in the single-decisionmaker case in Chap. 5. The derivatives $d\mathbf{f}^i/d\mathbf{y}$ are evaluated by the subsystems. Using the coordination algorithm of Fig. 5.2 this way, the DM_0 selects the best possible direction for the interaction variables, given the choices made by the local decisionmakers. For a similar algorithm in regional planning, for the linear case, see Rietveld (1980).

A counterpart of this feasible method in the nonfeasible method would be a scheme where the DM_0 decides the trade-offs between the subsystems (λ^{ri}) (see Fig. 5.3). This option is a very indirect way of pursuing some goals. The DM_0 becomes a referee, λ^{ri}, more or less, determining how much weight the objectives of subsystem i get in comparison with the other subsystems.

A weakness of Stackelberg schemes is that they do not include negotiations. In a general case, it may happen that the solution is not Pareto optimal.

6.3 NEGOTIATIONS WITH INDIFFERENCE TRADE-OFFS

Consider the system depicted in Fig. 6.1. The subsystem objectives are directly overall objectives. (In this figure, the formalism of the previous chapters is used.) Let each subsystem have its own decisionmaker. In other words, though the separate decisionmakers each have their own goals, they are dependent on one another by virtue of physical connections.

In this case, negotiations are needed. The situation is complex, owing to multiple objectives. As a suitable means of negotiation, indifference trade-offs between the subsystems may be useful.

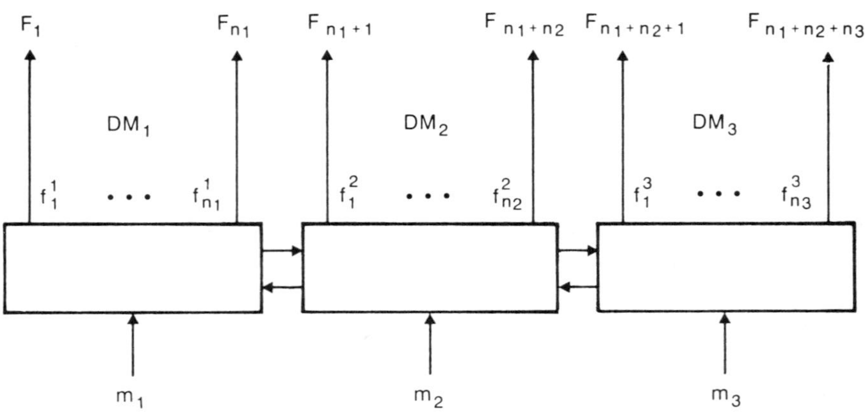

Figure 6.1 Case with multiple decisionmakers.

Assume that the decisionmakers agree on some preliminary values for λ^{ri}'s ($i \neq r$). After such a consensus, a feasible or nonfeasible scheme can be used to specify the corresponding interaction variables. In each subproblem, the decisionmaker uses his or her own preference. After convergence, or possibly before it, the decisionmakers renegotiate the λ^{ri}'s, if necessary. Note that, even if λ^{ri} is a trade-off between two objectives, all values of objectives are taken into account to determine the indifference value. Using the renegotiated λ^{ri}'s, the hierarchical algorithm is repeated as many times as necessary until there are no renegotiations.

The same principle of negotiation with indifference trade-offs can, in principle, be applied to any situation. We may have an objective structure (as in Fig. 4.6) and, in addition to local decisionmakers, we may have a decisionmaker interested in some F_i's that are not directly local objectives. That is, we may always think that the decisionmakers agree on the $\lambda_1, \ldots, \lambda_n$ trade-offs, and that these indifference trade-offs are used in the same way as in problems with a single decisionmaker. In these negotiations, the relations between trade-offs at the upper level and the lower-level subsystems given in Eqs. (5.27) and (5.28) may be used.

This approach has, however, its practical limits, which are reached when the number of trade-offs and negotiations gets too high.

6.4 NEGOTIATIONS WITH TRADE-OFFS

6.4.1 Introduction and Problem Formulation

In the previous chapter and at the beginning of this chapter, indifference trade-offs were used in the derivations, except for the discussion in Sect. 5.9, where it was shown how trade-offs—obtained by, for example, the surrogate worth trade-off (SWT) method (see Chap. 2)—can be used in the derived schemes.

Here, an iterative method corresponding to the schemes discussed in Sect. 6.3 will be derived whereby trade-offs between the subsystems can be calculated. For decisionmakers, it is usually easier to react to trade-offs than to agree on indifference trade-offs.

In Sect. 6.4.2, a computationally efficient way of calculating trade-offs between subsystems hierarchically will be considered. Then, in Sect. 6.4.3, the negotiation scheme will be presented, followed by an example in Sect. 6.4.4.

The scheme is derived using the following mathematical models for each decisionmaker i ($i = 1, \ldots, N$):

$$\min \begin{bmatrix} f_1^i(\mathbf{x}_i, \mathbf{m}_i, \mathbf{y}_i) \\ \vdots \\ f_{n_i}^i(\mathbf{x}_i, \mathbf{m}_i, \mathbf{y}_i) \end{bmatrix} \tag{6.1}$$

subject to

$$\mathbf{y}_i = \mathbf{H}_i(\mathbf{x}_i, \mathbf{m}_i) \tag{6.2}$$

$$\mathbf{g}_i(\mathbf{x}_i, \mathbf{m}_i, \mathbf{y}_i) \leq \mathbf{0} \tag{6.3}$$

$$\mathbf{x}_i = \sum_{j=1}^{N} C_{ij}\mathbf{y}_j \tag{6.4}$$

For an explanation of the terms in the model and of the form of the formulation, see Sect. 4.4. The systems considered are depicted in Fig. 6.1 (for $N = 3$).

In the iterative negotiation schemes proposed in Sect. 6.4.3, the local information given to the decisionmakers at each iteration is given in the form of trade-offs among the decisionmakers. Mathematically, these trade-offs are defined in the same way as in the single-decisionmaker case. That is, a trade-off between the objectives f_k^i (of DM_i) and f_l^j (of DM_j) ($j \neq i$), denoted by λ_{kl}^{ij}, is the amount of f_l^j DM_j gains when DM_i loses one unit in f_k^i (the other objectives remaining constant). The trade-off between f_k^i and f_l^i is denoted by λ_{kl}^i.

In some cases, the objectives are coupled so that it is not possible to keep the other objectives constant when two objectives change. That is, in these cases a unit change in f_k^i will inevitably change two or more objectives. In these cases, a trade-off must be given as a vector of numbers indicating corresponding changes in the other objectives (a so-called total trade-off; see Chankong and Haimes, 1983).

In the special case of sequential couplings, one may use the term *impact* for trade-off. The impact notion, as well as its practical calculation, are thoroughly discussed in Gomide (1982) and Gomide and Haimes (1983).

6.4.2 Numerical Calculation of Trade-Offs

To keep the mathematical treatment as tractable as possible, let us consider the case $N = 2$ (two connected systems and two decisionmakers). A generalization to an arbitrary value of N is trivial.

In every phase of the negotiations, we are interested only in alternatives that are Pareto optimal. It is suggested that the Pareto optimal points be generated by the constraint (ϵ-constraint) method (see Sect. 4.3.2) because it does not require convexity assumptions and because trade-offs can be calculated in connection with it.

If we are interested in the trade-off between one objective, say, f_1^1, and another at $\mathbf{f}^1 = \boldsymbol{\epsilon}^1 = (\epsilon_1^1, \ldots, \epsilon_n^1)^T$, $\mathbf{f}^2 = \boldsymbol{\epsilon}^2 = (\epsilon_1^2, \ldots, \epsilon_{n2}^2)^T$, we set up the following ordinary optimization problem:

$$\min f_1^1(x_1, m_1, y_1) \tag{6.5}$$

subject to

$$f_j^1(x_1, m_1, y_1) \leqslant \epsilon_j^1 \qquad j \neq 1 \tag{6.6}$$

$$f_j^2(x_2, m_2, y_2) \leqslant \epsilon_j^2 \qquad j = 1, \ldots, n_2 \tag{6.7}$$

$$\mathbf{y}_i = \mathbf{H}_i(\mathbf{x}_i, \mathbf{m}_i) \qquad \mathbf{g}_i(\mathbf{x}_i, \mathbf{m}_i, \mathbf{y}_i) \leqslant \mathbf{0} \qquad \mathbf{x}_i = \sum_{j=1}^{N} C_{ij}\mathbf{y}_j \qquad i = 1, \ldots, N \tag{6.8}$$

If, at a unique solution of the above problem, we have $\mathbf{f}^1 = \boldsymbol{\epsilon}^1$ and $\mathbf{f}^2 = \boldsymbol{\epsilon}^2$, this point is indeed a Pareto optimal alternative (see Sect. 4.3.2). Furthermore, strictly positive Lagrange multipliers associated with the constraints in Eqs. (6.6) and (6.7) are, under some technical conditions, trade-offs between f_1^1 and the corresponding objectives (for details, see Chankong and Haimes, 1983).

In some cases, the optimization algorithm that solves the problem given by Eqs. (6.5)–(6.8) also calculates the Lagrangian multipliers. Usually, however, the Lagrangian multipliers must be calculated by resolving the problem with perturbated ϵ values. In this section, we suggest that an alternative technique based on the augmented Lagrangian be used to generate the Lagrange multipliers. That is, the problem given in Eqs. (6.5)–(6.8) is transformed into a series of problems of the following form (k denotes the iteration index) (see Bertsekas, 1976):

$$\min \left\{ f_1^1(x_1, m_1, y_1) + \sum_{j=2}^{n_1} (\lambda_{1j}^1)^k [f_j^1(x_1, m_1, y_1) - \epsilon_j^1 + (u_j^1)^2] \right.$$
$$+ \sum_{j=1}^{n_2} (\lambda_{1j}^{12})^k [f_j^2(x_2, m_2, y_2) - \epsilon_j^2 + (u_j^2)^2] + \frac{1}{2} c \left(\sum_{j=2}^{n_1} [f_j^1(x_1, m_1, y_1) - \epsilon_j^1 \right. \tag{6.9}$$
$$\left. \left. + (u_j^1)^2]^2 + \sum_{j=1}^{n_2} [f_j^2(x_2, m_2, y_2) - \epsilon_j^2 + (u_j^2)^2]^2 \right) \right\}$$

subject to the original constraints given by Eqs. (6.2)–(6.4). Here, c is a positive parameter and the u_j^i's are slack variables associated with the augmented Lagrangian method. The Lagrange multipliers $(\lambda_{1j}^1)^k$ $(j = 2, \ldots, n_1)$, $(\lambda_{1j}^{12})^k$ $(j = 1, \ldots, n_2)$ are updated according to the equations given in Fig. 6.2 until they converge (for details see Bertsekas, 1976).

The augmented Lagrangian technique is beneficial not only for the calculation of trade-offs and the elimination of the constraints given by Eqs. (6.6) and (6.7) but also for the decomposition of the problem. In this problem where the overall system consists of coupled systems, a hierarchical solution method may be useful. [The problem given by Eqs. (6.5)–(6.8) does not lend itself to decomposition because the function to be minimized depends on the objective of only one subsystem.] Since we have quadratic terms in Eq. (6.9), almost any hierarchical decomposition method can be used to solve the problem given by Eqs. (6.9) and (6.8). In Fig. 6.2, the input prediction and balance method based on the augmented Lagrangian is assumed, due to its fast convergence (Gomide, 1982; Findeisen et al., 1980).

6.4.3 Negotiation Procedure and Its Convergence

We first present a negotiation scheme involving two decisionmakers and then consider its convergence. Finally, we address a general case. Consider a problem such as that given by Eqs. (6.1)–(6.4) where $N = 2$ (two subsystems, two decisionmakers).

Master Program

Are $y_1^k - x_2^k = 0$, $y_2^k - x_1^k = 0$, $\left.\frac{dF_1}{dx_1}\right|_k - (p_1^k)^T + c(x_1^k - y_2^k)^T = 0$, $\left.\frac{dF_2}{dx_2}\right|_k - (p_2^k)^T + c(x_2^k - y_1^k)^T = 0$,

$d_j^1 \triangleq f_j^1\,(x_1^k, m^k(x_1^k), y_1^k) - \epsilon_j^1 + ((u_j^1)^k)^2 = 0 \quad (j = 2,\ldots,n_1)$, $d_j^2 \triangleq f_j^2\,(x_2^k, m_2^k(x^k), y_2^k) - \epsilon_j^2 + ((u_j^2)^k)^2 = 0 \quad (j = 1,\ldots,n_2)$

If yes, stop; otherwise:

$p_1^{k+1} = p_1^k + c(y_2^k - x_1^k)$, $p_2^{k+1} = p_2^k + c(y_1^k - x_2^k)$,

$x_1^{k+1} = y_2^k + (1/c)(p_1^k - (dF_1/dx_1)|_k)$, $x_2^{k+1} = y_1^k + (1/c)(p_2^k - (dF_2/dx_2)|_k)$,

$(\lambda_{1j}^1)^{k+1} = (\lambda_{1j}^1)^k + cd_j^1 \quad (j = 2,\ldots,n_1)$, $(\lambda_{1j}^{12})^{k+1} = (\lambda_{1j}^{12})^k + cd_j^2 \quad (j = 1,\ldots,n_2)$.

Master Program → Subproblem 1: $x_1^k, x_2^k, (\lambda_{1j}^1)^k, p_2^k$

Subproblem 1 → Master Program: $y_1^k, m_1^k(x^k), (u_j^1)^k$

Subproblem 2 → Master Program: $y_2^k, m_2^k(x^k), (u_j^2)^k$

Master Program → Subproblem 2: $x_1^k, x_2^k, (\lambda_{1j}^{12})^k, p_1^k$

Subproblem 1

$$\min_{y_2,\, m_2,\, u_j^1} \{F_1(x_1^k, m_1, y_1; (\lambda_1^1)^k, \epsilon^1, u^1) + (p_2^k)^T(y_1 - x_2^k) + \frac{1}{2}c\,\|x_2^k - y_1\|^2\}$$

subject to $y_1 = H_1(x_1^k, m_1)$, $g_1(x_1^k, m_1, y_1) \leq 0$,

where $F_1(x_1^k, m_1, y_1; (\lambda_1^1)^k, \epsilon^1, u^1) =$

$$f_1^1\,(x_1^k, m_1, y_1) + \sum_{j=2}^{n_2} (\lambda_{1j}^1)^k (f_j^1\,(x_1^k, m_1, y_1) - \epsilon_j^1 + (u_j^1)^2)$$

$$+ \frac{1}{2}c \sum_{j=2}^{n_2} (f_j^1\,(x_1^k, m_1, y_1) - \epsilon_j^1 + (u_j^1)^2)^2.$$

Denote solution by $y_1^k, m_1^k(x^k), (u_j^1)^k$

Subproblem 2

$$\min_{y_2,\, m_2,\, u_j^2} \{F_2(x_2^k, m_2, y_2; (\lambda_1^{12})^k, \epsilon^2, u^2) + (p_1^k)^T(y_2 - x_1^k) + \frac{1}{2}c\,\|x_1^k - y_2\|^2\}$$

subject to $y_2 = H_2(x_2^k, m_2)$, $g_2(x_2^k, m_2, y_2) \leq 0$,

where $F_2(x_2^k, m_2, y; (\lambda_1^{12}), \epsilon^2, u^2) =$

$$\sum_{j=1}^{n_2} (\lambda_{1j}^{12})^k (f_j^2\,(x_2^k, m_2, y_2) - \epsilon_j^2 + (u_j^2)^2)$$

$$+ \frac{1}{2}c \sum_{j=1}^{n_2} (f_j^2\,(x_2^k, m_2, y_2) - \epsilon_j^2 + (u_j^2)^2)^2.$$

Denote solution by $y_2^k, m_2^k(x^k), (u_j^2)^k$

Figure 6.2 Calculation of trade-offs.

In this problem, the decisionmakers are not interested in each other's preference structure per se; that is, when the couplings are fixed, the decisionmakers are free to trade off their respective objectives in their subsystems.

However, the decisionmakers are dependent on each other via the couplings, whose values must be agreed upon. For that purpose, we suggest that the trade-off between the first objectives of the subsystems be used to give information to the decisionmakers on their interdependence. Thus, the following five-step scheme is proposed:

Step 1: The analyst fixes some values for the coupling variables ($y_1 = x_2$, $y_2 = x_1$).

Step 2: The decisionmakers optimize their subsystems.

Step 3: The analyst calculates the trade-off between f_1^1 and f_1^2, λ_{11}^{12}, by using, for example, the method in Sect. 6.4.2.

Step 4: Based on the trade-off information, the decisionmakers agree on the direction and amount of change in f_1^1 and f_1^2 (using their own decision rules).

Step 5: The analyst calculates the corresponding new values for the coupling variables. (Steps 2 through 5 are repeated, until no significant changes occur.)

Here, the trade-off between the first objectives has been used as a simple measure of the interdependence of the decisionmakers. Note that each decisionmaker's reaction to that trade-off may also depend naturally on the levels of all objectives, since these determine the utilities of the decisionmakers. In fact, at the beginning, the objectives of, for example, DM_1 may be so high that both parties agree to lower f_1^1, whatever the trade-off λ_{11}^{12} is.

In step 4, the changes in f_1^1 and f_1^2 may be based on the trade-off λ_{11}^{21} (i.e., $\Delta f_1^1 = -\lambda_{11}^{12}\Delta f_1^2$, where Δf_1^1 and Δf_2^1 are changes agreed upon). In this case, the analyst calculates the appropriate exact Pareto optimal point (corresponding to $\epsilon_1^2 = f_1^2 + \Delta f_1^2$) and the values for the coupling variables in step 5. Alternatively, the analyst determines (by using the method in Sect. 6.4.2) Pareto optimal pairs (f_1^1, f_1^2), from which set the decisionmakers select one. In this case, step 5 is included in step 4, and step 3 is skipped.

Next, consider the convergence of the scheme. An iterative negotiation scheme converges if the objective functions converge to some limits when the process is continued by consistent decisionmakers, and if the limit values of the objectives are a best compromise among all feasible objective values for the decisionmakers.

We will consider the convergence of the above scheme, making the following assumptions: (A1) a best-compromise solution exists, and (A2) an (not necessarily known) additive differentiable group utility function exists (for group utility functions, see Sage, 1977).

To see the major necessary condition for the convergence of the proposed scheme, let us further make the following assumption: (A3) all trade-offs between the objectives f_j^i ($i = 1, 2; j = 1, \ldots, n_i$) exist and all trade-offs in the subsystem optimizations in Step 2 exist. We do not assume that the decisionmakers use multicriteria optimization methods based on trade-offs; trade-offs are considered here for theoretical reasons. We also assume that (A4) the best-compromise solution is the only point where the trade-offs equal the indifference trade-offs of the group utility function (which ensures a local maximum to be a global one). It is easy to see that assumptions A1 and A2 guarantee that, in the negotiation process, the values of objectives will approach some limits (note that the scheme improves group utility in every phase).

Now, at the limit values of objectives, we know that the trade-off between f_1^1 and f_1^2 is equal to the indifference trade-off of the group utility function, and that the trade-offs in subsystem multicriteria optimizations equal also the corresponding indifference trade-offs. But note that, in subsystem optimizations, the couplings were kept constant. That is, trade-offs in subsystem optimizations are not necessarily trade-offs for the whole system. A simple assumption guarantees that trade-offs associated with the subsystem optimizations are also trade-offs for the whole system: (A5) when determining any trade-off between objectives of the same subsystem, the fixed objectives fix the coupling variables associated with that particular subsystem.

If assumption A5 holds, then the trade-offs in subsystem optimization in step 2 are also trade-offs for the whole system. Then, based on assumption A4, the negotiation scheme convergences.

To see an interpretation of A5, which is the most crucial assumption, let us put it in a weaker form and call it A5′: fixing the levels of the objectives of $N - 1$ subsystems fixes the coupling variables of the remaining subsystem.

Assumption A5′ is an independence condition for the subsystems, meaning that, when the levels of the objectives of $N - 1$ subsystems are fixed, the decision variables of these $N - 1$ subsystems do not affect the trade-offs of the remaining subsystem via changing the coupling variables of the particular subsystem.

One general class of systems where assumption A5′ usually holds consists of those cases where couplings between the subsystems are due to only common resources. In these cases, fixing the levels of the objectives of $N - 1$ subsystems usually fixes the amount of resources these $N - 1$ subsystems require. Thus, the amount of resources (for couplings, see Sect. 4.4) for the remaining subsystem is fixed.

Above we have suggested a simple iterative negotiation scheme and considered its convergence requirements. It is clear that assumption A5 (or A5′) is the most crucial one. In fact, one can relax, for example, assumption A3; it is not necessary to have trade-offs between all objectives. A case with total trade-offs is encountered in the example in the next section.

It should be noted that the iterative decisionmaking scheme suggested here can be used also by a single decisionmaker for decomposing a decision problem into smaller subproblems.

It is easy to see how the case $N = 2$ can be generalized for a general number of decisionmakers. The only change occurs in step 3, where now the trade-offs between the first objectives of all subsystems are calculated. That is, the analyst determines, for example, the trade-offs λ_{11}^{12}, λ_{11}^{13}, ..., λ_{11}^{1N}. Based on this information, the decisionmakers must agree on changes in their first objectives, Δf_1^1, $\ldots$, Δf_1^N which are related:

$$\Delta f_1^1 = \sum_{j=2}^{N} \lambda_{11}^{1j} \Delta f_1^j \tag{6.10}$$

6.4.4 Example

Consider the hypothetical Bow River Valley water quality management problem described by, for example, Hwang and Masud (1979). The valley with a cannery, two cities, and a state park is depicted in Fig. 6.3.

The objectives are (notation precedes the discussion on subsystems):

f_1^1 = Addition to the tax rate at Bowville [percent (%) additional tax rate]
f_2^1 = Dissolved oxygen (DO) level at Bowville [milligrams per liter (mg/L)]
f_3^1 = Percentage return on equity at Pierce-Hall Cannery (% earnings)
f_1^2 = DO level at Robin State Park (mg/L)
f_2^2 = DO level at Plympton (mg/L)
f_3^2 = Addition to the tax rate at Plympton (% additional tax rate)

On the relevant range of values, the objective equations given by Hwang and

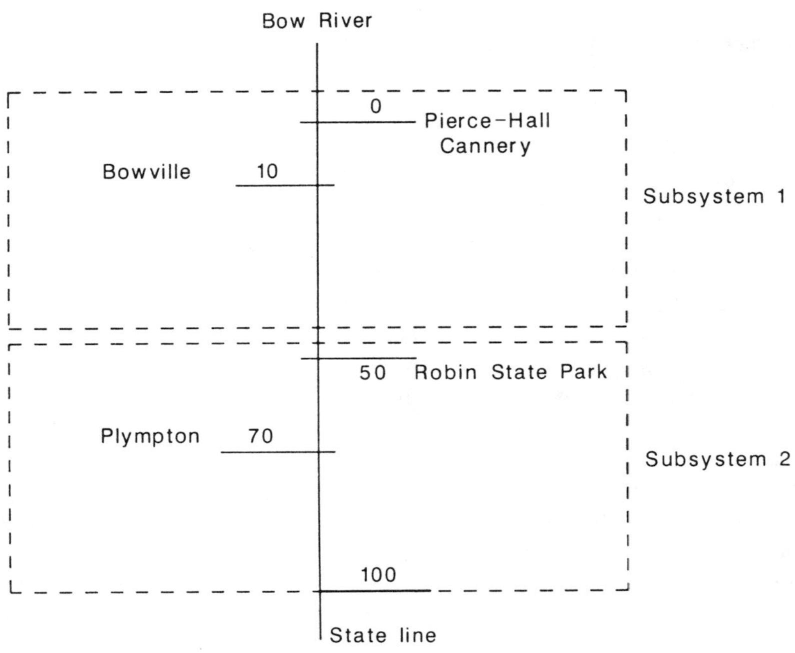

Figure 6.3 Bow River valley.

Masud (1979) can, for practical purposes, be simplified to the following equations:

$$f_1^1 = 1.8 \times 10^{-3}\left(\frac{532}{1.09 - x_2^2} - 532\right) \tag{6.11}$$

$$f_2^1 = 4.75 + 2.27(x_1 - 0.3) \tag{6.12}$$

$$f_3^1 = 7.5 - 0.012\left(\frac{59}{1.09 - x_1^2} - 59\right) \tag{6.13}$$

$$f_1^2 = 2.0 + 0.524(x_1 - 0.3) + 2.79(x_2 - 0.3) + 0.882\left(\frac{0.39}{1.39 - x_1^2} - 0.3\right) + 2.65\left(\frac{0.39}{1.39 - x_2^2} - 0.3\right) \tag{6.14}$$

$$f_2^2 = 4.48 + 0.319 f_1^2 \tag{6.15}$$

$$f_3^2 = 2.5 \times 10^{-3}\left(\frac{450}{1.09 - x_3^2} - 450\right) \tag{6.16}$$

where x_1, x_2, and x_3 are decision variables [proportionate reductions in carbonaceous biological oxygen demand (BOD) loads at Cannery, Bowville, and Plympton, respectively].

In fact, only x_1 and x_2 are decision variables, because, in practice, x_3 is determined via the following equation (minimum DO level at the state line):

$$0.187 + 0.166 f_2^2 + 3.34(x_3 - 0.3) + 2.62\left(\frac{0.39}{1.39 - x_3^2} - 0.3\right) = 3.5 \tag{6.17}$$

The other constraints are $0.3 \leq x_i \leq 1$ $(i = 1, 2, 3)$.

For the case, we could imagine four decisionmakers at the four locations. In the following discussion, we assume two decisionmakers or two groups of decisionmakers, DM_1 and DM_2. DM_1 represents interests at the cannery and Bowville (which are dependent on each other, because the work force of the cannery lives in Bowville). DM_2 represents the interests of Robin State Park and Plympton (whose interests are parallel: to reduce the upstream pollution). Hence, we consider this case as consisting of two subsystems, as depicted in Fig. 6.3.

The coupling variable, in this case, is the DO content at the boundary of the two respective regions, that is, the DO at the park (f_1^2). In this case, a coupling variable equals an objective.

As in many real-life problems, the objectives are always in conflict. That is, all feasible alternatives are Pareto optimal. The objective functions are also simple, so that there is no need here to use the techniques developed in Sect. 6.4.2. Instead, the trade-offs in the following discussion are calculated analytically by using the definition of trade-offs.

Let us apply the interactive negotiation scheme that was proposed in the previous section, step by step, assuming, for our simulation purposes, a group utility function:

$$U(\mathbf{f}^1, \mathbf{f}^2) = -0.7f_1^1 + f_2^1 + 0.4f_3^1 + f_1^2 + f_2^2 - 0.7f_3^2$$

The subsystem independence assumption A5′ is satisfied: that is, if the objectives of subsystem 1 (the upstream system) are fixed, they fix also the coupling variable (the DO at the boundary) (see the systems equations); if the objectives of subsystem 2 are fixed, the coupling variable, equaling an objective in subsystem 2, is fixed.

Step 1: The analyst specifies a value for the coupling variable, for example, $(y_1 = x_2 =)\ f_1^2 = 4$.

Step 2: The decisionmakers optimize their subsystems. DM$_1$ trades off his or her objectives f_1^1, f_2^1, and f_3^1, keeping $f_1^2 = 4$ [Eq. (6.14)]. DM$_1$ chooses, by (implicitly) maximizing function $-0.7f_1^1 + f_2^1 + 0.4f_3^1$,

$$f_1^1 = 0.65, f_2^1 = 5.97, f_3^1 = 6.37$$

For DM$_2$, the optimization problem is trivial, because the fixed value of f_1^2 determines unique $f_1^2 = 4$, $f_2^2 = 5.76$, and $f_3^2 = 1.39$ [Eqs. (6.15), (6.17), and (6.16)].

Steps 3, 4, and 5: The decisionmakers make their trade-off. There is no (partial) trade-off between f_1^1 (Bowville's tax) and f_1^2 (the park's DO), because a change in f_1^1 will have an impact on f_1^2, f_2^2, and f_3^2 at the same time. Hence, the decisionmakers simply have to deal with a trade-off between f_1^1 and (f_1^2, f_2^2, f_3^2). In any case, this is simpler than dealing with all six objectives. In this simulation, it is assumed that a set of values of $(f_1^1; f_1^2, f_2^2, f_3^2)$ is given to the decisionmakers and they agree on the following values: $f_1^1 = 1.74$, $f_1^2 = 4.86$, $f_2^2 = 6.03$, $f_3^2 = 1.54$. These values maximize the group utility function, when f_2^1 and f_3^1 are constant, as determined in step 1.

After this, the steps are repeated from the beginning. Table 6.1 shows how the iterative negotiation proceeds.

Table 6.1 Results of the simulated negotiations[a]

Iteration		f_1^1	f_2^1	f_3^1	f_1^2	f_2^2	f_3^2	Decision variables		
								x_1	x_2	x_3
1.	Steps 1,2	0.65	5.97	6.37	4.00	5.76	1.39	0.839	0.702	0.802
	Steps 3–5	1.74	5.97	6.37	4.86	6.03	1.54	0.839	0.857	0.817
2.	Step 2	1.60	6.07	5.99	4.86	6.03	1.54	0.878	0.846	0.817
	Steps 3–5	1.78	6.07	5.99	4.95	6.06	1.52	0.878	0.859	0.815
3.	Step 2	1.76	6.07	5.95	4.95	6.06	1.52	0.881	0.859	0.815
	Steps 3–5	1.77	6.07	5.95	4.96	6.06	1.52	0.881	0.860	0.815

[a]Explanations of f variables are given in the text.

The iterative negotiation scheme converges rapidly. It can be used also by a single decisionmaker. Compared with the solution technique (SEMOPS) of a single decisionmaker presented by Hwang and Masud (1979) for this problem, the above iterative procedure is much simpler in numerical calculation. Furthermore, here the trade-offs (impacts) between the decisionmakers are explicitly known to the decisionmakers, which contributes to their confidence in the negotiations.

REFERENCES

Bertsekas, D. P., Multiplier Methods: A Survey, *Automatica,* vol. 12, pp. 133–145, 1976.

Chankong, V., and Haimes, Y. Y., *Multiobjective Decision Making: Theory and Methodology,* North-Holland, New York, 1983.

Cruz, J. B., Jr., Leader-follower Strategies for Multilevel Systems, *IEEE Trans. Automat. Contr.,* vol. AC-23, no. 2, 1978.

Findeisen, W., Brdys, M., Malinowski, K., Tatjevski, P., and Wozniak, A., *Control and Coordination in Hierarchical Systems,* Wiley, New York, 1980.

Gomide, F. A. C., *Hierarchical Multistage, Multiobjective Impact Analysis,* Ph.D. dissertation, Department of Systems Engineering, Case Western Reserve University, Cleveland, Ohio, 1982.

Gomide, F. A. C., and Haimes, Y. Y., Multiobjective, Multistage Impact Analysis Method: Methodology and Examples, *ORSA/TIMS Joint Nat. Meeting,* Chicago, Ill., April 25–27, 1983.

Hwang, C.-L., and Masud, A. S. M., *Multiple Objective Decision-Making Methods and Applications,* Springer-Verlag, New York, 1979.

Rietveld, P., *Multiple Objective Decision Methods and Regional Planning,* North-Holland, Amsterdam, 1980.

Sage, A. P., *Methodology for Large-Scale Systems,* McGraw-Hill, New York, 1977.

Stackelberg, H. von, *The Theory of the Market Economy,* Oxford University Press, London, 1952.

CHAPTER

SEVEN

DISCRETE DYNAMIC MODELS

7.1 INTRODUCTION

This chapter discusses decisionmaking methods that deal with discrete dynamic problems having multiple criteria. We exclude methods based on dynamic programming, which are presented in Chap. 8. Discrete dynamic multicriteria problems do not differ mathematically from the static ones dealt with in the previous chapters; they are simply a special case in which the subsystems are connected in series. The results obtained so far are applicable to them. However, because of their series structure, special algorithms for dynamic problems emerge, such as the two-point boundary method presented in Sect. 7.4

7.2 PROBLEM FORMULATION

The discrete dynamic problem dealt with in the following problem formulation is referred to as a discrete-time dynamic problem. That is, we are considering a system that evolves with time. Note that the mathematical formulation applies directly to a static case where the subsystems are connected in series.

Consider the discrete-time system depicted, in block diagram form, in Fig. 7.1. In the ith stage $[i = 0, \ldots, (N - 1)]$ let

$\mathbf{x}_i$ = state vector
$\mathbf{f}^i = (f^i_1, \ldots, f^i_{n_i})^T$ = vector of objectives
n_i = the number of objectives
$\mathbf{m}_i$ = decision vector

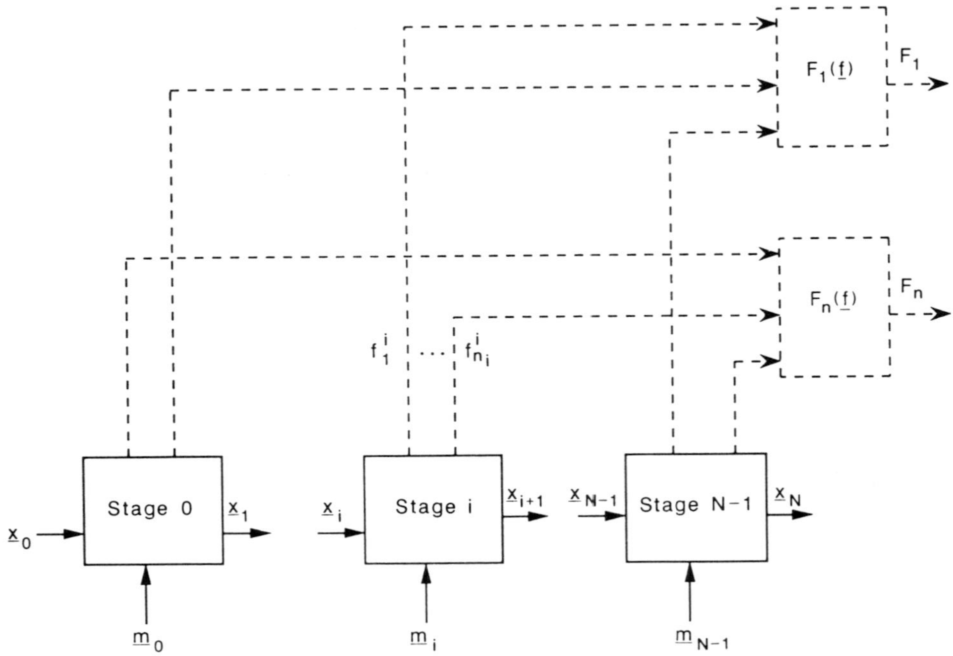

Figure 7.1 Discrete-time dynamic problem: structure of the system (solid line) and objectives (broken line).

Furthermore, let the state change at stage i according to

$$\mathbf{x}_{i+1} = \mathbf{H}_i(\mathbf{x}_i, \mathbf{m}_i) \qquad i = 0, \ldots, (N-1) \qquad \mathbf{x}_0 \text{ given} \tag{7.1}$$

The objectives of the stages are assumed to be functions of the stage variables:

$$f_j^i = f_j^i(\mathbf{x}_i, \mathbf{m}_i) \qquad i = 0, \ldots, (N-1) \qquad j = 1, \ldots, n_i \tag{7.2}$$

For brevity, let

$$\mathbf{x} = (\mathbf{x}_0^T, \ldots, \mathbf{x}_{N-1}^T)^T$$

$$\mathbf{f} = [(\mathbf{f}^0)^T, \ldots, (\mathbf{f}^{N-1})^T]^T = \text{vector of all stage objectives} \tag{7.3a}$$

$$\mathbf{m} = (\mathbf{m}_0^T, \ldots, \mathbf{m}_{N-1}^T)^T = \text{vector of all decision variables}$$

If the dependence of $\mathbf{x}$ on $\mathbf{x}_0$ is suppressed, then $\mathbf{x} = \mathbf{x}(\mathbf{m})$ [determined by Eq. (7.1)]. Finally, let the decision vectors $\mathbf{m}_i$ be constrained by the following equations:

$$\mathbf{g}_i(\mathbf{x}_i, \mathbf{m}_i) \leq \mathbf{0} \qquad i = 0, \ldots, (N-1) \tag{7.3b}$$

The overall objectives $F_1, \ldots, F_n$ that interest the decisionmaker are, in general, some functions of $\mathbf{f}$. A generalization to the case where the F_i's depend also on the final state $\mathbf{x}_N$ is straightforward. We can now state the multicriteria discrete-time problem as follows:

$$\min_{\mathbf{m}} \{F_1(\mathbf{f}^0, \ldots, \mathbf{f}^{N-1}), \ldots, F_n(\mathbf{f}^0, \ldots, \mathbf{f}^{N-1})\} \tag{7.4a}$$

with

$$\mathbf{f}^i = [f_1^i(\mathbf{x}_i, \mathbf{m}_i), \ldots, f_{n_i}^i(\mathbf{x}_i, \mathbf{m}_i)]^T \qquad i = 0, \ldots, (N-1)$$

subject to

$$\mathbf{x}_{i+1} = \mathbf{H}_i(\mathbf{x}_i, \mathbf{m}_i) \qquad i = 0, \ldots, (N-1) \qquad \mathbf{x}_0 \text{ given} \tag{7.4b}$$

$$\mathbf{g}_i(\mathbf{x}_i, \mathbf{m}_i) \leq \mathbf{0} \qquad i = 0, \ldots, (N-1) \tag{7.4c}$$

For brevity, let $\mathbf{F} = (F_1, \ldots, F_n)^T$.

The problem formulation is rather general, allowing, for example, objectives that may vary with time. In one common case of interest, we have the same number of objectives in the stages:

$$F_j(\mathbf{f}^0, \ldots, \mathbf{f}^{N-1}) = \sum_{i=0}^{N-1} f_j^i \tag{7.5}$$

That is, the decisionmaker is interested in the total (or average) of the objectives over the optimization period.

7.3 STRAIGHTFORWARD APPROACH

We consider three methods using a straightforward approach: the weighting method, the surrogate worth trade-off (SWT) method, and the gradient method.

7.3.1 Weighting Method

In the weighting method (see Sect. 4.3.3) the problem given by Eq. (7.4) is converted into a series of single-objective problems of the following form (subject to the same constraints):

$$\min \sum_{i=1}^{n} w_i F_i \tag{7.6}$$

where

$$w_i \geq 0 \qquad \text{and} \qquad \sum_{i=1}^{n} w_i = 1$$

By varying the weights, different Pareto optimal points result. All subproblems of the form given in Eq. (7.6) are single-objective discrete-time problems. They can be solved by any appropriate algorithm developed for discrete-time dynamic problems. There are plenty of techniques available, especially when we have objectives F_i of the form given in Eq. (7.5) (see, e.g., Bryson and Ho, 1969).

The weighting method, although efficient computationally, has a severe limitation: in order to generate all Pareto optimal points, the Pareto optimal surface in the objective space must be convex (and even strictly convex, to avoid additional complications). For more details, see Haimes et al. (1975), in which the continuous case is treated. As can readily be seen, there is no essential difference between the discrete-time and the continuous-time cases.

7.3.2 Surrogate Worth Trade-Off (SWT) Method

When the SWT method is applied to discrete-time dynamic problems, the ϵ-constraint method is used, where one objective function constitutes the primary objective and all other objectives are constraints (see Chap. 2). Pareto optimal solutions for the problem in Eq. (7.4), are generated, using the following ϵ-constraint form:

$$\min F_1 \tag{7.7}$$

subject to

$$\begin{aligned} & F_2 \leqslant \epsilon_2 \\ & \vdots \\ & F_n \leqslant \epsilon_n \\ & \mathbf{x}_{i+1} = \mathbf{H}_i(\mathbf{x}_i, \mathbf{m}_i) \qquad i = 0, \ldots, (N-1) \qquad \mathbf{x}_0 \text{ given} \\ & \mathbf{g}_i(\mathbf{x}_i, \mathbf{m}_i) \leqslant \mathbf{0} \qquad i = 0, \ldots, (N-1) \end{aligned} \tag{7.8}$$

By varying the ϵ's, different Pareto optimal points are generated.

The SWT and interactive SWT methods, which are based on the ϵ-constraint method, do not suffer from the convexity limitations of the weighting method. A drawback is the additional computational work caused by the additional constraints. For more detail, see Haimes et al. (1975), in which continuous-time cases are treated. Also see Haimes and Hall (1974), Haimes (1977), and Chankong and Haimes (1978). The subproblems are solved using Pontryagin's principle; the objectives are of integral form equivalent to Eq. (7.5) in the continuous case.

7.3.3 Gradient Method

The method discussed here is termed the gradient method because it reduces to the usual gradient method when the number of objectives is one, and because the starting point to derive it is the application of the gradient method to the utility function. Geoffrion et al. (1972) discuss a corresponding method in the static case.

Consider the problem given by Eqs. (7.4a) and (7.4b) [without the inequality constraints in Eq. (7.4c)]. Assume that there exists a differentiable underlying loss (negative utility) function $L(F_1, \ldots, F_n)$: $R^n \to R$ for the decisionmaker's preferences. The loss function is not directly used in the method. Instead, indifference trade-offs (marginal rates of substitution) given by the decisionmaker are used. The following relationship between the loss function and these trade-offs is used (see Sect. 5.2):

$$\frac{\partial L(F_1, \ldots, F_n)/\partial F_j}{\partial L(F_1, \ldots, F_n)/\partial F_i} = \lambda_{ij}^*(F_1, \ldots, F_n) \tag{7.9a}$$

where $\lambda_{ij}^*(F_1, \ldots, F_n)$ is the decisionmaker's indifference trade-off between the ith and jth objectives when the objectives have values $F_1, \ldots, F_n$.

For notational convenience, we define λ_{ij}^* for $i = j$. From Eq. (7.9a), we get $\lambda_{ii}^* = 1$, which we use as a definition of λ_{ii}^* for all i.

The method is an iterative one; the iteration index is denoted by the superscript k. At the kth iteration we have a decision vector $\mathbf{m}^k$ and $\mathbf{x}^k \triangleq \mathbf{x}(\mathbf{m}^k)$ with a corresponding value $\mathbf{f}^k$ for $\mathbf{f}$ and with a corresponding value $\mathbf{F}^k$ for the overall objective vector $\mathbf{F}$. To serve as a new direction, $\Delta\mathbf{m}^k$, for seeking a better value of $\mathbf{m}$, the direction in which the loss function decreases fastest is selected. Hence, we evaluate the total derivative of L with respect to $\mathbf{m}$ at the values of the kth iteration:

$$\begin{aligned}
&\frac{d}{d\mathbf{m}} L\{F_1[\mathbf{f}(\mathbf{x}(\mathbf{m}), \mathbf{m})], \ldots, F_n[\mathbf{f}(\mathbf{x}(\mathbf{m}), \mathbf{m})]\}\Big|_{m=m^k} \\
&= \sum_{j=1}^{n} \frac{\partial L}{\partial F_j}(\mathbf{F}^k) \frac{d}{d\mathbf{m}} F_j[\mathbf{f}(\mathbf{x}(\mathbf{m}), \mathbf{m})]\Big|_{m=m^k} \\
&= \frac{\partial L}{\partial F_1}(\mathbf{F}^k) \sum_{j=1}^{n} \left[\frac{\partial L(\mathbf{F}^k)/\partial F_j}{\partial L(\mathbf{F}^k)/\partial F_1}\right] \frac{d}{d\mathbf{m}} F_j[\mathbf{f}(\mathbf{x}(\mathbf{m}), \mathbf{m})]\Big|_{\mathbf{m}=\mathbf{m}^k} \\
&= \frac{\partial L}{\partial F_1}(\mathbf{F}^k) \sum_{j=1}^{n} \lambda_{1j}^{*k} \frac{d}{d\mathbf{m}} F_j[\mathbf{f}(\mathbf{x}(\mathbf{m}), \mathbf{m})]\Big|_{\mathbf{m}=\mathbf{m}^k} \\
&= \frac{\partial L}{\partial F_1}(\mathbf{F}^k) \left\{\frac{d}{d\mathbf{m}} \sum_{j=1}^{n} \lambda_{1j}^{*k} F_j[\mathbf{f}(\mathbf{x}(\mathbf{m}), \mathbf{m})]\right\}\Big|_{\mathbf{m}=\mathbf{m}^k}
\end{aligned} \tag{7.9b}$$

Here, in the first equality, the chain rule of differentiation is used. In the second equality, $\partial L/\partial F_1$ is assumed to be nonzero (and it is strictly positive because we are minimizing objectives). The third equality results from the use of Eq. (7.9a). In the fourth equality, simple transfer of constants in the derivation is made.

The term $\partial L/\partial F_1$ in Eq. (7.9b) is a positive constant, and it can be left out in determining the direction of the gradient. To evaluate the gradient by making use of the system's dynamisms, consider first a general function $E[\mathbf{x}(\mathbf{m}), \mathbf{m}]$. This function is of the same form as that in Eq. (7.9b). For notational simplicity, we evaluate the gradient of E, and then apply the result to Eq. (7.9b).

For brevity, let

$$\mathbf{m}_{i,j} = (\mathbf{m}_i, \mathbf{m}_{i+1}, \ldots, \mathbf{m}_{j-1}, \mathbf{m}_j) \qquad i, j = 0, \ldots, (N-1) \qquad i \leqslant j$$

$\mathbf{x}_i(\mathbf{x}_j, \mathbf{m}_{j,i-1})$ = the value of $\mathbf{x}_i$, as given by the system's equations in Eq. (7.1), when starting from an initial state $\mathbf{x}_j$ and applying a decision sequence $\mathbf{m}_{j,i-1}$; $i = 1, \ldots, N$; $j = 0, \ldots, (N-1)$; $i > j$.

In accordance with the above shorthand notation $\mathbf{x}(\mathbf{m})$ (where $\mathbf{x}_0$ is suppressed), the initial state $\mathbf{x}_0$ is left out in the expressions $\mathbf{x}_i(\mathbf{x}_0, \mathbf{m}_{0,i-1})$ in the following discussion. Furthermore, define, for $i = 0, \ldots, N$,

$$\begin{aligned} E^i(\mathbf{x}_i, \mathbf{m}) = E\{[&\mathbf{x}_0, \mathbf{x}_1(\mathbf{m}_0), \mathbf{x}_2(\mathbf{m}_{0,1}), \ldots, \mathbf{x}_{i-1}(\mathbf{m}_{0,i-2}), \mathbf{x}_i, \\ &\mathbf{x}_{i+1}(\mathbf{x}_i, \mathbf{m}_i), \mathbf{x}_{i+2}(\mathbf{x}_i, \mathbf{m}_{i,i+1}), \ldots, \\ &\mathbf{x}_N(\mathbf{x}_i, \mathbf{m}_{i,N-1})], \mathbf{m}\} \end{aligned} \tag{7.10a}$$

Note that, on the basis of the systems equations in Eq. (7.1),

$$E[\mathbf{x}(\mathbf{m}), \mathbf{m}] = E^i[\mathbf{x}_i(\mathbf{m}_{0,i-1}), \mathbf{m}] \qquad i = 1, \ldots, N \tag{7.10b}$$

and $E[\mathbf{x}(\mathbf{m}), \mathbf{m}] = E^0(\mathbf{x}_0, \mathbf{m})$.

For the following derivations, note the following: $dE[\mathbf{x}(\mathbf{m}), \mathbf{m}]/d\mathbf{m}$ means the (total) derivative of $E[\mathbf{x}(\mathbf{m}), \mathbf{m}]$ as function of m with respect to $\mathbf{m}$; $\partial E^i[\mathbf{x}(\mathbf{m}), \mathbf{m}]/\partial \mathbf{x}$ means the (partial) derivative with respect to the E^i's first vector argument $[\mathbf{x}(\mathbf{m})]$; and $\partial E^i[\mathbf{x}(\mathbf{m}), \mathbf{m}]/\partial \mathbf{m}$ is the (partial) derivative with respect to E^i's second argument $(\mathbf{m})$. Furthermore, note that the derivative of a scalar with respect to a vector is a row vector, and a derivative of a vector with respect to a vector is a matrix where the ith row is the derivative of the ith component.

By using Eq. (7.10b) we calculate, for $i = 0, \ldots, (N-1)$, at $\mathbf{m}^k$,

$$\begin{aligned} \left.\frac{dE[\mathbf{x}(\mathbf{m}), \mathbf{m}]}{d\mathbf{m}_i}\right|_{\mathbf{m}=\mathbf{m}^k} &= \left.\frac{dE^{i+1}[\mathbf{x}_{i+1}(\mathbf{m}_{0,i}), \mathbf{m}]}{d\mathbf{m}_i}\right|_{\mathbf{m}=\mathbf{m}^k} \\ &= \left.\frac{\partial E^{i+1}[\mathbf{x}_{i+1}(\mathbf{m}_{0,i}), \mathbf{m}]}{\partial \mathbf{m}} \frac{d\mathbf{m}}{d\mathbf{m}_i}\right|_{\mathbf{m}=\mathbf{m}^k} \\ &\quad + \left.\frac{\partial E^{i+1}[\mathbf{x}_{i+1}(\mathbf{m}_{0,i}), \mathbf{m}]}{\partial \mathbf{x}_{i+1}} \frac{d\mathbf{x}_{i+1}(\mathbf{m}_{0,i})}{d\mathbf{m}_i}\right|_{\mathbf{m}=\mathbf{m}^k} \\ &= \left.\frac{\partial E^{i+1}[\mathbf{x}_{i+1}(\mathbf{m}_{0,i}), \mathbf{m}]}{\partial \mathbf{m}} \frac{d\mathbf{m}}{d\mathbf{m}_i}\right|_{\mathbf{m}=\mathbf{m}^k} \\ &\quad + (\boldsymbol{\delta}_{i+1}^k)^T \frac{\partial H_i(\mathbf{x}_i^k, \mathbf{m}_i^k)}{\partial \mathbf{m}_i} \end{aligned} \tag{7.10c}$$

where the last equation follows directly from the following definition:

$$(\boldsymbol{\delta}_{i+1}^k)^T = \left.\frac{\partial E^{i+1}}{\partial \mathbf{x}_{i+1}}\right|_{\mathbf{m}=\mathbf{m}^k} \qquad i = 0, \ldots, (N-1) \tag{7.10d}$$

and from the systems equations in Eq. (7.1).

Note that, if $\mathbf{x}_N$ does not appear in $E[\mathbf{x}(\mathbf{m}), \mathbf{m}]$—as in our case—then we have [see Eq. (7.10d)]

$$\boldsymbol{\delta}_N^k = \mathbf{0} \tag{7.10e}$$

For the $\boldsymbol{\delta}_i^k$'s we obtain a recursion formula by noting first that [see Eq. (7.10a)]

$$\begin{aligned} E^i(\mathbf{x}_i, \mathbf{m}) = E\{&\mathbf{x}_0, \mathbf{x}_1(\mathbf{m}_0), \ldots, \mathbf{x}_{i-1}(\mathbf{m}_{0,i-2}), \mathbf{x}_i, \\ &\mathbf{x}_{i+1}(\mathbf{x}_i, \mathbf{m}_i), \mathbf{x}_{i+2}(\mathbf{x}_{i+1}(\mathbf{x}_i, \mathbf{m}_i), \mathbf{m}_{i-1}), \\ &\mathbf{x}_{i+3}(\mathbf{x}_{i+1}(\mathbf{x}_i, \mathbf{m}_i), \mathbf{m}_{i-1,i-2}), \ldots, \\ &\mathbf{x}_N(\mathbf{x}_{i+1}(\mathbf{x}_i, \mathbf{m}_i), \mathbf{m}_{i-1,N-1}), \mathbf{m}\} \end{aligned}$$

Hence (note the definition of E^{i+1}), for $i = 0, \ldots, (N-1)$,

$$\begin{aligned} \boldsymbol{\delta}_i^k = \left.\left(\frac{\partial E^i}{\partial \mathbf{x}_i}\right)^T\right|_{\mathbf{m}=\mathbf{m}^k} = \left.\left(\frac{\partial E}{\partial \mathbf{x}_i}\right)^T\right|_{\mathbf{m}=\mathbf{m}^k} \\ + \left[\left.\frac{\partial E^{i+1}}{\partial \mathbf{x}_{i+1}}\right|_{\mathbf{m}=\mathbf{m}^k} \left.\frac{d\mathbf{x}_{i+1}(\mathbf{x}_i, \mathbf{m}_i)}{d\mathbf{x}_i}\right|_{\mathbf{m}=\mathbf{m}^k}\right]^T \end{aligned}$$

Or, by taking the systems equations in Eq. (7.1) and the definition of $\boldsymbol{\delta}_{i+1}^k$ into account,

$$\boldsymbol{\delta}_i^k = \left.\left(\frac{\partial E}{\partial \mathbf{x}_i}\right)^T\right|_{\mathbf{m}=\mathbf{m}^k} + \left[\frac{\partial \mathbf{H}_i(\mathbf{x}_i, \mathbf{m})}{\partial \mathbf{x}_i}\right]^T \boldsymbol{\delta}_{i+1} \qquad i = 0, \ldots, (N-1) \tag{7.10f}$$

Applying the above results [Eqs. (7.10c), (7.10e), and (7.10f)] to our case, where

$$E[\mathbf{x}(\mathbf{m}), \mathbf{m}] = \sum_{j=1}^{n} \lambda_{1j}^{*k} F_j[f(\mathbf{x}(\mathbf{m}), \mathbf{m})]$$

yields the following equations (where $\boldsymbol{\delta}$ has been replaced by $\hat{\boldsymbol{\delta}}$):

$$\begin{aligned} \left.\frac{d}{\mathbf{m}_i} L\right|_{\mathbf{m}=\mathbf{m}^k} \simeq \sum_{j=1}^{n} \lambda_{1j}^{*k} \frac{\partial F_j[\mathbf{f}(\mathbf{x}^k, \mathbf{m}^k)]}{\partial \mathbf{f}^i} \frac{\partial \mathbf{f}^i(\mathbf{x}_i^k, \mathbf{m}_i^k)}{\partial \mathbf{m}_i} \\ + \hat{\boldsymbol{\delta}}_{i+1} \frac{\partial \mathbf{H}_i(\mathbf{x}_i^k, \mathbf{m}_i^k)}{\partial \mathbf{m}_i} \qquad i = 0, \ldots, (N-1) \end{aligned} \tag{7.11}$$

where $\hat{\boldsymbol{\delta}}_i^k$ $(i = 1, \ldots, N)$ is calculated (backward) from the following difference equation:

$$\begin{aligned} \hat{\boldsymbol{\delta}}_i^k = \frac{\partial \mathbf{H}_i(\mathbf{x}_i^k, \mathbf{m}_i^k)^T}{\partial \mathbf{x}_i} \hat{\boldsymbol{\delta}}_{i+1}^k + \left\{\sum_{j=1}^{n} \lambda_{1j}^{*k} \frac{\partial F_j[\mathbf{f}(\mathbf{x}^k, \mathbf{m}^k)]}{\partial \mathbf{f}^i} \frac{\partial \mathbf{f}(\mathbf{x}_i^k, \mathbf{m}_i^k)}{\partial \mathbf{x}_i}\right\} \\ i = 1, \ldots, (N-1) \qquad \hat{\boldsymbol{\delta}}_N^k = 0 \end{aligned} \tag{7.12}$$

The gradient can be calculated via Eqs. (7.11) and (7.12). Then we can proceed as proposed in Geoffrion et al. (1972); that is, we give the decisionmaker the values of the objectives along the direction of the negative gradient (note: we have a minimization task here). The decisionmaker decides how far to go in this direction. He then determines the indifference trade-offs. Then a new gradient is calculated, using Eqs. (7.12) and (7.11), and so on, until the preferred solution is found.

Computationally, solving the subproblems of this algorithm is no more tedious than solving those in a single-objective problem. Note that it was not necessary to make any assumptions concerning the structure of the objectives [such as Eq. (7.5)]. Note also that no constraints, such as those found in Eq. (7.4c), were assumed in the development. Constraints on the decision vector of the form $\mathbf{g}_i(\mathbf{m}_i) \leq \mathbf{0}$ are usually easily incorporated into a gradient method, but constraints on the state variable $\mathbf{x}$ are generally not easy to handle. As can be readily seen, a corresponding algorithm for the continuous-time problem can be derived.

7.4 TWO-POINT BOUNDARY VALUE METHOD

Consider another version of the problem given in Sect. 7.2, this time without inequality constraints; that is,

$$\min_{\mathbf{m}} \{F_1(\mathbf{f}^0, \ldots, \mathbf{f}^{N-1}), \ldots, F_n(\mathbf{f}^0, \ldots, \mathbf{f}^{N-1})\} \tag{7.13a}$$

with

$$\mathbf{f}^i = [f_1^i(\mathbf{x}_i, \mathbf{m}_i), \ldots, f_{n_i}^i(\mathbf{x}_i, \mathbf{m}_i)]^T \qquad i = 0, \ldots, (N-1)$$

subject to

$$\mathbf{H}_i(\mathbf{x}_i, \mathbf{m}_i) - \mathbf{x}_{i+1} = \mathbf{0} \qquad i = 0, \ldots, (N-1) \qquad \mathbf{x}_0 \text{ given} \tag{7.13b}$$

In the following discussion, a multiobjective counterpart to single-objective two-point boundary value methods is derived. For the problem given in Eqs. (7.13a) and (7.13b), denote the decisionmakers' indifference trade-offs by $\lambda_{12}^*(F), \ldots, \lambda_{1n}^*(F)$. Applying the necessary conditions for preferred solutions (see Sect. 5.2) to the problem in Eqs. (7.13a) and (7.13b) yields

$$\sum_{k=1}^{n} \lambda_{1k}^*(\mathbf{F}) \frac{\partial F_k}{\partial \mathbf{x}_i} + \sum_{j=1}^{N} \hat{\boldsymbol{\delta}}_j^T \frac{\partial}{\partial \mathbf{x}_i} [-\mathbf{x}_j + \mathbf{H}_{j-1}(\mathbf{x}_{j-1}, \mathbf{m}_{j-1})] = \mathbf{0} \tag{7.14}$$

$$i = 1, \ldots, N$$

$$\sum_{k=1}^{n} \lambda_{1k}^*(\mathbf{F}) \frac{\partial F_k}{\partial \mathbf{m}_i} + \sum_{j=1}^{N} \hat{\boldsymbol{\delta}}_j^T \frac{\partial}{\partial \mathbf{m}_i} [-\mathbf{x}_j + \mathbf{H}_{j-1}(\mathbf{x}_{j-1}, \mathbf{m}_{j-1})] = \mathbf{0} \tag{7.15}$$

$$i = 0, \ldots, (N-1)$$

Developing the first terms in Eqs. (7.14) and (7.15), as in Eq. (5.26), and leaving out terms that are not functions of $\mathbf{x}_i$ or $\mathbf{m}_i$ in the second sums, yields

$$\sum_{j=1}^{n_i} \hat{\lambda}_{1k}^{i*}(\mathbf{f})\frac{\partial f_j^i}{\partial \mathbf{x}_i} - \hat{\boldsymbol{\delta}}_i^T + \hat{\boldsymbol{\delta}}_{i+1}^T \frac{\partial \mathbf{H}_i(\mathbf{x}_i, \mathbf{m}_i)}{\partial \mathbf{x}_i} = \mathbf{0} \qquad i = 1, \ldots, (N-1) \tag{7.16a}$$

$$\hat{\boldsymbol{\delta}}_N = \mathbf{0} \tag{7.16b}$$

$$\sum_{j=1}^{n_i} \hat{\lambda}_{1k}^{i*}(\mathbf{f})\frac{\partial f_j^i}{\partial \mathbf{m}_i} + \hat{\boldsymbol{\delta}}_{i+1}^T \frac{\partial \mathbf{H}_i(\mathbf{x}, \mathbf{m}_i)}{\partial \mathbf{m}_i} = \mathbf{0} \qquad i = 0, \ldots, (N-1) \tag{7.17}$$

where [see Eq. (5.27)]

$$\hat{\lambda}_{1j}^{i*}(\mathbf{f}) = \sum_{k=1}^{n} \lambda_{1k}^{*}(\mathbf{F}(\mathbf{f}))\frac{\partial F_k}{\partial f_j^i} \qquad j = 1, \ldots, n_i \tag{7.18}$$

Dividing Eqs. (7.16) and (7.17) by $\hat{\lambda}_{11}^{i*}$ (which is assumed to be nonzero) yields

$$\sum_{j=1}^{n_i} \lambda_{1j}^i(\mathbf{f})\frac{\partial f_j^i}{\partial \mathbf{x}_i} - \boldsymbol{\delta}_i^T + \lambda^{i(i+1)}(\mathbf{f})\,\boldsymbol{\delta}_{i+1}^T \frac{\partial \mathbf{H}_i(\mathbf{x}_i, \mathbf{m}_i)}{\partial \mathbf{x}_i} = \mathbf{0}$$

$$i = 1, \ldots, (N-1) \tag{7.19a}$$

$$\boldsymbol{\delta}_N = \mathbf{0} \tag{7.19b}$$

$$\sum_{j=1}^{n_i} \lambda_{1j}^i(\mathbf{f})\frac{\partial f_j^i}{\partial \mathbf{m}_i} + \lambda^{i(i+1)}(\mathbf{f})\boldsymbol{\delta}_{i+1}^T \frac{\partial \mathbf{H}_i(\mathbf{x}_i, \mathbf{m}_i)}{\partial \mathbf{m}_i} = 0 \qquad i = 0, \ldots, (N-1) \tag{7.20}$$

where [see Eqs. (5.28), (5.34), and (5.29)]

$$\lambda_{1j}^i(\mathbf{f}) = \frac{\hat{\lambda}_{1j}^{i*}(\mathbf{f})}{\hat{\lambda}_{11}^{i*}(\mathbf{f})} \tag{7.21}$$

$$\boldsymbol{\delta}_i = \frac{\hat{\boldsymbol{\delta}}_i}{\hat{\lambda}_{11}^{i*}(\mathbf{f})} \tag{7.22}$$

$$\lambda^{k1}(\mathbf{f}) = \frac{\lambda_{11}^{1*}(\mathbf{f})}{\lambda_{11}^{k*}(\mathbf{f})} \tag{7.23}$$

As in Eq. (5.28), let $\lambda_s^i = (1, \lambda_{12}^i, \lambda_{13}^i, \ldots, \lambda_{1n_i}^i)^T$.

Note that Eq. (7.20) is, by the general necessary condition (Sect. 5.2), a necessary condition for the following multicriteria problems [$i = 0, \ldots, (N-1)$]:

$$\min_{\mathbf{m}_i} \begin{bmatrix} f_1^i(\mathbf{x}_i, \mathbf{m}_i) + \lambda^{i(i+1)}\,\boldsymbol{\delta}_{i+1}^T\mathbf{H}_i(\mathbf{x}_i, \mathbf{m}_i) \\ f_2^i(\mathbf{x}_i, \mathbf{m}_i) \\ \vdots \\ f_{n_i}^i(\mathbf{x}_i, \mathbf{m}_i) \end{bmatrix} \tag{7.24}$$

with indifference trade-offs λ_{1j}^i ($j = 2, \ldots, n_i$) and an indifference trade-off between stages, $\lambda^{i(i+1)}$.

Furthermore, note that Eqs. (7.19a) and (7.19b) are recursion formulas for $\boldsymbol{\delta}$ (to be solved backward):

$$\boldsymbol{\delta}_i = \lambda^{i(i+1)}(\mathbf{f}) \left(\frac{\partial \mathbf{H}_i}{\partial \mathbf{x}_i}\right)^T \boldsymbol{\delta}_{i+1} + \sum_{j=1}^{n_i} \lambda_{1j}^i(\mathbf{f}) \left(\frac{\partial f_j^i}{\partial \mathbf{x}_j}\right)^T \tag{7.25a}$$

$$\boldsymbol{\delta}_N = \mathbf{0} \tag{7.25b}$$

Assume, for a moment, that Eq. (7.20) is also a sufficient condition for the problem in Eq. (7.24). Then, altogether we have elements for a two-point boundary value problem, as we have in the single-objective case: the system equation [Eq. (7.13b)] with an initial value $\mathbf{x}_0$, an auxiliary variable ($\boldsymbol{\delta}$) with a difference equation with a boundary value $\boldsymbol{\delta}_N = \mathbf{0}$, and an optimization problem [Eq. (7.24)] for each stage.

One way the single-objective case is different is the fact that here, at each stage of a general case, there is a dependence on values of objectives in other stages ($\mathbf{f}$). A natural way to overcome this difficulty is to use the relaxation method. That is, first a value for the vector $\mathbf{f}$ is guessed. At the beginning, in solving the problem in Eq. (7.24) or evaluating $\boldsymbol{\delta}_i$ [Eq. (7.25)] at a stage i, the guessed values of the objectives in other stages are used when needed. Every time a new value for a component of $\mathbf{f}$ is evaluated in solving the problem in Eq. (7.24), the new value is used in the other stages when needed.

Several techniques developed in the single-objective case can now be directly modified to handle the above two-point boundary value problem. As an example, consider the relaxation method. Its basic idea is first to guess a trajectory for the auxiliary variable ($\boldsymbol{\delta}$, here), then to integrate the state equation, solving at each stage the decision variable using the guessed value for $\boldsymbol{\delta}$. Then, the auxiliary variable is solved backward, using the obtained values for $\mathbf{x}$ and $\mathbf{m}$, and so on, in turn.

As in the single-objective case, a strong optimization result (such as Pontryagin's theorem) does not hold in this multiobjective problem. That is, a solution of the original problem satisfies Eq. (7.20) but constitutes only stationary solutions for the problems in Eq. (7.24). Therefore, convexity requirements are needed that guarantee unique solutions to Eq. (7.20) and the problems in Eq. (7.24). Here, convexity specifications depend also on trade-offs that are not known beforehand. Hence, in general, the method must undergo a trial to see if it works.

To get some insight into the convexity requirements, consider Eq. (7.20) in the special case where all λ's and $\boldsymbol{\delta}$'s are positive constants. A sufficient condition for a unique $\mathbf{m}_i$ solution is the strict convexity of the f_j^i's ($j = 1, \ldots, n_i$) and $\mathbf{H}_i$ with respect to $\mathbf{m}_i$. Note that, as in the gradient method of Sect. 7.3.3, the objective structure may be general and constraints of the form $\mathbf{g}_i(\mathbf{m}_i) \leq 0$ are easily incorporated, but constraints on the $\mathbf{x}_i$'s are difficult to handle.

7.5 HIERARCHICAL METHODS

In this section, the results given in Chap. 5 will be applied to the problem given in Sect. 7.2, using the feasible and the nonfeasible methods. Note that in hierarchical schemes, the stages are solved in parallel.

7.5.1 Feasible Method

Consider the feasible method with partial derivatives that was given in Sect. 5.5.1. Using the identity

$$\mathbf{y}_i = \mathbf{x}_{i+1} \qquad i = 0, \ldots, (N-1)$$

we obtain the scheme seen in Fig. 7.2. Here, the subproblems are given by Eq. (7.26):

$$\min_{\mathbf{m}_i} \begin{bmatrix} f^i_1\,(\mathbf{x}^k_i, \mathbf{m}_i) \\ \vdots \\ f^i_{n_i}\,(\mathbf{x}^k_i, \mathbf{m}_i) \end{bmatrix} \tag{7.26a}$$

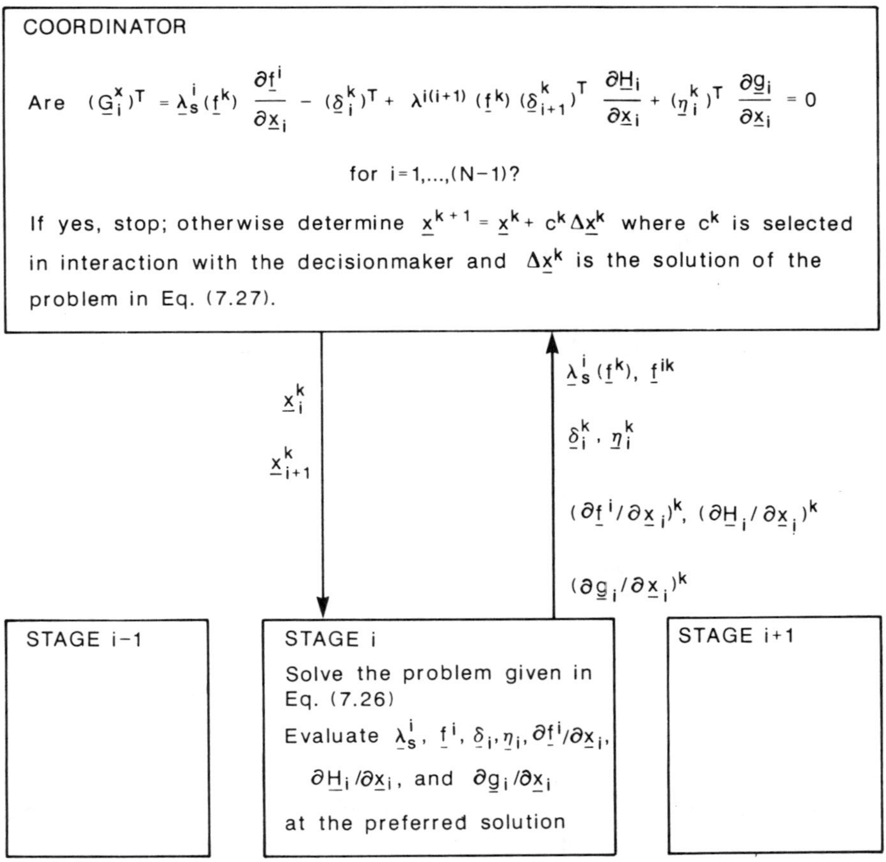

Figure 7.2 Feasible scheme.

subject to

$$\mathbf{x}_{i+1}^k - \mathbf{H}_i(\mathbf{x}_i^k, \mathbf{m}_i) = \mathbf{0} \tag{7.26b}$$

$$\mathbf{g}_i(\mathbf{x}_i^k, \mathbf{m}_i) \leqslant \mathbf{0} \tag{7.26c}$$

with indifference trade-offs given by $\boldsymbol{\lambda}_s^i$.

The linear programming problem for the determination of a new search direction $\Delta\mathbf{x}$ is now the following problem [see Eqs. (5.71), (5.61)–(5.65), and (5.73)]:

$$\min \sum_{i=1}^{N-1} \lambda^{ri}(\mathbf{f}^k) \left\{ (\mathbf{G}_i^x)^T \Delta\mathbf{x}_i - (\eta_i^k)^T \left(\frac{\partial \mathbf{g}_i}{\partial \mathbf{x}_i} \Delta\mathbf{x}_i + \frac{\partial \mathbf{g}_i}{\partial \mathbf{m}_i} \Delta\mathbf{m}_i \right) \right\} \tag{7.27}$$

subject to

$$-1 \leqslant \Delta\mathbf{x}_i \leqslant 1 \qquad \text{(componentwise)}$$

$$\Delta\mathbf{x}_{i+1} = (\partial\mathbf{H}_i)/(\partial\mathbf{x}_i)\Delta\mathbf{x}_i + (\partial\mathbf{H}_i)/(\partial\mathbf{m}_i)\Delta\mathbf{m}_i$$

$$(\partial(\mathbf{g}_i)_j/\partial\mathbf{x}_i)\Delta\mathbf{x}_i + (\partial(\mathbf{g}_i)_j/\partial\mathbf{m}_i)\Delta\mathbf{m}_i \leqslant \mathbf{0} \qquad \text{for } j \in \{j|(\mathbf{g}_i(\mathbf{x}_i^k, \mathbf{m}_i^k))_j = \mathbf{0}\}$$

$$i = 0, \ldots, (N-1)$$

$$\Delta\mathbf{x}_0 = \mathbf{0}$$

where

$$(\mathbf{G}_i^x)^T = \boldsymbol{\lambda}_s^i(\mathbf{f}^k) \frac{\partial \mathbf{f}^i}{\partial \mathbf{x}_i} - (\boldsymbol{\delta}_i^k)^T + \lambda^{i(i+1)}(\mathbf{f}^k)(\boldsymbol{\delta}_{i+1}^k)^T \frac{\partial \mathbf{H}_i}{\partial \mathbf{x}_i} + (\boldsymbol{\eta}_i^k)^T \frac{\partial \mathbf{g}_i}{\partial \mathbf{x}_i}$$

In contrast to the previous methods, the $\mathbf{g}_i(\mathbf{x}_i, \mathbf{m}_i) \leqslant \mathbf{0}$ constraints do not present a principal difficulty. But computationally, these are naturally not always easy.

7.5.2 Nonfeasible Method

In the nonfeasible method, the connections between the stages are relaxed. Consequently, we make the following changes in the problem formulation. Let $\mathbf{y}_i$ denote the final state of stage i, $i = 0, \ldots, (N-1)$. The system equation [Eq. (7.1)] is replaced by

$$\mathbf{y}_i = \mathbf{H}_i(\mathbf{x}_i, \mathbf{m}_i) \qquad i = 0, \ldots, (N-1) \tag{7.28}$$

And, as additional constraints, we take

$$\mathbf{y}_i - \mathbf{x}_{i+1} = \mathbf{0} \qquad i = 0, \ldots, (N-2) \tag{7.29}$$

[Note that $\mathbf{x}_N$ does not appear in the objectives; we do not have to write Eq. (7.29) for $i = N - 1$).] Otherwise, the problem formulation remains the same. Note that the change in the formulation is a technical one; the problem remains the same. The scheme in Sect. 5.6 is now directly applicable. We obtain the scheme given in Fig. 7.3.

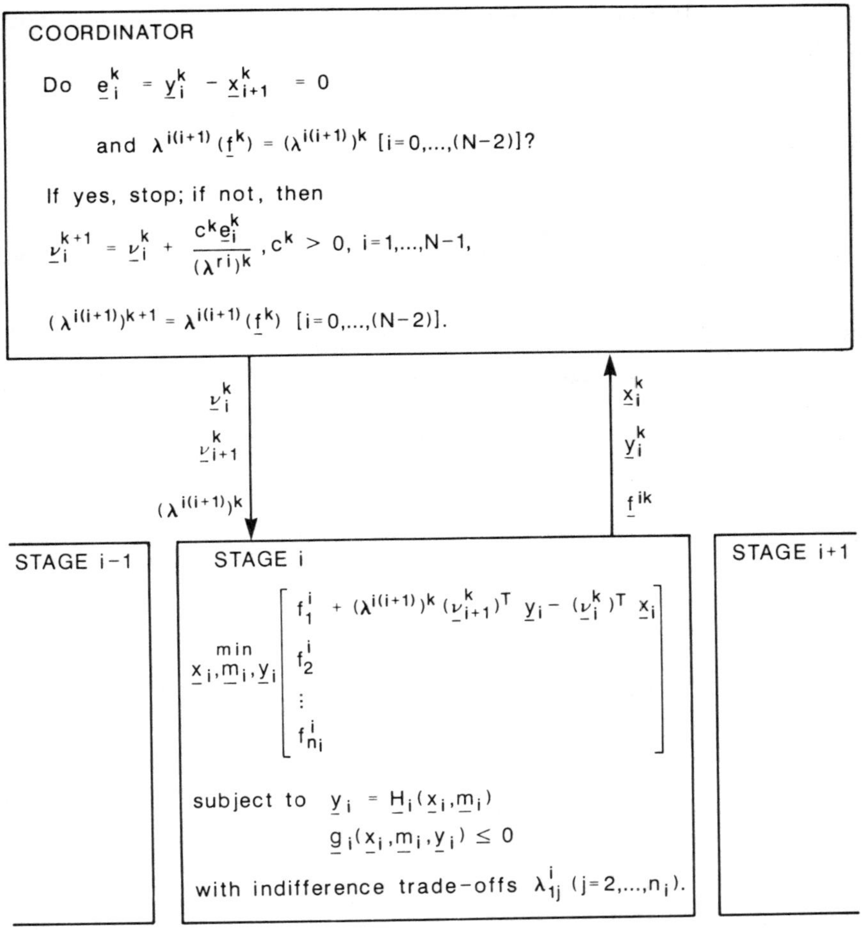

Figure 7.3 Nonfeasible scheme.

7.6 DISCUSSION OF THE METHODS

As in the static case, methods that use the straightforward approach probably demand less interaction with the decisionmaker than the other schemes which are suitable for developing multiple-decisionmaker schemes. Selection of an algorithm for a particular case depends on the properties of the problem: is it a discrete- or continuous-time problem; do the objectives change in time or are they constant; of what type are the constraints and functions? As in single-objective problems, there is no best method: every method discussed has its best application area.

In practice, there are several important problems where there is also a hierarchy with respect to the time spans for which the decisions are made. For these kinds of problems, it is more difficult to derive general mathematical results. For an example of this kind of problem, see Tarvainen et al. (1983).

REFERENCES

Bryson, A. E., and Ho, Y. C., *Applied Optimal Control*, Blaisdell, Mass., 1969.

Chankong, V., and Haimes, Y. Y., The Interactive Surrogate Worth Trade-off (ISWT) Method for Multiobjective Decision-making, in *Multicriteria Problem Solving*, ed. S. Zoints, Springer, New York, 1978.

Geoffrion, A. M., Dyer, J., and Feinberg, A., An Interactive Approach for Multicriterion Optimization with an Application to the Operation of an Academic Department, *Management Sci.*, vol. 19, pp. 357–368, 1972.

Haimes, Y. Y., and Hall, W. A., Multiobjectives in Water Resources System Analysis: The Surrogate Worth Trade-off Method, *Water Resources Res.*, vol. 10, p. 614, 1974.

Haimes, Y. Y., Hall, W. A., and Freedman, H. T., *Multiobjective Optimization in Water Resources Systems: The Surrogate Worth Trade-off Method*, Elsevier, Amsterdam, 1975.

Haimes, Y. Y., *Hierarchical Analyses of Water Resources Systems: Modeling and Optimization of Large-scale systems*, McGraw-Hill, New York, 1977.

Tarvainen, K., Haimes, Y. Y., and Lefkowitz, I., Decomposition Methods in Multiobjective Discrete-time Dynamic Problems, *Automatica*, vol. 19, pp. 15–28, 1983.

CHAPTER

EIGHT

MULTIOBJECTIVE DYNAMIC PROGRAMMING

8.1 INTRODUCTION

This chapter considers multicriteria decisionmaking methods based on dynamic programming. Section 8.2 presents the mathematical problem formulation. In Sect. 8.3, algorithms for finding preferred solutions are derived, and in Sect. 8.4, algorithms for determining Pareto optimal alternatives are considered.

8.2 PROBLEM FORMULATION

Consider the problem in Eq. (7.4) with a separable objective structure; that is,

$$\min_{\mathbf{m}_0,\ldots,\mathbf{m}_{N-1}} \begin{bmatrix} \sum_{i=0}^{N-1} f_1^i(\mathbf{x}_i, \mathbf{m}_i) \\ \vdots \\ \sum_{i=0}^{N-1} f_n^i(\mathbf{x}_i, \mathbf{m}_i) \end{bmatrix} \triangleq (*) \tag{8.1a}$$

subject to

$$\mathbf{x}_{i+1} = \mathbf{H}_i(\mathbf{x}_i, \mathbf{m}_i) \qquad i = 0, \ldots, (N-1) \qquad \mathbf{x}_0 \text{ given} \tag{8.1b}$$

$$\mathbf{g}_i(\mathbf{x}_i, \mathbf{m}_i) \leqslant \mathbf{0} \qquad i = 0, \ldots, (N-1) \tag{8.1c}$$

In Sect. 8.3, the solution of the above problem is a preferred solution determined by a decisionmaker who is interested in the problem. An algorithm based on Tarvainen (1981) will be derived where the problem is decomposed into N different stages.

In Sect. 8.4, the solution concept is that of Pareto optimality. The basis of the derivation of schemes is found in Sect. 8.3. (The dynamic programming algorithm for Pareto optimal alternatives was first derived by Brown and Strauch, 1965.)

8.3 ALGORITHMS FOR DETERMINING PREFERRED SOLUTIONS

8.3.1 Straightforward Approach

In a straightforward approach (see Sects. 4.9 and 5.1), we have the following subproblems to be solved for different values of the ϵ_j's ($j = 2, \ldots, n$) in the interactive SWT or SWT method:

$$\min \sum_{i=0}^{N-1} f_1^i (\mathbf{x}_i, \mathbf{m}_i)$$

subject to

$$\sum_{i=0}^{N-1} f_j^i(\mathbf{x}_i, \mathbf{m}_i) \leqslant \epsilon_j \qquad j = 2, \ldots, n$$

$$\mathbf{x}_{i+1} = \mathbf{H}_i(\mathbf{x}_i, \mathbf{m}_i) \qquad i = 0, \ldots, (N-1) \qquad \mathbf{x}_0 \text{ given}$$

$$\mathbf{g}_i(\mathbf{x}_i, \mathbf{m}_i) \leqslant \mathbf{0} \qquad i = 0, \ldots, (N-1)$$

These subproblems are solved by dynamic programming (Bellman, 1957). Note that every constraint of the form $F_j \leqslant \epsilon_j$ ($j = 2, \ldots, n$) increases the dimension of the state space by one. For further specifications, see Tauxe et al. (1979a, b; 1980). However, under certain conditions in problem formulation, the curse of dimensionality can be overcome (see Chankong et al., 1981).

8.3.2 Consistency Condition

In the single-objective case ($n = 1$), dynamic programming is based on equalities of the following form:

$$\min_{\mathbf{x},\mathbf{y}} f(\mathbf{x}, \mathbf{y}) = \min_{\mathbf{x}} \{\min_{\mathbf{y}} f(\mathbf{x}, \mathbf{y})\} \tag{8.2}$$

This equation is readily proven mathematically. In the multiobjective case, a corresponding decomposition condition is needed and must be postulated as a consistency requirement for the decisionmaker.

Consider a general multicriteria problem:

$$\min_{\mathbf{f} \in G} \begin{bmatrix} f_1 \\ \vdots \\ f_n \end{bmatrix} \tag{8.3}$$

where $\mathbf{f} = (f_1, \ldots, f_n)^T$, and G is a feasible set in the objective space. Let the feasible set G be covered by subsets $G_{\mathbf{p}}$ (parameterized by $\mathbf{p}$) such that

$$\cup_{\mathbf{p}} G_{\mathbf{p}} = G \tag{8.4}$$

The following consistency condition is postulated:

$$\min_{\mathbf{f} \in G} \begin{bmatrix} f_1 \\ \vdots \\ f_n \end{bmatrix} = \min_{\mathbf{p}} \left\{ \min_{\mathbf{f} \in G\mathbf{p}} \begin{bmatrix} f_1 \\ \vdots \\ f_n \end{bmatrix} \right\} \tag{8.5}$$

That is, it is required that, if the decisionmaker first selects a preferred solution from each subset $G_{\mathbf{p}}$ of the feasible set G and then selects a preferred one among these preferred solutions of the subsets, then the decisionmaker obtains the same result as when he or she chooses a preferred solution directly from G. This is a natural requirement, but in empirical studies, inconsistencies certainly occur when tests of consistency conditions are made. In a special case where the parameter $\mathbf{p}$ appears explicitly as a decision variable, the consistency condition can be presented in, for example, the following form:

$$\begin{matrix} \min\limits_{\mathbf{p}, \mathbf{y}} \\ \text{subject to} \\ (\mathbf{p}, \mathbf{t}) \in G_1 \\ \mathbf{p} \in G_2 \end{matrix} \begin{bmatrix} f_1(\mathbf{p}, \mathbf{t}) \\ \vdots \\ \\ f_n(\mathbf{p}, \mathbf{t}) \end{bmatrix} = \begin{matrix} \min\limits_{\mathbf{p}} \\ \text{subject to} \\ \mathbf{p} \in G_2 \\ \\ \end{matrix} \left\{ \begin{matrix} \min\limits_{\mathbf{t}} \\ \text{subject to} \\ (\mathbf{p}, \mathbf{t}) \in G_1 \\ \\ \end{matrix} \begin{bmatrix} f_1(\mathbf{p}, \mathbf{t}) \\ \vdots \\ \\ f_n(\mathbf{p}, \mathbf{t}) \end{bmatrix} \right\}$$

8.3.3 Recursion Formulas with Multiobjective Subproblems

To derive the dynamic programming procedure given below, it has been assumed that the condition given in Eq. (8.5) holds for the whole problem [Eqs. (8.1a)–(8.1c)] and that it holds also for every subproblem obtained in using the consistency condition and for the arbitrary subset cover $\{G_p\}$ that satisfies Eq. (8.4).

Consider, first, a parameterization with a parameter $\mathbf{s}_{N-1}$:

$$\mathbf{s}_{N-1} = \begin{bmatrix} \mathbf{x}_{N-1} \\ F_1^{N-1} \\ \vdots \\ F_n^{N-1} \end{bmatrix} \tag{8.6}$$

where $F_j^{N-1} = \Sigma_{t=0}^{N-2} f_j^t$ (the cumulative value of the jth objective up to state $N - 1$), $j = 1, \ldots, n$.

Every $G_{\mathbf{s}_{N-1}}$ consists of those objective vectors that can be generated so that the related $\mathbf{x}_{N-1}$, F_1^{N-1}, ..., and F_n^{N-1} are the specified ones. Mathematically, for an $\mathbf{s}_{N-1}$ given by Eq. (8.6),

$$\begin{aligned} G_{\mathbf{s}_{N-1}} = \Bigg\{ \left(\sum_{i=0}^{N-1} f_1^i, \ldots, \sum_{i=0}^{N-1} f_n^i \right)^T \Bigg| & (\exists \mathbf{m}, \mathbf{x})[\mathbf{x}_{N-1}(\mathbf{m}) = \mathbf{x}_{N-1} \quad \text{and} \\ & (\forall j \in \{1, \ldots, n\}) \left(\sum_{t=0}^{N-2} f_j^t = F_j^{N-1} \right) \\ & \text{and the systems equations (8.1b) and (8.1c) hold]} \Bigg\} \\ = \{(F_1^{N-1} + f_1^{N-1}(\mathbf{x}_{N-1}, \mathbf{m}_{N-1}), \ldots, & F_n^{N-1} \\ + f_n^{N-1}(\mathbf{x}_{N-1}, \mathbf{m}_{N-1}))^T \mid & \text{previous statement}\} \end{aligned} \tag{8.7}$$

where the last equality is a simple identity.

The range of $\mathbf{s}_{N-1}$ is all values for which $G_{\mathbf{s}_{N-1}}$ is nonempty. An explicit definition of the range is not needed, as is seen below. Clearly, the relation $\cup_{\mathbf{p}} G_{\mathbf{p}} = G$ [Eq. (8.4)] holds in this case. Applying the consistency condition yields [note from Eq. (8.7) that a given $G_{\mathbf{s}_{N-1}}$ is generated by the $\mathbf{m}_{N-1}$'s that satisfy Eq. (8.1c)]

$$(*) = \min_{\mathbf{s}_{N-1}} \left[\begin{array}{c} \min\limits_{\mathbf{m}_{N-1}} \\ \\ \text{subject to} \\ \\ \mathbf{g}_{N-1}(\mathbf{x}_{N-1}, \mathbf{m}_{N-1}) \le 0 \end{array} \begin{bmatrix} F_1^{N-1} + f_1^{N-1}(\mathbf{x}_{N-1}, \mathbf{m}_{N-1}) \\ \vdots \\ F_n^{N-1} + f_n^{N-1}(\mathbf{x}_{N-1}, \mathbf{m}_{N-1}) \end{bmatrix} \right] \tag{8.8}$$

Let the preferred solution for $\mathbf{m}_{N-1}$ in the inner minimization of Eq. (8.8), for a given $\mathbf{s}_{N-1}$, be denoted by $\mathbf{m}_{N-1}^0(\mathbf{s}_{N-1})$. For $j = 1, \ldots, n$, let $P_j^{N-1}(\mathbf{s}_{N-1}) = f_j^{N-1}[\mathbf{x}_{N-1}, \mathbf{m}_{N-1}^0(\mathbf{s}_{N-1})]$ (the preferred value of f_j^{N-1} for a given $\mathbf{s}_{N-1}$).

Next, consider a parameterization of the original problem with a parameter

$$\mathbf{s}_{N-2} = \begin{bmatrix} \mathbf{x}_{N-2} \\ F_1^{N-2} \\ \vdots \\ F_n^{N-2} \end{bmatrix} \tag{8.9}$$

where $F_j^{N-2} = \sum_{t=0}^{N-3} f_j^t$ (the cumulative value of the jth objective up to the stage $N-2$), $j = 1, \ldots, n$. In the same way as Eq. (8.8) was derived, we have

$$(*) = \min_{\mathbf{s}_{N-2}} \left[\begin{array}{l} \min\limits_{\mathbf{m}_{N-2}, \mathbf{m}_{N-1}} \\ \text{subject to} \\ \mathbf{g}_{N-2}(\mathbf{x}_{N-2}, \mathbf{m}_{N-2}) \le \mathbf{0} \\ \\ \mathbf{g}_{N-1}(\mathbf{x}_{N-1}, \mathbf{m}_{N-1}) \le \mathbf{0} \\ \mathbf{x}_{N-1} = \mathbf{H}_{N-2}(\mathbf{x}_{N-2}, \mathbf{m}_{N-2}) \end{array} \begin{bmatrix} F_1^{N-2} + f_1^{N-2}(\mathbf{x}_{N-2}, \mathbf{m}_{N-2}) \\ \quad + f_1^{N-1}(\mathbf{x}_{N-1}, \mathbf{m}_{N-1}) \\ \vdots \\ F_n^{N-2} + f_n^{N-2}(\mathbf{x}_{N-2}, \mathbf{m}_{N-2}) \\ \quad + f_n^{N-1}(\mathbf{x}_{N-1}, \mathbf{m}_{N-1}) \end{bmatrix} \right] \tag{8.10}$$

Consider the inner minimization in Eq. (8.10). Parameterize it with $\mathbf{m}_{N-2}$, which readily yields [see Eq. (8.5)]

$$(*) = \min_{\mathbf{s}_{N-2}} \left[\begin{array}{l} \min\limits_{\mathbf{m}_{N-2}} \\ \text{subject to} \\ \mathbf{g}_{N-2}(\mathbf{x}_{N-2}, \mathbf{m}_{N-2}) \leq \mathbf{0} \end{array} \left\{ \begin{array}{l} \min\limits_{\mathbf{m}_{N-1}} \\ \text{subject to} \\ \mathbf{g}_{N-1}(\mathbf{x}_{N-1}, \mathbf{m}_{N-1}) \leq \mathbf{0} \\ \mathbf{x}_{N-1} = \mathbf{H}_{N-2}(\mathbf{x}_{N-2}, \mathbf{m}_{N-2}) \end{array} \begin{bmatrix} F_1^{N-2} + f_1^{N-2}(\mathbf{x}_{N-2}, \mathbf{m}_{N-2}) \\ + f_1^{N-1}(\mathbf{x}_{N-1}, \mathbf{m}_{N-1}) \\ F_n^{N-2} + f_n^{N-2}(\mathbf{x}_{N-2}, \mathbf{m}_{N-2}) \\ + f_n^{N-1}(\mathbf{x}_{N-1}, \mathbf{m}_{N-1}) \end{bmatrix} \right\} \right] \tag{8.11a}$$

Consider the innermost minimization in Eq. (8.11a). Note that it is an inner minimization in Eq. (8.8), with

$$\mathbf{x}_{N-1} = \mathbf{H}_{N-2}(\mathbf{x}_{N-2}, \mathbf{m}_{N-2})$$

$$F_j^{N-1} = F_j^{N-2} + F_j^{N-2}(\mathbf{x}_{N-2}, \mathbf{m}_{N-2}) \qquad j = 1, \ldots, n$$

or, for brevity, with

$$\mathbf{s}_{N-1}(\mathbf{s}_{N-2}, \mathbf{m}_{N-2}) = [\mathbf{H}_{N-2}(\mathbf{x}_{N-2}, \mathbf{m}_{N-2}), F_1^{N-2} + f_1^{N-2}(\mathbf{x}_{N-2}, \mathbf{m}_{N-2}), \ldots, F_n^{N-2} + f_n^{N-2}(\mathbf{x}_{N-2}, \mathbf{m}_{N-2})]^T$$

If the results of the inner minimization in Eq. (8.8) are used, Eq. (8.11a) will take the following form:

$$\begin{array}{l} \min\limits_{\mathbf{x}_{N-2}} \\ (*) = F_1^{N-2} \\ \vdots \\ F_n^{N-2} \end{array} \left[\begin{array}{l} \min\limits_{\mathbf{m}_{N-2}} \\ \text{subject to} \\ \mathbf{g}_{N-2}(\mathbf{x}_{N-2}, \mathbf{m}_{N-2}) \leq \mathbf{0} \end{array} \begin{bmatrix} F_1^{N-2} + f_1^{N-2}(\mathbf{x}_{N-2}, \mathbf{m}_{N-2}) \\ + P_1^{N-1}[\mathbf{s}_{N-1}(\mathbf{s}_{N-2}, \mathbf{m}_{N-2})] \\ \vdots \\ F_n^{N-2} + f_n^{N-2}(\mathbf{x}_{N-2}, \mathbf{m}_{N-2}) \\ + p_n^{N-1}[\mathbf{s}_{N-1}(\mathbf{s}_{N-2}, \mathbf{m}_{N-2})] \end{bmatrix} \right] \tag{8.11b}$$

Let a solution to the inner optimization be denoted by $\mathbf{m}_{N-2}^0(\mathbf{s}_{N-2})$. For $j = 1, \ldots, n$, let

$$P_j^{N-2}(\mathbf{s}_{N-2}) = f_j^{N-2}[\mathbf{x}_{N-2}, \mathbf{m}_{N-2}^0(\mathbf{s}_{N-2})] + P_j^{N-1}\{\mathbf{s}_{N-1}[\mathbf{s}_{N-2}, \mathbf{m}_{N-2}^0(\mathbf{s}_{N-2})]\}$$

(the optimal cumulative value of the jth objective from the stage $N - 2$ on, for a given $\mathbf{s}_{N-2}$).

The optimization with respect to $\mathbf{m}_{N-2}$ and $\mathbf{m}_{N-1}$ in Eq. (8.10) has been reduced to two consecutive optimizations in Eq. (8.11b). Next, we consider a parameterization with respect to $(\mathbf{x}_{N-3}, F_1^{N-3}, \ldots, F_n^{N-3})$ and do a subparameterization with respect to $\mathbf{m}_{N-3}$. In the innermost minimization with respect to $\mathbf{m}_{N-2}$ and

$\mathbf{m}_{N-1}$, we can use the results from the inner minimization in Eq. (8.11b), and so on. A general formulation for the $N - 1$ problems to be solved can be presented as follows $[i = 0, \ldots, (N - 1)]$; $\mathbf{s}_i = (\mathbf{x}, F_1^i, \ldots, F_n^i)$:

$$\begin{array}{ll} \min\limits_{\mathbf{m}_i} & \begin{bmatrix} F_1^i + f_1^i(\mathbf{x}_i, \mathbf{m}_i) + P_1^{i+1}(\mathbf{s}_{i+1}(\mathbf{s}_i, \mathbf{m}_i)) \\ \\ \vdots \\ \\ F_n^i + f_n^i(\mathbf{x}_i, \mathbf{m}_i) + P_n^{i+1}(\mathbf{s}_{i+1}(\mathbf{s}_i, \mathbf{m}_i)) \end{bmatrix} \\ \text{subject to} & \\ \mathbf{g}_i(\mathbf{x}_i, \mathbf{m}_i) \leq \mathbf{0} & \end{array} \tag{8.12}$$

where

$$\mathbf{s}_{i+1}(\mathbf{s}_i, \mathbf{m}_i) = [\mathbf{H}_i(\mathbf{x}_i, \mathbf{m}_i), F_1^i + f_1^i(\mathbf{x}_i, \mathbf{m}_i), \ldots, F_n^i + f_n^i(\mathbf{x}_i, \mathbf{m}_i)]^T$$

$$i = 1, \ldots, (N - 1) \tag{8.13}$$

$$\mathbf{s}_o = [\mathbf{x}_o, 0, \ldots, 0]^T \tag{8.14}$$

$$P_j^{i+1}(\mathbf{s}_{i+1}) = f_j^{i+1}[\mathbf{x}_{i+1}, \mathbf{m}_{i+1}^0(\mathbf{s}_{i+1})] + P_j^{i-2}\{\mathbf{s}_{i+2}[\mathbf{s}_{i+1}, \mathbf{m}_{i+1}^0(\mathbf{s}_{i+1})]\}$$

$$j = 1, \ldots, n, \qquad i = 0, \ldots, (N - 2) \tag{8.15}$$

$$P_j^N(\mathbf{s}_N) = 0 \qquad j = 1, \ldots, n \tag{8.16}$$

The problems in Eq. (8.12) are solved backward, starting with $i = N - 1$. As in ordinary dynamic programming, each problem in Eq. (8.12) has to be solved for several values of $\mathbf{s}_i$. When solving the ith problem ($i \geq 1$), the feasible set of $\mathbf{s}_i$ is not known. Hence, the ith problem is possibly solved also for some infeasible values of $\mathbf{s}_i$. This does not matter, because when we solve the last problem ($i = 0$), there is a known initial value for $\mathbf{s}_0$ [Eq. (8.14)]. The feasible initial condition and the system's equations that hold between the $\mathbf{s}_i$'s cause the infeasible $\mathbf{s}_i$'s ($i \geq 1$) to drop out in the final optimization phase.

Here, the only essential difference from the single-objective case is the inclusion of the past performance via the F_j^i's. This difference occurs because, in a general case, the preferences for future performance also depend on the past levels of objectives. Hence, the dimension of the state space increases by n (or, in the method of Sect. 8.3.1, by $n - 1$; on the other hand, every dynamic programming problem needs to be solved several times for different values of ϵ's).

Note that, if we set $n = 1$ (i.e., we have a single-objective problem), the F_1^i's in the corresponding problem in Eq. (8.12) can be omitted since they are constants, and we have a usual dynamic programming problem.

Compared to the method in Sect. 8.3.1, the above method assumes decisionmaking with respect to each stage, which sometimes may be a drawback. On the other hand, the above method produces a feedback law.

8.4 ALGORITHM FOR DETERMINING PARETO OPTIMAL ALTERNATIVES

The algorithm for preferred solutions derived in the previous section can be used as a starting point to develop a dynamic programming procedure for determining

Pareto optimal alternatives. The mathematical problem formulation is the same as in the previous consideration (Sect. 8.2). However, now the solution of the minimization given in Eq. (8.1a) and of all other minimization problems yields the set of Pareto optimal alternatives.

The consistency condition given in Eq. (8.5) can now be derived mathematically, provided that the problem [Eq. (8.3)] is *dominance bounded*. The concept of dominance boundedness was introduced by Yu and Seiford (1981), where they derive a dynamic programming algorithm for a more general case than here.

A multicriteria decision problem is said to be dominance bounded if, for each feasible alternative **f**, either **f** is nondominated (Pareto optimal) or it is dominated by a nondominated feasible alternative.

It can be shown that a sufficient condition for a problem to be nondominance bounded is the compactness of the set of feasible objective vectors (see Brown and Strauch, 1965). In most practical applications this set is compact (bounded and closed).

Lemma 8.1 Consider the problem in Eq. (8.3), where the solution of the minimization denotes the set of Pareto optimal alternatives, and consider the parameterization given in Eq. (8.4). If the problem is dominance bounded, then the consistency condition in Eq. (8.5) holds.

To prove the lemma, let

$$S = \{\mathbf{f}|\mathbf{f} \text{ is Pareto optimal in } G\} \tag{8.17}$$

$$S_{\mathbf{p}} = \{\mathbf{f}|\mathbf{f} \text{ is Pareto optimal in } G_{\mathbf{p}}\} \tag{8.18}$$

$$S_{\mathbf{pp}} = \{\mathbf{f}|\mathbf{f} \text{ is Pareto optimal in } \cup_{\mathbf{p}} S_{\mathbf{p}}\} \tag{8.19}$$

With this notation, the consistency condition given in Eq. (8.5) assumes the following form:

$$S = S_{\mathbf{pp}} \tag{8.20}$$

First, take $\mathbf{f} \in S$. Due to the parameterization equation (8.4), there is a $\bar{\mathbf{p}}$ such that $\mathbf{f} \in G_{\bar{\mathbf{p}}}$. Since **f** is Pareto optimal in G, it is Pareto optimal in $\mathbf{G}_{\bar{\mathbf{p}}}$ (which is a subset of G). That is, $\mathbf{f} \in S_{\bar{\mathbf{p}}}$. Therefore **f** is taken into account when $\cup_{\mathbf{p}} S_{\mathbf{p}}$ is considered for determining $S_{\mathbf{pp}}$. Since f is Pareto optimal in G, it is Pareto optimal in $\cup_{\mathbf{p}} S_{\mathbf{p}}$ (which is a subset of G). That is, $\bar{f} \in S_{\mathbf{pp}}$. Thus, $S \subset S_{\mathbf{pp}}$. (This relation holds without dominance boundedness.)

Second, take $\mathbf{f} \in S_{\mathbf{pp}}$. To prove that $\mathbf{f} \in S$, assume the contrary: $\mathbf{f} \notin S$. By nondominance boundedness, there exists $\mathbf{f}' \in S$ such that $\mathbf{f}'$ dominates **f**. By the reasoning of the previous paragraph, $\mathbf{f}' \in S_{\mathbf{pp}}$. That is, $\mathbf{f}, \mathbf{f}' \in S_{\mathbf{pp}}$, where $\mathbf{f}'$ dominates **f**. This is a contradiction: $S_{\mathbf{pp}}$ is a set of Pareto optimal alternatives. Hence, $\mathbf{f} \in S$ and $S_{\mathbf{pp}} \subset S$. In all, $S = S_{\mathbf{pp}}$. This completes the proof of Lemma 8.1.

Assuming that the consistency condition given in Eq. (8.5) holds for the problem of determining Pareto optimal alternatives, one can proceed exactly in the same manner as in the previous section concerning preferred solutions. The only technical difference is that now the solutions denoted by $\mathbf{m}_i^0(\mathbf{s}_i)$ and the corresponding P_j^i's are sets. This brings no principal difficulty in the minimizations in different stages. Hence, we arrive at the same equations as before [Eqs. (8.12)–(8.16)], where the sets are handled in a natural way (this will be formally explained later).

A major difference arises because in the minimizations given in Eq. (8.12), the F_j^i's are constants. Using the definition of Pareto optimality, one immediately sees that the Pareto optimal solution sets of $\mathbf{m}_i$ are not affected by constants. Therefore, the F_j^i's can be left out of Eq. (8.12) and in the definition of the state s_i. Hence, Eqs. (8.12)–(8.16) can now be presented in the following form:

$$\min_{\mathbf{m}_i} \quad \text{subject to} \quad \mathbf{g}_i(\mathbf{x}_i, \mathbf{m}_i) \leq 0 \quad \begin{bmatrix} f_1^i(\mathbf{x}_i, \mathbf{m}_i) + P_1^{i+1}[\mathbf{H}_i(\mathbf{x}_i, \mathbf{m}_i)] \\ \vdots \\ f_n^i(\mathbf{x}_i, \mathbf{m}_i) + P_n^{i+1}[\mathbf{H}_i(\mathbf{x}_i, \mathbf{m}_i)] \end{bmatrix} \qquad i = 0, \ldots, (N-1) \tag{8.21}$$

where

$$P_j^{i+1}(\mathbf{x}_{i+1}) = f_j^{i+1}[\mathbf{x}_{i+1}, \mathbf{m}_{i+1}^0(\mathbf{x}_{i+1})] + P_j^{i+2}\{\mathbf{x}_{i+2}[\mathbf{x}_{i+1}, \mathbf{m}_{i+1}^0(\mathbf{x}_{i+1})]\}$$
$$j = 1, \ldots, n, \qquad i = 0, \ldots, (N-2) \tag{8.22}$$

with

$$P_j^N(\mathbf{x}_N) = 0 \qquad j = 1, \ldots, n \tag{8.23}$$

$\mathbf{m}_i^0(\mathbf{x}_i)$ is the Pareto optimal solution set of the problem given in Eq. (8.21), $i = 0, \ldots, (N-1)$.

Here, a natural shortened notation concerning sets is used. That is, for example, when $m(x)$ is a set and f and p are functions, then $f(m(x)) + p(m(x))$ is the set $\{f(y) + p(y) | y \in m(x)\}$.

The dynamic programming algorithm for generating Pareto optimal alternatives may save a considerable amount of computation time, as in the single-criterion case.

REFERENCES

Bellman, R., *Dynamic Programming*, Princeton University Press, Princeton, N.J., 1957.

Brown, T. A., and Strauch, R. E., Dynamic Programming in Multiplicative Lattices, *J. Math. Anal. Appl.*, vol. 12, p. 364, 1965.

Chankong, V., Haimes, Y. Y., and Gemperline, D., A Multiobjective Dynamic Programming Method for Capacity Expansion, *IEEE Trans. Automat. Contr.*, vol. AC-26, no. 5, pp. 1195–1207, 1981.

Tarvainen, K., *Hierarchical-Multiobjective Optimization*, Ph.D. dissertation, Department of Systems Engineering, Case Western Reserve University, Cleveland, Ohio, 1981.

Tauxe, G. W., Inman, R. R., and Mades, D. M., Multiobjective Dynamic Programming: A Classic Problem Redressed, *Water Resources Res.*, vol. 15, p. 1398, 1979a.

Tauxe, G. W., Inman, R. R., and Mades, D. M., Multiobjective Dynamic Programming with Application to a Reservoir, *Water Resources Res.*, vol. 15, p. 1403, 1979b.

Tauxe, G. W., Mades, D. M., and Inman, R. R., Multiple Objectives in Reservoir Operation. *J. Water Resources Planning Management Div., Proc. Am. Soc. Civ. Eng.*, vol. 106, p. 225, 1980.

Yu, P. L., and Seiford, L., Multistage Decision Problems with Multiple Criteria, in *Multiple Criteria Analysis*, ed. P. Nijkamp and J. Spronk, Gower, Hampshire, England, 1981.

CHAPTER

NINE

HIERARCHICAL OVERLAPPING COORDINATION: SINGLE-OBJECTIVE MODELS

9.1 INTRODUCTION

In the modeling process of large-scale and complex systems, more than one mathematical model is likely to emerge; each of these models may focus on a specific aspect of the system, yet all may be regarded as acceptable representations of the system. This phenomenon is particularly common in hierarchical multilevel modeling, where more than one decomposition approach may be both feasible and desirable (Macko and Haimes, 1978).

This chapter considers the decomposition and coordination problems of large-scale and complex systems that have more than one hierarchical overlapping structure, and a coordination that is both hierarchical and overlapping is developed.

The concept and importance of the hierarchical overlapping coordination (HOC), is considered in Sect. 9.2. A mathematical formulation of an HOC that has two decompositions is presented in Sect. 9.3. Section 9.4 deals with the convergence properties of the HOC procedure, and computational examples are given in Sect. 9.5 to demonstrate how the HOC procedure works.

9.2 CONCEPT OF HOC

HOC as a concept was first proposed in a paper by Haimes and Macko (1973) and was extended in Macko and Haimes (1978). In this section, we consider the concept of HOC and demonstrate its usefulness with some examples.

Let us consider first a very simple example in order to understand HOC as a concept. Figure 9.1 shows the so-called matrix organization structure of a production system of an industrial corporation (Vancil, 1973). For illustrative purposes, this system is assumed to have a natural decomposition into a marketing division and a manufacturing division. Two sectors, which are concerned with product A and product B, are assumed to exist in the marketing division. Likewise, three plants, which are located in different areas, are assumed to exist in the manufacturing division. Each of the two product sectors has a manager and each of the three plants also has a manager. Let us call the decomposition of this structure into a marketing division the product decomposition and the decomposition into a manufacturing division the plant decomposition. Clearly, the sectors in the product decomposition overlap the plants in the plant decomposition. The product managers' decisions also overlap the plant managers' decisions. For example, a decision by the manager of product sector A overlaps the decisions of the managers of the three plants. The hierarchical representation of this overlapping organization structure of the production system is depicted in the two ways shown in Figs. 9.2 and 9.3.

Product managers are concerned with the system's profits and plant managers are concerned with the system's costs and the sum of those yields the net profit of the system (Vancil, 1973). That is, these two different decompositions deal with the different aspects of the system. The data bases of these two decompositions differ from each other and receive different information from inside and outside of the system. It is valuable to consider these different types of decomposition simultaneously. By thinking of different hierarchical structures together we can hope to achieve the synergism for the synergistic analysis of a large-scale, complex system. However, a method must be found that can deal in an organized way with the many and varied decompositions of a complex system. HOC has been developed to find such a method.

Suppose that the overall objective of the matrix production system represented in Fig. 9.1 is to maximize a given measure of net profit, with each manager cooperating in order to achieve it. Then, a desirable decisionmaking structure would be one in which (1) each individual manager's decisions are feasible in the

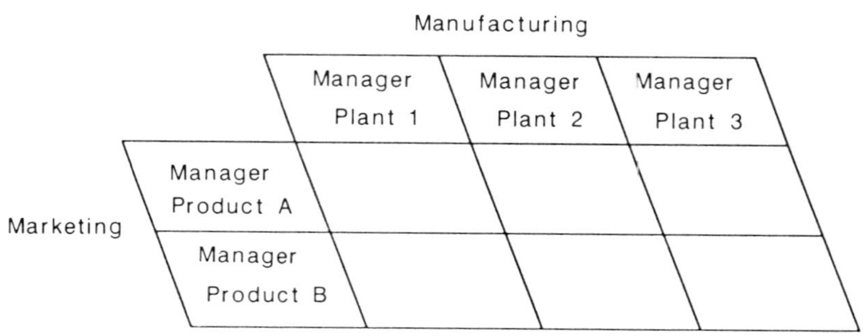

Figure 9.1 Matrix organization of a production system.

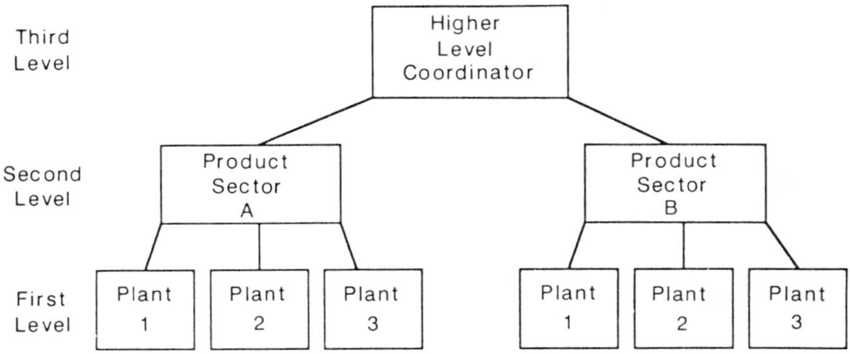

Figure 9.2 Product-plant decomposition.

overall system, and (2) the information exchange between the product managers and the plant managers leads to a sequence of decisions that produce an improving overall benefit that converges to the optimum (Macko and Haimes, 1978).

HOC has this decisionmaking structure. That is, the optimal solution is found by the coordination of different decompositions.

Let us now consider further the concept of HOC by looking at a more complicated example: the problem of planning and managing the Maumee River Basin water resources system (Haimes, 1977).

The Maumee River Basin spans parts of the states of Ohio, Indiana, and Michigan. It can be divided into five political-geographic planning subareas (PSAs) to make a political-geographic decomposition as shown in Fig. 9.4. It can also be divided into eight hydrological subareas, as shown in Fig. 9.5. Figures similar to the three-level hierarchical structures shown in Figs. 9.2 and 9.3 could be drawn up to depict the decomposition structure of the Maumee River Basin, with a higher-level coordinator at the third level, the political-geographic decomposition at the second level or at the lowest level, and the hydrological decomposition at the lowest level, or at the second level.

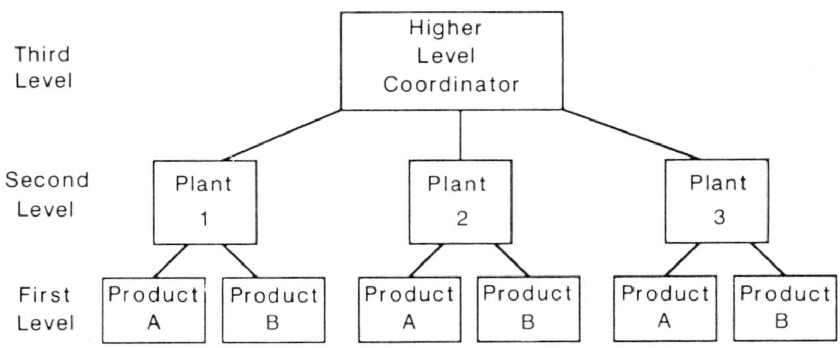

Figure 9.3 Plant-product decomposition.

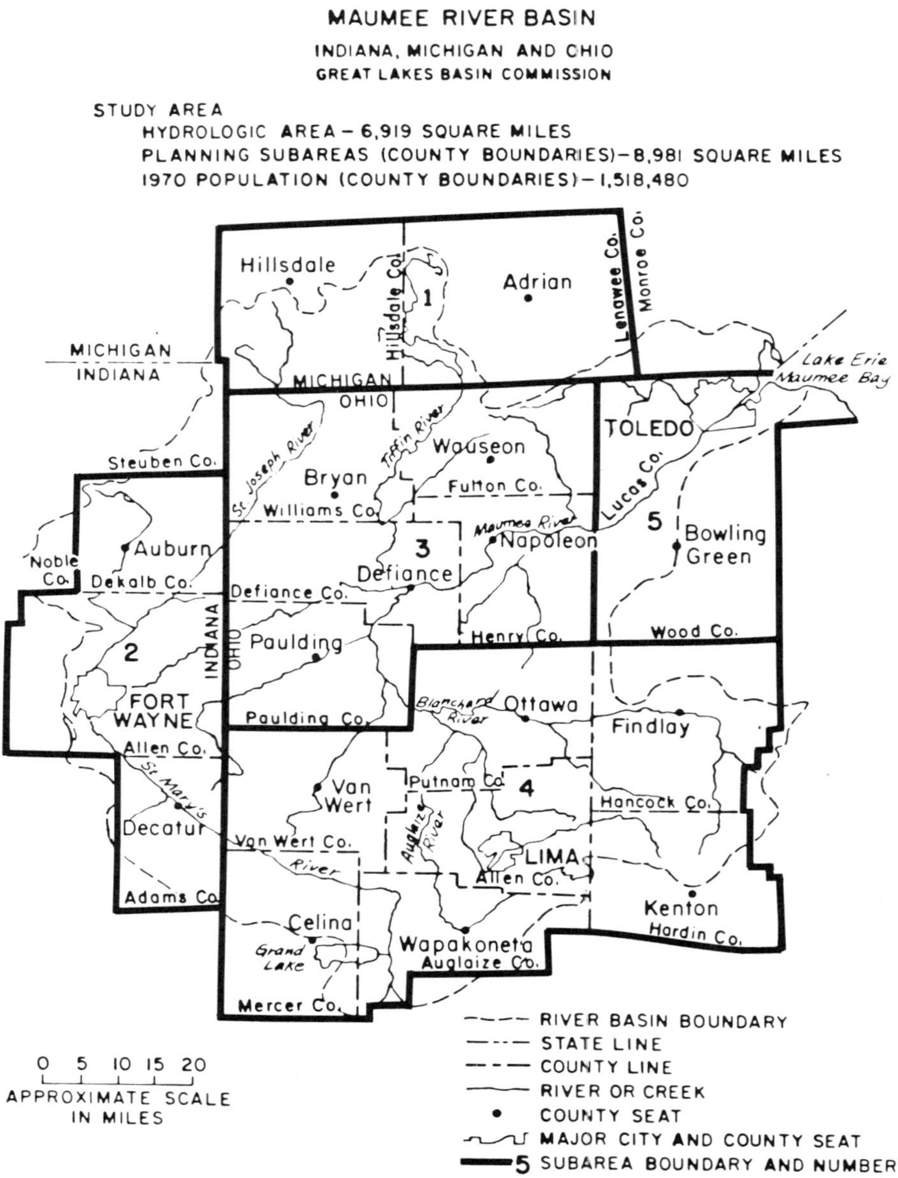

Figure 9.4 Political-geographic decomposition.

In the absence of some coordinating concept, such as HOC, modelers would be implicitly forced to choose between a hydrologically or politically based model or decomposition. Each decomposition represents and uncovers important aspects not available through the other. The availability and credibility of the data bases are particularly critical. The manipulation of data bases to serve and suit artifi-

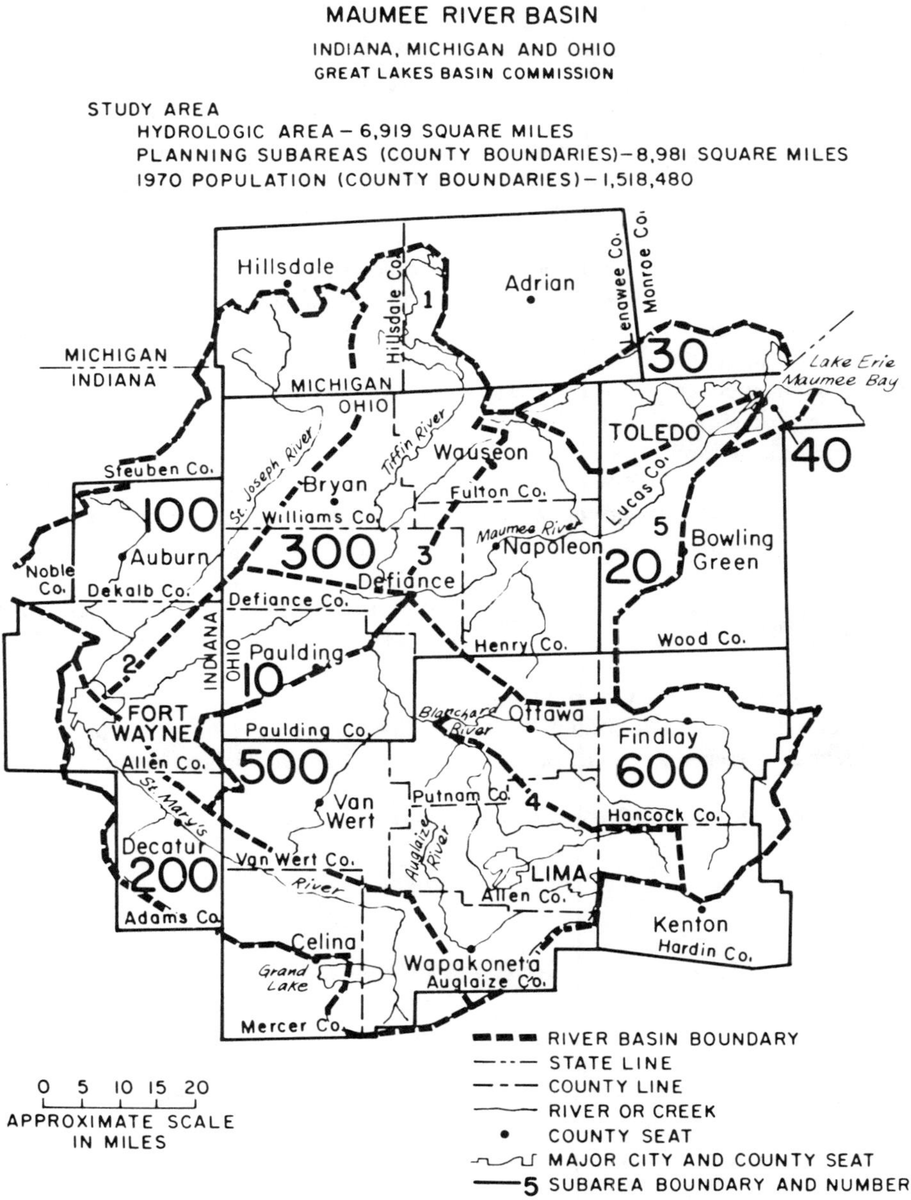

Figure 9.5 Hydrological decomposition.

cially imposed demands and constraints (artificial because they are imposed through the modeling process) necessitates compromise, and compromise can lead to an ultimate deterioration in model credibility. For example, data concerning streamflow, water quality, and floods are available on a hydrological basis and are collected by the U.S. Geological Survey, the U.S. Environmental Protection Agency, and the U.S. Army Corps of Engineers, respectively. On the other hand, data concerning population dynamics, employment, and other economic activities are available on political-geographic bases and are collected by agencies such as the U.S. Departments of Commerce, Labor, or Treasury. HOC enables the utmost utilization of these data bases with their minimum manipulation or misuse. This can be achieved by resorting to two simultaneous decompositions—hydrological and political-geographic—each of which might have a number of subsystems. In general, water resources systems (as well as many other large-scale systems) lend themselves to more than one decomposition or description. For instance, one could have a functional decomposition (water supply for and demand from the various sectors—agriculture, industry, municipality, aquatic life, etc.) and a temporal decomposition (long, intermediate and short term, on-line) as well as the hydrological and political-geographic decompositions already mentioned (Shima et al., 1981).

So far, for simplicity, we have been discussing systems with two different hierarchical structures (i.e., decompositions). However, large-scale systems sometimes have more than two hierarchical structures. The decisionmaking structure of HOC can be very effective in providing an optimal solution for such systems. Systems decomposed into N different decompositions were studied by Sung (1978) and Haftari (1981).

Finally, the advantages of HOC might be summarized as follows (Macko and Haimes, 1978):

1. It adds more flexibility in modeling.
2. It yields more realistic models by increasing the responsiveness of the model structures to organizational, political, technical, financial, and historical realities.
3. It increases the responsiveness of the model structures to the availability of data.
4. It provides a holographic view of the modeled system.

9.3 MATHEMATICAL FORMULATION OF HOC

In order to elucidate the basic concepts of HOC and to gain an understanding of its procedures, we will consider here a system that has linear equality equations (or linear equality constraints). HOC can, of course, be applicable to the more general problems in which a system has nonlinear equations (or constraints). More general cases can be found in Macko and Haimes (1978).

9.3.1 Overlapping Decompositions

Let the problem of the system be represented as follows:

Original Problem:

$$\max_{\mathbf{x}} f(\mathbf{x}) \tag{9.1}$$

subject to

$$A\mathbf{x} = \mathbf{c} \tag{9.2}$$

where

$f(\mathbf{x})$ = a real-valued concave function that is assumed to be differentiable and bounded
$\mathbf{x}$ = an $(n \times 1)$ decision vector
$\mathbf{c}$ = a $(p \times 1)$ fixed vector (given)
A = a $(p \times n)$ matrix.

$f(\mathbf{x})$ is an objective function of the system that assigns a measure of benefit to a decision vector $\mathbf{x}$.

For simplicity, let us suppose that the system has two potential decompositions. We will refer to these two decompositions as the α and the β decompositions. In terms of the examples in the previous section, the α decomposition may be a product or a political-geographic decomposition, and the β decomposition may be a plant or a hydrological decomposition.

Since each decomposition implies a different grouping of variables, the vectors $\mathbf{x}$ and $\mathbf{c}$ can be rearranged for each decomposition. Thus, the system represented by Eq. (9.2) may be equivalently represented by

$$A_\alpha \mathbf{x}_\alpha = \mathbf{c}_\alpha \tag{9.3}$$

$$A_\beta \mathbf{x}_\beta = \mathbf{c}_\beta \tag{9.4}$$

respectively, for each decomposition, where we define

$$\mathbf{x}_\alpha = P_\alpha \mathbf{x} \qquad \mathbf{c}_\alpha = Q_\alpha \mathbf{c} \qquad \mathbf{x}_\beta = P_\beta \mathbf{x} \qquad \mathbf{c}_\beta = Q_\beta \mathbf{c} \tag{9.5}$$

$$A_\alpha = Q_\alpha A P_\alpha^{-1} \qquad A_\beta = Q_\beta A P_\beta^{-1} \tag{9.6}$$

The nonsingular P and Q matrices mathematically represent the rearranging of the vectors $\mathbf{x}$ and $\mathbf{c}$. It is important to note that the decision vectors $\mathbf{x}_\alpha$ and $\mathbf{x}_\beta$ are related by the nonsingular transformation:

$$\mathbf{x}_\alpha = P_{\alpha\beta} \mathbf{x}_\beta \qquad \mathbf{x}_\beta = P_{\beta\alpha} \mathbf{x}_\alpha \tag{9.7}$$

where

$$P_{\alpha\beta} = P_\alpha P_\beta^{-1} \qquad P_{\beta\alpha} = P_\beta P_\alpha^{-1} \tag{9.8}$$

The original problem represented by Eqs. (9.1) and (9.2), when viewed for the α and β decomposition, becomes

α Problem:

$$\max_{x_\alpha} f(P_\alpha^{-1} \mathbf{x}_\alpha) \tag{9.9}$$

subject to

$$A_\alpha \mathbf{x}_\alpha = \mathbf{c}_\alpha \tag{9.10}$$

β Problem:

$$\max_{x_\beta} f(P_\beta^{-1} \mathbf{x}_\beta) \tag{9.11}$$

subject to

$$A_\beta \mathbf{x}_\beta = \mathbf{c}_\beta \tag{9.12}$$

Note that the α and β problems are the same as the original problem. Only the variable names are changed. The relations are diagrammed in Fig. 9.6.

Suppose the α problem is decomposed into n_α subsystems and the β problem is decomposed into n_β subsystems. The vectors $\mathbf{x}_\alpha$ and $\mathbf{c}_\alpha$ are assumed to be partitioned into n_α subvectors $\mathbf{x}_{\alpha i}$ and $\mathbf{c}_{\alpha i}$ $(i = 1, \ldots, n_\alpha)$:

$$\mathbf{x}_\alpha = (\mathbf{x}_{\alpha 1}^T, \mathbf{x}_{\alpha 2}^T, \ldots, \mathbf{x}_{\alpha n_\alpha}^T)^T \tag{9.13}$$

$$\mathbf{c}_\alpha = (\mathbf{c}_{\alpha 1}^T, \mathbf{c}_{\alpha 2}^T, \ldots, \mathbf{c}_{\alpha n_\alpha}^T)^T \tag{9.14}$$

So also are $\mathbf{x}_\beta$ and $\mathbf{c}_\beta$:

$$\mathbf{x}_\beta = (\mathbf{x}_{\beta 1}^T, \mathbf{x}_{\beta 2}^T, \ldots, \mathbf{x}_{\beta n_\beta}^T)^T \tag{9.15}$$

$$\mathbf{c}_\beta = (\mathbf{c}_{\beta 1}^T, \mathbf{c}_{\beta 2}^T, \ldots, \mathbf{c}_{\beta n_\beta}^T)^T \tag{9.16}$$

We assume the matrices A_α and A_β are decomposable as follows:

$$A_\alpha = B_\alpha + G_\alpha H_\alpha \tag{9.17}$$

$$A_\beta = B_\beta + G_\beta H_\beta \tag{9.18}$$

where B and H are block diagonal matrices compatible with the partitioning of the vectors by subsystems, and G denotes the set of matrices of elements of A that are not on the block diagonal.

Thus, the α and β problems become

$$\max f_\alpha(\mathbf{x}_\alpha, \mathbf{y}_\alpha) \tag{9.19}$$

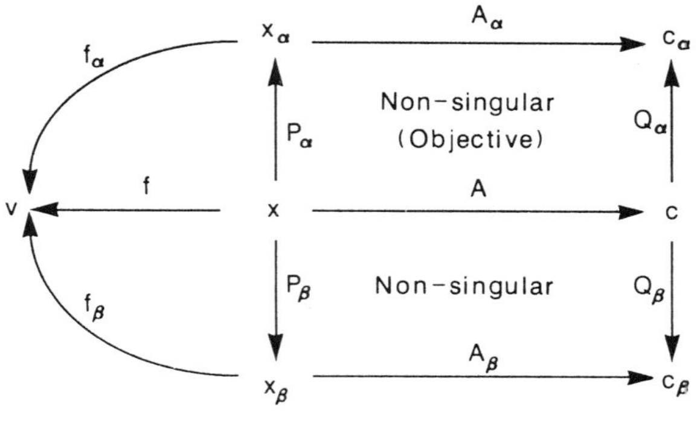

Figure 9.6 Relations of the variables [where $v \overset{\Delta}{=} f(x)$].

subject to

$$B_\alpha \mathbf{x}_\alpha + G_\alpha \mathbf{y}_\alpha = \mathbf{c}_\alpha \tag{9.20}$$

$$H_\alpha \mathbf{x}_\alpha = \mathbf{y}_\alpha \tag{9.21}$$

and

$$\max f_\beta(\mathbf{x}_\beta, \mathbf{y}_\beta) \tag{9.22}$$

subject to

$$B_\beta \mathbf{x}_\beta + G_\beta \mathbf{y}_\beta = \mathbf{c}_\beta \tag{9.23}$$

$$H_\beta \mathbf{x}_\beta = \mathbf{y}_\beta \tag{9.24}$$

where

$$f(P_\alpha^{-1}\mathbf{x}_\alpha) = f_\alpha(\mathbf{x}_\alpha, H_\alpha \mathbf{x}_\alpha) = f_\alpha(\mathbf{x}_\alpha, \mathbf{y}_\alpha) \tag{9.25}$$

$$f(P_\beta^{-1}\mathbf{x}_\beta) = f_\beta(\mathbf{x}_\beta, H_\beta \mathbf{x}_\beta) = f_\beta(\mathbf{x}_\beta, \mathbf{y}_\beta) \tag{9.26}$$

Also, $\mathbf{y}_\alpha$ is a vector of interaction or coupling variables of the subsystems in the α decomposition and $\mathbf{y}_\beta$ is a vector of the interaction or coupling variables of the subsystems in the β decomposition. $\mathbf{y}_\alpha$ and $\mathbf{y}_\beta$ are also assumed to be partitioned into n_α and n_β subvectors, respectively, as follows:

$$\mathbf{y}_\alpha = (\mathbf{y}_{\alpha 1}^T, \ldots, \mathbf{y}_{\alpha i}^T, \ldots, \mathbf{y}_{\alpha n_\alpha}^T)^T \tag{9.27}$$

$$\mathbf{y}_\beta = (\mathbf{y}_{\beta 1}^T, \ldots, \mathbf{y}_{\beta i}^T, \ldots, \mathbf{y}_{\beta n_\beta}^T)^T \tag{9.28}$$

From Eq. (9.7) and the definitions found in Eqs. (9.21) and (9.24), $\mathbf{y}_\alpha$ and $\mathbf{y}_\beta$ have the following relationships, respectively:

$$\mathbf{y}_\alpha = H_\alpha P_{\alpha\beta} \mathbf{x}_\beta \tag{9.29}$$

and

$$\mathbf{y}_\beta = H_\beta P_{\beta\alpha} \mathbf{x}_\alpha \tag{9.30}$$

Suppose the functions f_α and f_β are additively separable with respect to the subvectors $\mathbf{x}_{\alpha i}$ and $\mathbf{x}_{\beta i}$, respectively. Since the matrices B_α, H_α, B_β, and H_β are block diagonal and are compatible with the partitioning of the vectors, the α and β problems may be rewritten as follows, explicitly indicating the subsystems.

α Decomposition Problem:

$$\max f_\alpha(\mathbf{x}_\alpha, \mathbf{y}_\alpha) = \sum_{i=1}^{n_\alpha} f_{\alpha i}(\mathbf{x}_{\alpha i}, \mathbf{y}_{\alpha 1}, \ldots, \mathbf{y}_{\alpha n_\alpha}) \tag{9.31}$$

subject to

$$B_{\alpha i} \mathbf{x}_{\alpha i} + \sum_{j=1}^{n_\alpha} G_{\alpha ij} \mathbf{y}_{\alpha j} = \mathbf{c}_{\alpha i} \tag{9.32}$$

$$H_{\alpha i}\mathbf{x}_{\alpha i} = \mathbf{y}_{\alpha i} \qquad (i = 1, 2, \ldots, n_\alpha) \tag{9.33}$$

β Decomposition Problem:

$$\max f_\beta(\mathbf{x}_\beta, \mathbf{y}_\beta) = \sum_{i=1}^{n_\beta} f_{\beta i}(\mathbf{x}_{\beta i}, \mathbf{y}_{\beta 1}, \ldots, \mathbf{y}_{\beta n_\beta}) \tag{9.34}$$

subject to

$$B_{\beta i}\mathbf{x}_{\beta i} + \sum_{j=1}^{n_\beta} G_{\beta ij}\mathbf{y}_{\beta j} = \mathbf{c}_{\beta i} \tag{9.35}$$

$$H_{\beta i}\mathbf{x}_{\beta i} = \mathbf{y}_{\beta i} \qquad (i = 1, 2, \ldots, n_\beta) \tag{9.36}$$

where

n_α = the number of subsystems in the α decomposition
$\mathbf{x}_{\alpha i}$ = a decision vector of the ith subsystem in the α decomposition
$\mathbf{y}_{\alpha i}$ = an "output" of the ith subsystem affecting other subsystems in the α decomposition
$f_{\alpha i}$ = a benefit of the ith subsystem in the α decomposition

Similar definitions are used for n_β, $\mathbf{x}_{\beta i}$, $\mathbf{y}_{\beta i}$, and $f_{\beta i}$ in the β decomposition.

The overall problem can be reduced to independent uncoupled subproblems by fixing the interaction vectors in either decomposition. The ith subsystems in the α and β decompositions are represented as follows:

αi Subsystem, i = 1, …, n_α:

$$\max f_{\alpha i}(\mathbf{x}_{\alpha i}, \mathbf{y}_\alpha) \tag{9.37}$$

subject to

$$B_{\alpha i}\mathbf{x}_{\alpha i} + G_{\alpha i}\mathbf{y}_\alpha = \mathbf{c}_{\alpha i} \tag{9.38}$$

$$H_{\alpha i}\mathbf{x}_{\alpha i} = \mathbf{y}_{\alpha i} \tag{9.39}$$

βi Subsystem, i = 1, …, n_β:

$$\max f_{\beta i}(\mathbf{x}_{\beta i}, \mathbf{y}_\beta) \tag{9.40}$$

subject to

$$B_{\beta i}\mathbf{x}_{\beta i} + G_{\beta i}\mathbf{y}_\beta = \mathbf{c}_{\beta i} \tag{9.41}$$

$$H_{\beta i}\mathbf{x}_{\beta i} = \mathbf{y}_{\beta i} \tag{9.42}$$

where

$$G_{\alpha i}\mathbf{y}_\alpha = \sum_{j=1}^{n_\alpha} G_{\alpha ij}\mathbf{y}_{\alpha j}$$

$$G_{\beta i}\mathbf{y}_\beta = \sum_{j=1}^{n_\beta} G_{\beta ij}\mathbf{y}_{\beta j}$$

The original problem has now been decomposed into subproblems of the α and β decompositions. Before explaining the coordination procedure, we will make several summarizing remarks here.

1. The problem resulting from the α decomposition [Eqs. (9.31)–(9.33)] and the β decomposition [Eqs. (9.34)–(9.36)] are equivalent to each other and to the original problem. The decision variables and the fixed variables are rearranged in the decomposition problems.
2. Fixing the interaction vectors $\mathbf{y}_\alpha$ and $\mathbf{y}_\beta$ in either decomposition yields uncoupled subproblems.
3. These decompositions are feasible decompositions in the sense that any decision vector $\boldsymbol{\xi}_\alpha$ or $\boldsymbol{\xi}_\beta$ satisfies the constraints for fixed feasible values of $\mathbf{y}_\alpha$ and $\mathbf{y}_\beta$:

$$B_\alpha \boldsymbol{\xi}_\alpha + G_\alpha \mathbf{y}_\alpha = \mathbf{c}_\alpha \qquad H_\alpha \boldsymbol{\xi}_\alpha = \mathbf{y}_\alpha$$

   and

$$B_\beta \boldsymbol{\xi}_\beta + G_\beta \mathbf{y}_\beta = \mathbf{c}_\beta \qquad H_\beta \boldsymbol{\xi}_\beta = \mathbf{y}_\beta$$

4. The interaction vectors $\mathbf{y}_\alpha$ and $\mathbf{y}_\beta$ have the following relationships:

$$\mathbf{y}_\alpha = H_\alpha \mathbf{x}_\alpha = H_\alpha P_{\alpha\beta} \mathbf{x}_\beta$$

$$\mathbf{y}_\beta = H_\beta \mathbf{x}_\beta = H_\beta P_{\beta\alpha} \mathbf{x}_\alpha$$

9.3.2 Coordination Procedure of HOC

In the previous section the overall system was decomposed in two ways. Those two different decompositions can be coordinated as follows:

Algorithm 1

Step 1: We fix $\mathbf{y}_\alpha^{(1)}$ initially. Let $n = 1$.
Step 2: We solve the α decomposition problem in each subsystem [Eqs. (9.37)–(9.39)]. Let the solution be $\mathbf{x}_\alpha^{(n)}$.
Step 3: $\mathbf{y}_\beta^{(n)}$ is given by $\mathbf{y}_\beta^{(n)} = H_\beta P_{\beta\alpha} \mathbf{x}_\alpha^{(n)}$.
Step 4: We solve the β decomposition problem in each subsystem [Eqs. (9.40)–(9.42)]. Let the solution be $\mathbf{x}_\beta^{(n)}$.
Step 5: $\mathbf{y}_\alpha^{(n)}$ is fixed by $\mathbf{y}_\alpha^{(n)} = H_\alpha P_{\alpha\beta} \mathbf{x}_\beta^{(n)}$. Let $n = n + 1$. Go to step 2, where (n) implies the iteration number.

This process terminates when the optimum is achieved. That is,

$$f_\alpha(\mathbf{x}_\alpha^{(n)}, \mathbf{y}_\alpha^{(n)}) = f_\beta(\mathbf{x}_\beta^{(n)}, \mathbf{y}_\beta^{(n)}) = \max f(\mathbf{x})$$

subject to $A\mathbf{x} = \mathbf{c}$.

It might be practical to use the following two equations that establish a criterion that the improvement from one stage to the next be less than a prescribed small number ϵ:

$$|f_\alpha(\mathbf{x}_\alpha^{(n+1)}, \mathbf{y}_\alpha^{(n+1)}) - f_\alpha(\mathbf{x}_\alpha^{(n)}, \mathbf{y}_\alpha^{(n)})| < \epsilon \tag{9.43}$$

$$|f_\beta(\mathbf{x}_\beta^{(n+1)}, \mathbf{y}_\beta^{(n+1)}) - f_\beta(\mathbf{x}_\beta^{(n)}, \mathbf{y}_\beta^{(n)})| < \epsilon \tag{9.44}$$

However, these equations may not be good for a test for optimality for some problems. For example, we cannot distinguish the solution that is obtained due to the cycling from an optimal solution. We discuss this problem in the next section.

In Algorithm 1, the α problem was solved first. However, the β problem could just as well have been solved first. It makes no difference which one is chosen for starting. Figure 9.7 shows the structure of the coordination procedure of HOC.

If we take the example shown in Fig. 9.1, the decisionmaking structure of this coordination procedure takes the following form. Suppose that the α decom-

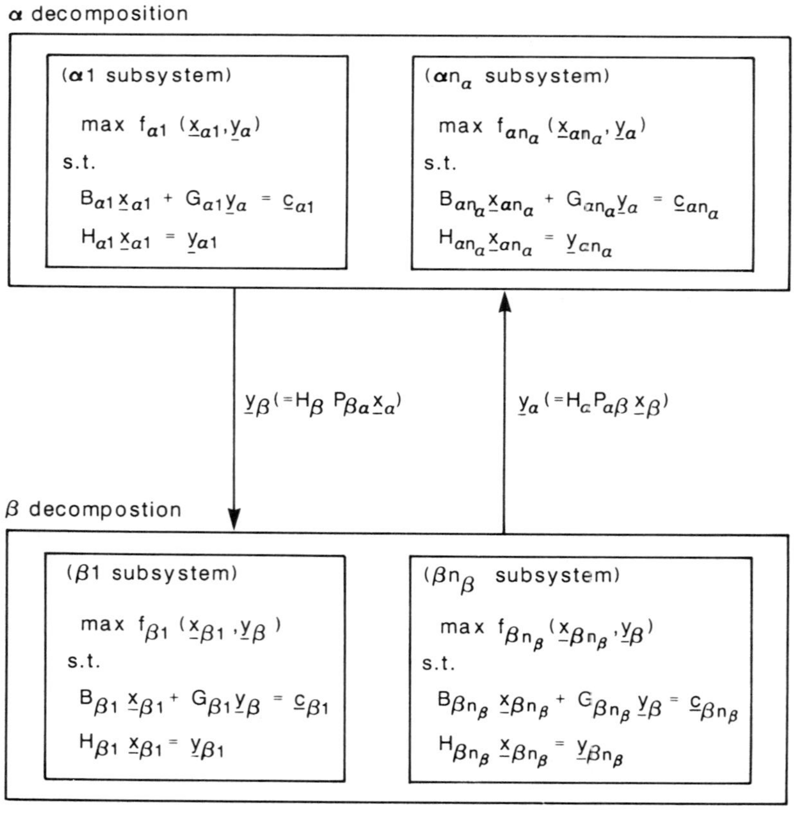

Figure 9.7 Structure of the overlapping coordination.

position is the product decomposition and the β decomposition is the plant decomposition. A manager in each product sector initially estimates the interaction vector $\mathbf{y}_\alpha$ of the product decomposition and calculates the optimal solution for the fixed $\mathbf{y}_\alpha$. Then the product managers propose to the plant managers an interaction vector $\mathbf{y}_\beta$ of the plant decomposition on the basis of their solution. The plant managers optimize their subsystems and find a solution based on the product manager's proposal. Then plant managers propose to the product managers a new interaction vector $\mathbf{y}_\alpha$ which is based on their optimal solution. The α problem is now solved again, and these procedures will be repeated until the product managers and the plant managers find an optimal solution.

In this section, the basic concept of the hierarchical overlapping coordination has been mathematically formulated by using a model that has linear equality constraints and a real-valued concave objective function. More general cases are formulated in Macko and Haimes (1978).

9.4 CONVERGENCE PROPERTIES OF HOC

There are some drawbacks to the HOC procedure; the principal one is the problem of convergence. Convergence to an optimal solution of the original problem is not guaranteed for some problems because cycling occurs. In this section, we will discuss the convergence properties of HOC and will present a test for optimality.

9.4.1 Convergence Properties of HOC for Problems with Linear Equality Constraints

We analyze the convergence properties of HOC for problems with linear constraints in finite-dimensional Euclidean space since this book concerns itself with finite-dimensional static problems. However, the reader should note that the original analysis was made in Hilbert space, and HOC is applicable not only to finite-dimensional static systems, but also to a much broader class of systems, namely, to finite- and infinite-dimensional static and dynamic systems (for details, see Macko and Haimes, 1978).

For the purpose of illustration, we will again present the original problem that was considered in the previous section:

$$\max f(\mathbf{x}) \tag{9.45}$$

subject to

$$A\mathbf{x} = \mathbf{c} \tag{9.46}$$

The convergence properties of HOC will be analyzed using the above model. Let us assume that (1) $f: R^n \rightarrow R^1$ (f is assumed to be concave, bounded, and differentiable); (2) A is a $(p \times n)$ matrix; and (3) H_α, H_β, P_α, and P_β are matrices of appropriate dimensions. Since the original problem is essentially equivalent to the α and β decompositions, a direct substitution of the relationships between variables shows that the problem

$$\max f(\mathbf{x})$$

subject to

$$A\mathbf{x} = \mathbf{c} \qquad H_\alpha P_\alpha \mathbf{x} = \mathbf{y}_\alpha^{(n)} \tag{9.47}$$

is equivalent under the transformation P_α to the problem in Eqs. (9.37)–(9.39) in the sense that

$$\mathbf{x} \text{ is } \begin{Bmatrix} \text{feasible} \\ \text{optimal} \end{Bmatrix}$$

in Eq. (9.47) if and only if

$$\mathbf{x}_\alpha = P_\alpha \mathbf{x} \text{ is } \begin{Bmatrix} \text{feasible} \\ \text{optimal} \end{Bmatrix}$$

in the problem in Eqs. (9.37)–(9.39).

Similarly, the problem

$$\max f(\mathbf{x})$$

subject to

$$A\mathbf{x} = \mathbf{c} \qquad H_\beta P_\beta \mathbf{x} = \mathbf{y}_\beta^{(n)} \tag{9.48}$$

is equivalent under the transformation P_β to the problem in Eqs. (9.40)–(9.42).

Suppose the sequences $\{\xi_{\alpha n}\}$ and $\{\xi_{\beta n}\}$, which correspond to the optimal $\mathbf{x}$ at each iteration in the α and β problems, are obtained by solving the problems in Eqs. (9.47) and (9.48) by using Algorithm 1. Then, for all $n > 0$,

$$\begin{aligned} &\text{i.} \quad \xi_{\alpha n} \text{ maximizes } f(\mathbf{x}) \text{ subject to } K_\alpha(\mathbf{x} - \xi_{\beta,n-1}) = 0 \\ &\text{ii.} \quad \xi_{\beta n} \text{ maximizes } f(\mathbf{x}) \text{ subject to } K_\beta(\mathbf{x} - \xi_{\alpha n}) = 0 \end{aligned} \tag{9.49}$$

where

$$K_\alpha \equiv \begin{pmatrix} A \\ H_\alpha P_\alpha \end{pmatrix} \qquad K_\beta \equiv \begin{pmatrix} A \\ H_\beta P_\beta \end{pmatrix}$$

and, moreover, Eq. (9.49) is valid for all $n > 0$ if and only if the sequences $\{P_\alpha \xi_{\alpha n}\}$ and $\{P_\beta \xi_{\beta n}\}$ are obtained by Algorithm 1. (Here it is implicitly assumed that the α and β problems are optimized with respect to every decision variable except the interaction variables. That is, only the decision variables that correspond to $\mathbf{y}_\alpha$ and $\mathbf{y}_\beta$ are fixed in the optimization problems.) The following properties are obtained.

Property 1 If Algorithm 1 is started with a feasible $\mathbf{x}$ (i.e., $A\mathbf{x} = \mathbf{c}$), or more precisely, a feasible $\mathbf{y}_\alpha$, then at each stage n of the process, the problems in Eqs. (9.47) and (9.48) will have nonempty feasible domains.

Property 2 Let $f^{\text{opt}} \equiv \max f(\mathbf{x}) : A\mathbf{x} = \mathbf{c}$ and suppose the sequences $\{\xi_{\alpha n}\}$ and $\{\xi_{\beta n}\}$ satisfy Eq. (9.49); then

$$\text{i.} \qquad f(\boldsymbol{\xi}_{\alpha n}) \leqslant f(\boldsymbol{\xi}_{\beta n}) \leqslant f(\boldsymbol{\xi}_{\alpha, n+1}) \tag{9.50}$$

$$\text{ii.} \qquad \lim_{n\to\infty} f(\boldsymbol{\xi}_{\alpha n}) = \lim_{n\to\infty} f(\boldsymbol{\xi}_{\beta n}) \equiv f^* \leqslant f^{\text{opt}} \tag{9.51}$$

Part i follows directly from Eq. (9.49) and part ii follows from part i and the fact that the bounded monotonically increasing sequences converge. (Recall that we assumed that f is bounded.)

The remaining question is under what conditions $f^* = f^{\text{opt}}$. The following property answers this question.

Property 3 Assume that f is concave and differentiable and that the constraints of both problems in Eqs. (9.47) and (9.48) are regular. Furthermore, assume that

$$N(A) = N(K_\alpha) + N(K_\beta) \tag{9.52}$$

where $N(\cdot)$ denotes the null space of $(\cdot)$. If the sequences $\{\boldsymbol{\xi}_{\alpha n}\}$ and $\{\boldsymbol{\xi}_{\beta n}\}$ are obtained by Algorithm 1, then any accumulation point $\boldsymbol{\xi}^*$ of either sequence solves the original problem; that is,

$$f(\boldsymbol{\xi}^*) = f^* = \max_{\mathbf{x}} f(\mathbf{x})$$

subject to

$$A\mathbf{x} = \mathbf{c}$$

PROOF Let ξ^* be an accumulation point of $\{\xi_{\alpha n}\}$ where the sequences $\{\xi_{\alpha n}\}$ and $\{\xi_{\beta n}\}$ are obtained by Algorithm 1. Define the real-valued functions F_α and F_β:

$$F_\alpha(\boldsymbol{\xi}) \equiv \max \{f(\mathbf{x}) : K_\alpha(\mathbf{x} - \boldsymbol{\xi}) = 0\}$$

$$F_\beta(\boldsymbol{\xi}) \equiv \max \{f(\mathbf{x}) : K_\beta(\mathbf{x} - \boldsymbol{\xi}) = 0\}$$

Then, the following statements hold:

1. F_α and F_β are lower semicontinuous.† (The reader is encouraged to prove this statement.)
2. By the definition, $F_\alpha(\boldsymbol{\xi}) \geqslant f(\boldsymbol{\xi})$ and $F_\beta(\boldsymbol{\xi}) \geqslant f(\boldsymbol{\xi})$ for all $\boldsymbol{\xi}$.
3. $f(\boldsymbol{\xi}^*) = f^* \geqslant f(\boldsymbol{\xi}_{\alpha n}) \geqslant f(\boldsymbol{\xi}_{\beta, n-1}) = F_\beta(\boldsymbol{\xi}_{\alpha, n-1})$ for all $n > 1$.
4. $f(\boldsymbol{\xi}^*) = f^* \geqslant f(\boldsymbol{\xi}_{\alpha n}) = F_\alpha(\boldsymbol{\xi}_{\alpha n})$ for all $n > 1$.

Since $\{\boldsymbol{\xi}_{\alpha n}\}$ has a subsequence $\{\boldsymbol{\xi}_i\}$ that converges to $\boldsymbol{\xi}^*$, and both F_α and F_β

†A numerical function θ defined on a set $\Gamma \subset R^n$ is said to be lower semicontinuous at $x^0 \in \Gamma$ if for each $\epsilon > 0$ there exists a $\delta > 0$ such that

$$\left\{ \begin{matrix} \|x - x^0\| < \delta \\ x \in \Gamma \end{matrix} \right\} \Rightarrow -\epsilon < \theta(x) - \theta(x^0)$$

(see Mangasarian, 1969, pp. 196, for more details).

are lower semicontinuous, for any $\epsilon > 0$ there exist $\delta > 0$ and $N > 0$ such that

$$i > N \Rightarrow \|\boldsymbol{\xi}^* - \boldsymbol{\xi}_i\| < \delta \Rightarrow F_\alpha(\boldsymbol{\xi}^*) < \epsilon + F_\alpha(\boldsymbol{\xi}_i)$$

and

$$F_\beta(\boldsymbol{\xi}^*) < \epsilon + F_\beta(\boldsymbol{\xi}_i)$$

Thus, statements 2–4 imply that, for all $\epsilon > 0$,

$$f(\boldsymbol{\xi}^*) \leqslant F_\alpha(\boldsymbol{\xi}^*) < \epsilon + F_\alpha(\boldsymbol{\xi}_i) \leqslant \epsilon + f(\boldsymbol{\xi}^*)$$

$$f(\boldsymbol{\xi}^*) \leqslant F_\beta(\boldsymbol{\xi}^*) < \epsilon + F_\beta(\boldsymbol{\xi}_i) \leqslant \epsilon + f(\boldsymbol{\xi}^*)$$

and hence

$$f(\boldsymbol{\xi}^*) = F_\alpha(\boldsymbol{\xi}^*) = F_\beta(\boldsymbol{\xi}^*)$$

Therefore, $\boldsymbol{\xi}^*$ maximizes $f(\mathbf{x})$ subject to $K_\alpha(\mathbf{x} - \boldsymbol{\xi}^*) = \mathbf{0}$ and $K_\beta(\mathbf{x} - \boldsymbol{\xi}^*) = \mathbf{0}$. Also, from the first-order condition,

$$\nabla f(\boldsymbol{\xi}^*)\delta\mathbf{x} = 0 \qquad \mathbf{K}_\alpha\delta\mathbf{x} = \mathbf{0}$$

$$\nabla f(\boldsymbol{\xi}^*)\delta\mathbf{x} = 0 \qquad \mathbf{K}_\beta\delta\mathbf{x} = \mathbf{0}$$

Thus, under the assumption of Eq. (9.52),

$$\nabla f(\boldsymbol{\xi}^*)\delta\mathbf{x} = 0 \qquad A\delta\mathbf{x} = \mathbf{0}$$

and $A\boldsymbol{\xi}^* = \mathbf{c}$. Therefore, since f is concave, $\boldsymbol{\xi}^*$ maximizes $f(\mathbf{x})$ subject to $A\mathbf{x} = \mathbf{c}$.

Q.E.D.

If Eq. (9.52) holds, we have the following property.

Property 4 Let A, H_α, and H_β be $(p \times n)$, $(q_\alpha \times n)$, and $(q_\beta \times n)$ matrices, respectively. Then,

$$N(A) = N(K_\alpha) + N(K_\beta) \Rightarrow q_\alpha + q_\beta \leqslant n - p$$

where $\text{rank}(A) = p$, $\text{rank}(H_\alpha) = q_\alpha$, and $\text{rank}(H_\beta) = q_\beta$.

PROOF By the definition of K_α and K_β, K_α and K_β are $[(p + q_\alpha) \times n]$ and $[(p + q_\beta) \times n]$ matrices. Thus,

$$\dim N(A) = \dim [N(K_\alpha) + N(K_\beta)] \leqslant \dim N(K_\alpha) + \dim (K_\beta)$$

Therefore,

$$n - p \leqslant n - p - q_\alpha + n - p - q_\beta \quad \text{and} \quad q_\alpha + q_\beta \leqslant n - p \tag{9.53}$$

Q.E.D.

Property 4 implies that the necessary condition for $N(A) = N(K_\alpha) + N(K_\beta)$

to hold is that the sum of the numbers of the interaction variables in the α and β decompositions does not exceed the number of independent solutions of $A\mathbf{x} = \mathbf{c}$.

The following property is very important for achieving convergence to the optimal solution.

Property 5

$$\{\mathbf{y}_\alpha\} \cap \{\mathbf{y}_\beta\} \neq \phi \Rightarrow N(A) \neq N(K_\alpha) + N(K_\beta)$$

where

$\{\mathbf{y}_\alpha\} \triangleq$ the set of the components of the vector $\mathbf{x}$ which satisfies $H_\alpha P_\alpha \mathbf{x} = \mathbf{y}_\alpha$

$\{\mathbf{y}_\beta\} \triangleq$ the set of the components of the vector $\mathbf{x}$ which satisfies $H_\beta P_\beta \mathbf{x} = \mathbf{y}_\beta$

PROOF Suppose $\{\mathbf{y}_\alpha\} \cap \{\mathbf{y}_\beta\} \neq \phi$. Then, there exists a variable x_i, which is a component of the vector $\mathbf{x}$, such that $x_i \in \{\mathbf{y}_\alpha\} \cap \{\mathbf{y}_\beta\}$. Then, δx_i, which is a small perturbation corresponding to x_i, is restricted to zero in both $N(K_\alpha)$ and $N(K_\beta)$. Therefore, $N(K_\alpha) + N(K_\beta) \neq N(A)$.

A more general case where the system is described by nonlinear equations is considered in Macko and Haimes (1978). The analysis there is made in Hilbert space, and we should recall that HOC can be applicable not only to the finite-dimensional static problems, but also to infinite-dimensional static and dynamic problems.

9.4.2 Convergence Properties of HOC for Problems with Linear Inequality Constraints

In this section, we restrict our analysis to finite-dimensional Euclidean space since the problems with which we are concerned are finite-dimensional static problems. We deal with the following mathematical model that has inequality constraints:

$$\max f(\mathbf{x}) \tag{9.54}$$

subject to

$$A\mathbf{x} \geqslant \mathbf{c} \tag{9.55}$$

$$\mathbf{x} \geqslant \mathbf{0} \tag{9.56}$$

where

f = a real-valued differentiable concave function and may be linear or nonlinear; we assume that the set $\{\mathbf{x} | A\mathbf{x} \geqslant c, \mathbf{x} \geqslant 0\}$ is compact if f is linear
A = a $(p \times n)$ matrix
$\mathbf{x}$ = an $(n \times 1)$ decision vector
$\mathbf{c}$ = a $(p \times 1)$ fixed vector

What is different here is that the constraint space is not the intersection of the hyperspaces (as it was in the equality constraint case) but is instead the intersection of the half spaces given by the constraint equations.

The convergence properties of HOC for problems with inequality constraints are considered in Shima (1982), where properties that are similar to those in the equality constraint cases are obtained. However, Eqs. (9.47) and (9.48) are rewritten as follows:

$$\max f(\mathbf{x})$$

subject to

$$A\mathbf{x} \geqslant \mathbf{c} \qquad H_\alpha P_\alpha \mathbf{x} = \mathbf{y}_\alpha^{(n)} \qquad \mathbf{x} \geqslant 0 \tag{9.57}$$

and

$$\max f(\mathbf{x})$$

subject to

$$A\mathbf{x} \geqslant \mathbf{c} \qquad H_\beta P_\beta \mathbf{x} = \mathbf{y}_\beta^{(n)} \qquad \mathbf{x} \geqslant 0 \tag{9.58}$$

Or, those problems are rewritten as follows, by introducing the surplus vector $\mathbf{x}_s$:

$$\max \hat{f}(\hat{\mathbf{x}})$$

subject to

$$\hat{A}\hat{\mathbf{x}} = \mathbf{c} \qquad \hat{H}_\alpha \hat{P}_\alpha \hat{\mathbf{x}} = \mathbf{y}_\alpha \qquad \hat{\mathbf{x}} \geqslant 0 \tag{9.59}$$

and

$$\max \hat{f}(\hat{\mathbf{x}})$$

subject to

$$\hat{A}\hat{\mathbf{x}} = \mathbf{c} \qquad \hat{H}_\beta \hat{P}_\beta \hat{\mathbf{x}} = \mathbf{y}_\beta \qquad \hat{\mathbf{x}} \geqslant 0 \tag{9.60}$$

where

$\hat{\mathbf{x}} = [\mathbf{x}^T, \mathbf{x}_s^T]^T$
$\hat{A} = [A, -I]$ (I is an identity matrix)
$\hat{H}$ and $\hat{P}$ are matrices modified to have appropriate dimensions.

In ways similar to the equality constraint case,

$$\mathbf{x} \text{ (or } \hat{\mathbf{x}}\text{) is } \begin{Bmatrix} \text{feasible} \\ \text{optimal} \end{Bmatrix}$$

in Eq. (9.57) or (9.59) if and only if

$$\mathbf{x}_\alpha = P_\alpha \mathbf{x} \text{ (or } \hat{\mathbf{x}}_\alpha = \hat{P}_\alpha \hat{\mathbf{x}}\text{) is } \begin{Bmatrix} \text{feasible} \\ \text{optimal} \end{Bmatrix}$$

in the problem that is decomposed in the subsystems in the α decomposition.

Likewise,

$$\mathbf{x} \text{ (or } \hat{\mathbf{x}}) \text{ is } \begin{Bmatrix} \text{feasible} \\ \text{optimal} \end{Bmatrix}$$

in Eq. (9.58) [or Eq. (9.60)] if and only if

$$\mathbf{x}_\beta = P_\beta \mathbf{x} \text{ (or } \hat{\mathbf{x}}_\beta = \hat{P}_\beta \hat{\mathbf{x}}) \text{ is } \begin{Bmatrix} \text{feasible} \\ \text{optimal} \end{Bmatrix}$$

in the problem that is decomposed in the subsystems in the β decomposition.

We will describe the properties that guarantee convergence to the optimal solution of the original problem (see Shima, 1982, for details).

The following properties are also obtained.

Property 1 If Algorithm 1 is started with any fixed feasible values of the interaction variables, then at each stage n of the process, the problems in Eq. (9.57) [or (9.59)] and Eq. (9.58) [or (9.60)] will have nonempty feasible domains.

Property 2 Let $f^{\text{opt}} = \max f(\mathbf{x}){:}A\mathbf{x} \geqslant \mathbf{c},\ \mathbf{x} \geqslant 0$, and suppose the sequences $\{\boldsymbol{\xi}_{\alpha n}\}$ and $\{\boldsymbol{\xi}_{\beta n}\}$ are obtained by Algorithm 1; then

$$\text{i.} \qquad f(\boldsymbol{\xi}_{\alpha n}) \leqslant f(\boldsymbol{\xi}_{\beta n}) \leqslant f(\boldsymbol{\xi}_{\alpha . n+1}) \tag{9.61}$$

$$\text{ii.} \qquad \lim_{n\to\infty} f(\boldsymbol{\xi}_{\alpha n}) = \lim_{n\to\infty} f(\boldsymbol{\xi}_{\beta n}) \equiv f^* \leqslant f^{\text{opt}}$$

The assumptions in this property are made in order to guarantee improvement of the solution each time the problem is solved in the α or β decompositions.

Property 3 Assume f is concave and differentiable and the constraints of the problems in Eqs. (9.57) and (9.58) are regular at accumulation point $\boldsymbol{\xi}^*$ and, furthermore, assume

$$X = X_\alpha + X_\beta \tag{9.62}$$

If the sequences $\{\boldsymbol{\xi}_{\alpha n}\}$ and $\{\boldsymbol{\xi}_{\beta n}\}$ are obtained by Algorithm 1, then any accumulation point $\boldsymbol{\xi}^*$ of either sequence solves the original problem; that is,

$$f(\boldsymbol{\xi}^*) = f^* = \max_{\mathbf{x}} f(\mathbf{x}) \qquad A\mathbf{x} \geqslant \mathbf{c} \qquad \mathbf{x} \geqslant 0$$

where

$$X = \{\mathbf{z} | (A\mathbf{z})_j \geqslant 0, j \in B_1, \mathbf{z}_j \geqslant 0, j \in B_2\} \tag{9.63}$$

$$X_\alpha = \{\mathbf{z} | (A\mathbf{z})_j \geqslant 0, j \in B_1, H_\alpha P_\alpha \mathbf{z} = 0, \mathbf{z}_j \geqslant 0, j \in B_2\} \tag{9.64}$$

$$X_\beta = \{\mathbf{z} | (A\mathbf{z})_j \geqslant 0, j \in B_1, H_\beta P_\beta \mathbf{z} = 0, \mathbf{z}_j \geqslant 0, j \in B_2\} \tag{9.65}$$

$$B_1 = \{j | (A\boldsymbol{\xi}^*)_j = \mathbf{c}_j\} \tag{9.66}$$

$$B_2 = \{j | \boldsymbol{\xi}_j^* = 0\} \tag{9.67}$$

$\boldsymbol{\xi}^*$ is an optimal point. $(A\mathbf{z})_j$ means the jth row of $A\mathbf{z}$. For more details, see Shima (1982) and Shima and Haimes (1982a,b; 1984).

It is difficult to check before solving the problem whether or not convergence to the optimal solution of the original problem is obtained by HOC. This is because the assumptions in Property 2 depend on the values of $\mathbf{y}_\alpha$ and $\mathbf{y}_\beta$ at each iteration, and Eq. (9.62) depends on the optimal point of the original problem.

In order to avoid a cycling and a waste of computational time, we recommend that the improvement in the value of the objective function be checked each time the α and β problem is solved, until the optimal solution is obtained.

The convergence properties of HOC for problems with inequality constraints are still not very clear. So far, very strong assumptions are obtained only in Shima (1982) and Shima and Haimes (1982b, 1984). The use of HOC is recommended if these problems have weakly coupled constraints. However, in some cases, cycling may occur. If convergence to the optimal solution cannot be obtained because of the cycling, then the problem is not suitable for HOC.

9.4.3 Tests for Optimality in HOC

In Sect. 9.2.2, Eqs. (9.43) and (9.44) were considered as the criteria for convergence to the optimal point. However, in some problems these criteria cannot distinguish a point where cycling is occurring from the optimal point. Other tests for optimality are considered in this section.

As we saw in the previous sections, a problem is optimized with respect to decision variables corresponding to interaction variables. More precisely, the α problem is not optimized with respect to the decision variables that correspond to the components of the vector $\mathbf{y}_\alpha$, and the β is not optimized with respect to the decision variables that correspond to the components of the vector $\mathbf{y}_\beta$. That is, the following conditions may not be satisfied in the α and β problems.

Linear Equality Constraint Case:

$$\frac{\partial f_\alpha}{\partial \mathbf{x}_{\alpha y}} + (\boldsymbol{\mu}_\alpha A_\alpha)_{\alpha y} = 0 \qquad \frac{\partial f_\beta}{\partial \mathbf{x}_{\beta y}} + (\boldsymbol{\mu}_\beta A_\beta)_{\beta y} = 0$$

Linear Inequality Constraint Case:

$$\frac{\partial f_\alpha}{\partial \mathbf{x}_{\alpha y}} + (\boldsymbol{\pi}_\alpha A_\alpha)_{\alpha y} \leqslant 0 \qquad \left[\frac{\partial f_\alpha}{\partial \mathbf{x}_{\alpha y}} + (\boldsymbol{\pi}_\alpha A_\alpha)_{\alpha y}\right]\mathbf{x}_{\alpha y} = 0$$

and

$$\frac{\partial f_\beta}{\partial \mathbf{x}_{\beta y}} + (\boldsymbol{\pi}_\beta A_\beta)_{\beta y} \leqslant 0 \qquad \left[\frac{\partial f_\beta}{\partial \mathbf{x}_{\beta y}} + (\boldsymbol{\pi}_\beta A_\beta)_{\beta y}\right]\mathbf{x}_{\beta y} = 0$$

where

$\mathbf{x}_{\alpha y} = H_\alpha P_\alpha \mathbf{x} = \mathbf{y}_\alpha$
$\mathbf{x}_{\beta y} = H_\beta P_\beta \mathbf{x} = \mathbf{y}_\beta$
$\boldsymbol{\mu}_\alpha$ = Lagrange multiplier vector of the constraint $A_\alpha \mathbf{x}_\alpha = \mathbf{c}_\alpha$; that is, $B_\alpha \mathbf{x}_\alpha + G_\alpha \mathbf{y}_\alpha = \mathbf{c}_\alpha$

$(\boldsymbol{\mu}_\alpha A_\alpha)\alpha y$ = coefficient vector corresponding to $\mathbf{x}_{\alpha y}$; $\boldsymbol{\mu}_\beta$ and $(\boldsymbol{\mu}_\beta A_\beta)_{\beta y}$ are defined similarly

$\boldsymbol{\pi}_\alpha$ = Kuhn-Tucker multiplier vector of the constraint $A_\alpha \mathbf{x}_\alpha \geqslant \mathbf{c}_\alpha$; that is, $B_\alpha \mathbf{x}_\alpha + G_\alpha \mathbf{y}_\alpha \geqslant \mathbf{c}_\alpha$

$(\boldsymbol{\pi}_\alpha A_\alpha)_{\alpha y}$ = coefficient vector corresponding to $\mathbf{x}_{\alpha y}$; $\boldsymbol{\pi}_\beta$ and $(\boldsymbol{\pi}_\beta A_\beta)_{\beta y}$ are also defined similarly.

Note that we assume the existence of $\boldsymbol{\mu}_\alpha$, $\boldsymbol{\mu}_\beta$, $\boldsymbol{\pi}_\alpha$, and $\boldsymbol{\pi}_\beta$ in this section.

Those conditions that are parts of the Kuhn-Tucker conditions of the α and β problems are not satisfied until the optimum is achieved. Those conditions can be utilized as the criteria for achievement of the optimum.

We can check whether those conditions are satisfied at each time the α or the β problem is solved. Then we go to the next step if those conditions are not satisfied. If those conditions are satisfied, we stop, since the optimal solution has been obtained.

In this section, tests for optimality are considered in the case where Lagrange multipliers and Kuhn-Tucker multipliers exist. If these do not exist, that is, if the optimal point for any value of $\mathbf{y}_\alpha$ or $\mathbf{y}_\beta$ is not regular, then the criteria of Eqs. (9.43) and (9.44) can be used as a stopping criterion.

9.5 EXAMPLE

9.5.1 Manufacturing Problem

Suppose an industrial company makes six subproducts (which are grouped into two main products with one common subproduct) by using manufacturing facilities in two plants. We will analyze the system from the viewpoints of two product managers (a product decomposition) and two plant managers (a plant or geographic decomposition). The objective of the system is to maximize the overall company profits. Data for this problem are presented in Table 9.1.

Table 9.1 Data for example problem

Plant	Capacity used per unit production rate, subproducts 1	2	3	4	5	6	Capacity available
Plant 1	1	2				4	8
		1					8
Plant 2			1	3	4		12
			1	1	3		6
			1			2	6
Unit profit ($\times 10^3$ dollars)	2	5	1	4	1	6	

Suppose that subproducts 1, 2, 6, and 3 are assembled into one product and are under the jurisdiction of one of the product managers. Subproducts 4, 5, and 3 are assembled into another product and are under the jurisdiction of the other manager. Subproduct 3 is utilized in both products, so that it acts as a coupling between the two products. Subproduct 6 is produced in both plants and, hence, acts as a coupling between the two plants.

Let the product decomposition be called the α decomposition and the plant decomposition be called the β decomposition.

Formulation Let x_i, $i = 1, 2, \ldots, 6$, denote the number of units to be produced of subproducts $i = 1, \ldots, 6$ and f be a resulting profit.

The overall problem is formulated as the following linear programming problem:

$$\max_{\mathbf{x}} f(\mathbf{x}) = 2x_1 + 5x_2 + x_3 + 4x_4 + x_5 + 6x_6 \tag{9.68}$$

subject to

$$x_1 + 2x_2 + 4x_6 \leqslant 8 \tag{9.69a}$$

$$x_2 \leqslant 8 \tag{9.69b}$$

$$x_3 + 3x_4 + 4x_5 \leqslant 12 \tag{9.69c}$$

$$x_3 + x_4 + 3x_5 \leqslant 6 \tag{9.69d}$$

$$x_3 + 2x_6 \leqslant 6 \tag{9.69e}$$

$$x_i \geqslant 0 \qquad (i = 1, \ldots, 6)$$

The constraints are represented by the matrix form:

$$A\mathbf{x} \leqslant \mathbf{c} \tag{9.70}$$

that is,

$$\begin{pmatrix} 1 & 2 & 0 & 0 & 0 & 4 \\ 0 & 1 & 0 & 0 & 0 & 0 \\ 0 & 0 & 1 & 3 & 4 & 0 \\ 0 & 0 & 1 & 1 & 3 & 0 \\ 0 & 0 & 1 & 0 & 0 & 2 \end{pmatrix} \begin{pmatrix} x_1 \\ x_2 \\ x_3 \\ x_4 \\ x_5 \\ x_6 \end{pmatrix} \leqslant \begin{pmatrix} 8 \\ 8 \\ 12 \\ 6 \\ 6 \end{pmatrix} \tag{9.71}$$

The rearrangement of the system of inequalities yields

$$A_\alpha x_\alpha \leqslant c_\alpha \tag{9.72}$$

given by

$$\begin{pmatrix} 1 & 2 & 4 & 0 & 0 & 0 \\ 0 & 1 & 0 & 0 & 0 & 0 \\ 0 & 0 & 2 & 0 & 0 & 1 \\ 0 & 0 & 0 & 3 & 4 & 1 \\ 0 & 0 & 0 & 1 & 3 & 1 \end{pmatrix} \begin{pmatrix} x_1 \\ x_2 \\ x_6 \\ x_4 \\ x_5 \\ x_3 \end{pmatrix} \leqslant \begin{pmatrix} 8 \\ 8 \\ 6 \\ 12 \\ 6 \end{pmatrix} \tag{9.73}$$

The transformation P_α is given by

$$\mathbf{x}_\alpha = P_\alpha \mathbf{x} \tag{9.74}$$

that is,

$$\begin{pmatrix} x_1 \\ x_2 \\ x_6 \\ x_4 \\ x_5 \\ x_3 \end{pmatrix} = \begin{pmatrix} 1 & 0 & 0 & 0 & 0 & 0 \\ 0 & 1 & 0 & 0 & 0 & 0 \\ 0 & 0 & 0 & 0 & 0 & 1 \\ 0 & 0 & 0 & 1 & 0 & 0 \\ 0 & 0 & 0 & 0 & 1 & 0 \\ 0 & 0 & 1 & 0 & 0 & 0 \end{pmatrix} \begin{pmatrix} x_1 \\ x_2 \\ x_3 \\ x_4 \\ x_5 \\ x_6 \end{pmatrix} \tag{9.75}$$

A transformation Q_α, which transforms $\mathbf{c}$ into $\mathbf{c}_\alpha$, can be obtained similarly. The matrix A_α can be decomposed as follows.

$$A_\alpha = B_\alpha + G_\alpha H_\alpha \tag{9.76}$$

that is,

$$\begin{pmatrix} 1 & 2 & 4 & 0 & 0 & 0 \\ 0 & 1 & 0 & 0 & 0 & 0 \\ 0 & 0 & 2 & 0 & 0 & 1 \\ 0 & 0 & 0 & 3 & 4 & 1 \\ 0 & 0 & 0 & 1 & 3 & 1 \end{pmatrix} = \left(\begin{array}{ccc:ccc} 1 & 2 & 4 & 0 & 0 & 0 \\ 0 & 1 & 0 & 0 & 0 & 0 \\ 0 & 0 & 2 & 0 & 0 & 0 \\ \hdashline 0 & 0 & 0 & 3 & 4 & 1 \\ 0 & 0 & 0 & 1 & 3 & 1 \end{array}\right) + \left(\begin{array}{c} 0 \\ 0 \\ 1 \\ \hline 0 \\ 0 \end{array}\right) [0\ 0\ 0\ 0\ 0\ 1] \tag{9.77}$$

Equation (9.73) is now decomposed to explicitly indicate the two product subsystems and the coupling between them.

***α* Decomposition (product decomposition)**

α1 Subsystem (Product 1):

$$\max f_{\alpha 1} = 2x_1 + 5x_2 + 6x_6 \tag{9.78}$$

subject to

$$B_{\alpha 1}\mathbf{x}_{\alpha 1} + G_{\alpha 1}\mathbf{y}_\alpha \leqslant \mathbf{c}_{\alpha 1} \tag{9.79}$$

$$H_{\alpha 1}\mathbf{x}_{\alpha 1} = \mathbf{y}_{\alpha 1} \tag{9.80}$$

$$\mathbf{x}_{\alpha 1} \geqslant 0$$

where

$$B_{\alpha 1} = \begin{pmatrix} 1 & 2 & 4 \\ 0 & 1 & 0 \\ 0 & 0 & 2 \end{pmatrix}$$

$$\mathbf{x}_{\alpha 1} = \begin{pmatrix} x_1 \\ x_2 \\ x_6 \end{pmatrix} \qquad G_{\alpha 1} = \begin{pmatrix} 0 \\ 0 \\ 1 \end{pmatrix} \qquad \mathbf{y}_\alpha = [x_3]$$

$$\mathbf{c}_{\alpha 1} = \begin{pmatrix} 8 \\ 8 \\ 6 \end{pmatrix} \qquad H_{\alpha 1} = [0, 0, 0] \qquad \mathbf{y}_{\alpha 1} = [0]$$

α2 Subsystem (Product 2):

$$\max f_{\alpha 2} = x_3 + 4x_4 + x_5 \tag{9.81}$$

subject to

$$B_{\alpha 2}\mathbf{x}_{\alpha 2} + G_{\alpha 2}\mathbf{y}_{\alpha} \leqslant \mathbf{c}_{\alpha 2} \tag{9.82}$$

$$H_{\alpha 2}\mathbf{x}_{\alpha 2} = \mathbf{y}_{\alpha 2} \qquad \mathbf{x}_{\alpha 2} \geqslant 0 \tag{9.83}$$

where

$$B_{\alpha 2} = \begin{pmatrix} 3 & 4 & 1 \\ 1 & 3 & 1 \end{pmatrix} \qquad \mathbf{x}_{\alpha 2} = \begin{pmatrix} x_4 \\ x_5 \\ x_3 \end{pmatrix}$$

$$G_{\alpha 2} = \begin{pmatrix} 0 \\ 0 \end{pmatrix} \qquad \mathbf{y}_{\alpha} = [x_3]$$

$$\mathbf{c}_{\alpha 2} = \begin{pmatrix} 12 \\ 6 \end{pmatrix} \qquad H_{\alpha 2} = [0 \;\; 0 \;\; 1] \qquad \mathbf{y}_{\alpha 2} = \mathbf{y}_{\alpha} = [x_3]$$

Let us now consider the β decomposition. Equations (9.70) and (9.71) are already in an order amenable to plantwise decomposition (β decomposition) into two subsystems. Hence, A corresponds to A_β and $\mathbf{x}$ corresponds to $\mathbf{x}_\beta$. Therefore, the transformations P_β and Q_β are identity matrices.

The problem is decomposed to explicitly indicate the two plant subsystems and couplings as follows.

β Decomposition (plant decomposition)

β1 Subsystem (Plant 1):

$$\max f_{\beta 1} = 2x_1 + 5x_2 \tag{9.84}$$

subject to

$$B_{\beta 1}\mathbf{x}_{\beta 1} + G_{\beta 1}\mathbf{y}_{\beta} \leqslant \mathbf{c}_{\beta 1} \tag{9.85}$$

$$H_{\beta 1}\mathbf{x}_{\beta 1} = \mathbf{y}_{\beta 1} \tag{9.86}$$

where

$$B_{\beta 1} = \begin{pmatrix} 1 & 2 \\ 0 & 1 \end{pmatrix} \qquad \mathbf{x}_{\beta 1} = \begin{pmatrix} x_1 \\ x_2 \end{pmatrix}$$

$$G_{\beta 1} = \begin{pmatrix} 4 \\ 0 \end{pmatrix} \qquad \mathbf{y}_{\beta} = [x_6]$$

$$\mathbf{c}_{\beta 1} = \begin{pmatrix} 8 \\ 8 \end{pmatrix} \qquad H_{\beta 1} = [0, 0] \qquad \mathbf{y}_{\beta 1} = [0]$$

β2 Subsystem (Plant 2):

$$\max f_{\beta 2} = x_3 + 4x_4 + x_5 + 6x_6 \tag{9.87}$$

subject to

$$B_{\beta 2}\mathbf{x}_{\beta 2} + G_{\beta 2}\mathbf{y}_{\beta} \leqslant \mathbf{c}_{\beta 2} \tag{9.88}$$

$$H_{\beta 2}\mathbf{x}_{\beta 2} = \mathbf{y}_{\beta 2} \tag{9.89}$$

$$B_{\beta 2} = \begin{pmatrix} 1 & 3 & 4 & 0 \\ 1 & 1 & 3 & 0 \\ 1 & 0 & 0 & 2 \end{pmatrix} \quad \mathbf{x}_{\beta 2} = \begin{pmatrix} x_3 \\ x_4 \\ x_5 \\ x_6 \end{pmatrix}$$

$$G_{\beta 2} = \begin{pmatrix} 0 \\ 0 \\ 0 \end{pmatrix} \qquad \mathbf{y}_{\beta} = [x_6] \qquad \mathbf{c}_{\beta 2} = \begin{pmatrix} 12 \\ 6 \\ 6 \end{pmatrix}$$

$$H_{\beta 2} = [0 \;\; 0 \;\; 0 \;\; 1] \qquad \mathbf{y}_{\beta 2} = \mathbf{y}_{\beta} = [x_6]$$

The coupling variable x_3 in the α decomposition (the product decomposition) is related to the decision vector $\mathbf{x}_\beta$ of the β decomposition as follows:

$$\mathbf{y}_\alpha = H_\alpha P_{\alpha\beta}\mathbf{x}_\beta \tag{9.90}$$

that is,

$$[x_3] = [0 \;\; 0 \;\; 0 \;\; 0 \;\; 0 \;\; 1] \begin{pmatrix} 1 & 0 & 0 & 0 & 0 & 0 \\ 0 & 1 & 0 & 0 & 0 & 0 \\ 0 & 0 & 0 & 0 & 0 & 1 \\ 0 & 0 & 0 & 1 & 0 & 0 \\ 0 & 0 & 0 & 0 & 1 & 0 \\ 0 & 0 & 1 & 0 & 0 & 0 \end{pmatrix} \begin{pmatrix} x_1 \\ x_2 \\ x_3 \\ x_4 \\ x_5 \\ x_6 \end{pmatrix} \tag{9.91}$$

Similarly, the coupling variable x_5 in the β decomposition (the plant decomposition) is related to the decision vector $\mathbf{x}_\alpha$ of the α decomposition (the product decomposition) as follows:

$$\mathbf{y}_\beta = H_\beta P_{\beta\alpha}\mathbf{x}_\alpha \tag{9.92}$$

that is,

$$[x_6] = [0 \;\; 0 \;\; 0 \;\; 0 \;\; 0 \;\; 1] \begin{pmatrix} 1 & 0 & 0 & 0 & 0 & 0 \\ 0 & 1 & 0 & 0 & 0 & 0 \\ 0 & 0 & 0 & 0 & 0 & 1 \\ 0 & 0 & 0 & 1 & 0 & 0 \\ 0 & 0 & 0 & 0 & 1 & 0 \\ 0 & 0 & 1 & 0 & 0 & 0 \end{pmatrix} \begin{pmatrix} x_1 \\ x_2 \\ x_6 \\ x_4 \\ x_5 \\ x_3 \end{pmatrix} \tag{9.93}$$

Figure 9.8 depicts the overlapping structure of this problem.

9.5.2 Test for Optimality

The α problem has not yet been optimized with respect to x_3 and the β problem has not yet been optimized with respect to x_5. The following conditions can be

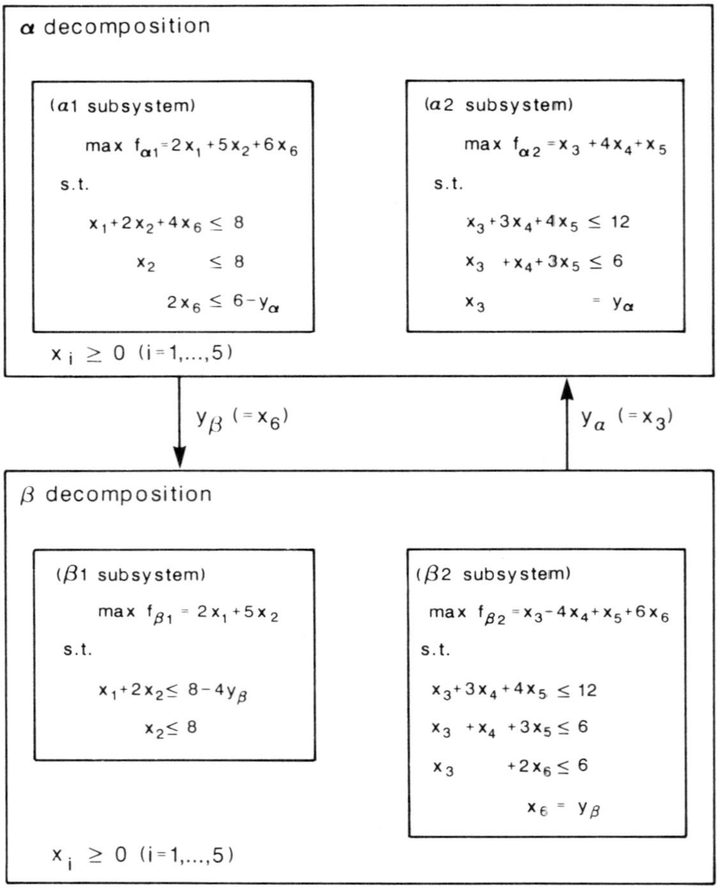

Figure 9.8 Structure of the example problem.

used as a test for optimality because the optimum will be achieved if these conditions are satisfied.

For the α problem,

$$1 - \pi_{\alpha 3} - \pi_{\alpha 4} - \pi_{\alpha 5} \leqslant 0 \tag{9.94}$$

$$(1 - \pi_{\alpha 3} - \pi_{\alpha 4} - \pi_{\alpha 5})x_3 = 0 \tag{9.95}$$

For the β problem,

$$6 - 4\pi_{\beta 1} - 2\pi_{\beta 5} \leqslant 0 \tag{9.96}$$

$$(6 - 4\pi_{\beta 1} - 2\pi_{\beta 5})x_6 = 0 \tag{9.97}$$

where $\pi_{\alpha 3}$, $\pi_{\alpha 4}$, and $\pi_{\alpha 5}$ are, respectively, the Kuhn-Tucker multipliers of the constraints $2x_6 \leqslant 6 - y_\alpha$, $x_3 + 3x_4 + 4x_5 \leqslant 12$, and $x_3 + x_4 + 3x_5 \leqslant 6$ of the α decomposition. $\pi_{\beta 1}$ and $\pi_{\beta 5}$ are, respectively, the Kuhn-Tucker multipliers of the constraints $x_1 + 2x_2 \leq 8 - 4y_\beta$ and $x_3 + 2x_6 \leq 6$ of the β decomposition.

Now we will solve the problem.

9.5.2.1 Solution procedure

Step 1. The feasible value of y_α is initially fixed. Let $y_\alpha^{(1)} = 1$. The α problem is solved in each subsystem. By solving the α1 subsystem, we can get $\mathbf{x}_{\alpha 1}^{(1)} = (x_1, x_2, x_6)^T = (0, 4, 0)^T$,

$$\pi_{\alpha 3} = 0 \qquad f_{\alpha_1}^{(1)} = 20$$

Likewise, by solving the α2 subsystem, we can get

$$\mathbf{x}_2^{(1)} = (x_4, x_5, x_3)^T = \left(\frac{11}{3}, 0, 1\right)^T$$

$$\pi_{\alpha 4} = \frac{4}{3} \qquad \pi_{\alpha 5} = 0 \qquad f_{\alpha 2}^{(1)} = 15$$

Therefore, $f_\alpha^{(1)} = 20 + 15 = 35$.

We check the conditions given in Eqs. (9.94) and (9.95):

$$1 - \pi_{\alpha 3} - \pi_{\alpha 4} - \pi_{\alpha 5} = -\frac{1}{3} < 0$$

$$(1 - \pi_{\alpha 3} - \pi_{\alpha 4} - \pi_{\alpha 5})x_3 = -\frac{1}{3} \neq 0$$

Equation (9.95) is not satisfied.

Step 2. Let $y_\beta^{(1)} = x_6 = 0$. The β problem is solved in each subsystem. By solving the β1 subsystem, we get

$$x_{\beta 1}^{(1)} = (x_1, x_2)^T = (0, 4)^T$$

$$\pi_{\beta 1} = 5/2 \qquad f_{\beta 1}^{(1)} = 20$$

Likewise, by solving the β2 subsystem, we get

$$\mathbf{x}_{\beta 2}^{(1)} = (x_3, x_4, x_5, x_6)^T = (0, 4, 0, 0)^T$$

$$\pi_{\beta 5} = 0 \qquad f_{\beta 2}^{(1)} = 16.$$

Therefore, $f_\alpha^{(1)} < f_\beta^{(1)}$.

We check the conditions given in Eqs. (9.96) and (9.97):

$$6 - 4\pi_{\beta 1} - 2\pi_{\beta 5} = -4 < 0$$

$$(6 - 4\pi_{\beta 1} - 2\pi_{\beta 5})x_6 = 0$$

Both conditions are satisfied. Therefore, this solution is optimal.

If we solve the α problem again for a given $y_\alpha^{(2)} = x_3 = 0$, then we will have exactly the same solution and the same value of the objective function. Therefore, the solution is

$$\mathbf{x}^{\text{opt}} = (x_1, x_2, x_3, x_4, x_5, x_6)^T$$
$$= (0, 4, 0, 4, 0, 0)^T$$
$$f^{\text{opt}} = 36$$

After becoming familiar with the HOC, one may be able to directly write the subproblems in Eqs. (9.78)–(9.83) in the α decomposition and the subproblems in Eqs. (9.84)–(9.89) in the β decomposition. It then will not be necessary to consider the transformation matrices. For more examples, see Mendu et al. (1980) and Sect. 11.2.3 of Chap. 11.

Exercise 1 Let the overall problem be

$$\min f(\mathbf{x}) = 4x_1 + x_2 + 5x_3 + 3x_4 + 5x_5 + 2x_6 + 7x_7$$

subject to

$$\begin{aligned} x_1 + x_2 + 4x_3 &\geq 8 \\ x_2 &\leq 5 \\ 2x_3 + x_4 &\leq 1.5 \\ x_5 + 2x_6 + 3x_7 &\geq 7 \\ 5x_4 + 2x_7 &\leq 1.2 \\ x_i \geq 0 \qquad i = 1, 2, \ldots, 7 \end{aligned}$$

Let $\mathbf{x}_{\alpha 1} = (x_1, x_2, x_3)^T$, $\mathbf{x}_{\alpha 2} = (x_4, x_5, x_6, x_7)^T$, $y_\alpha = x_3$, $\mathbf{x}_{\beta 1} = (x_1, x_2, x_3, x_4)^T$, $\mathbf{x}_{\beta 2} = (x_5, x_6, x_7)^T$, and $y_\beta = x_7$. Solve the problem by HOC.

Answer: $\mathbf{x}^{\text{opt}} = (0, 5, 0.75, 0, 0, 2.6, 0.6)^T$. (This exercise has been obtained by modifying Example 1 in Sung, 1978.)

REFERENCES

Haftari, A., *Integration of Overlapping Coordination and Multiobjective Optimization*, Ph.D. dissertation, Case Western Reserve University, Cleveland, Ohio, 1981.

Haimes, Y. Y., *Hierarchical Analyses of Water Resources Systems: Modeling and Optimization of Large-Scale Systems*, McGraw-Hill, New York, 1977.

Haimes, Y. Y., and Macko, D., Hierarchical Structures in Water Resources Systems Management, *IEEE Trans. Syst. Man Cybern.*, vol. SMC-3, no. 4, pp. 396–402, 1973.

Macko, D., and Haimes, Y. Y., Overlapping Coordination of Hierarchical Structures, *IEEE Trans. Syst. Man Cybern.*, vol. SMC-8, no. 10, pp. 745–751, 1978.

Mangasarian, O. L., *Nonlinear Programming*, McGraw-Hill, New York, 1969.

Mendu, S. R., Macko, D., and Haimes, Y. Y., Computational Aspects of Overlapping Coordination Methodology for Linear Hierarchical Systems, *IEEE Trans. Syst. Man Cybern.*, vol. SMC-10, no. 2, pp. 68–78, 1980.

Shima, T., *Hierarchical Overlapping Coordination of Large Scale Systems with Multiple Objectives*, Ph.D. dissertation, Case Western Reserve University, Cleveland, Ohio, 1982.

Shima, T., and Haimes, Y. Y., The Convergence Properties of Hierarchical Overlapping Coordination: Part I, The Equality Constraint Case, Technical Paper No. LSS&PA-82-4, Center for Large Scale Systems and Policy Analysis, Case Western Reserve University, Cleveland, Ohio, 1982a.

Shima, T., and Haimes, Y. Y., The Convergence Properties of Hierarchical Overlapping Coordination: Part II, The Inequality Constraint Case, Technical Paper No. LSS&PA-82-5, Center for Large Scale Systems and Policy Analysis, Case Western Reserve University, Cleveland, Ohio, 1982b.

Shima, T., and Haimes, Y. Y., The Convergence Properties of Hierarchical Overlapping Coordination, *IEEE Trans. Syst. Man Cybern.*, vol. SMC-14, 74–87, 1984.

Shima, T., Tarvainen, K., and Haimes, Y. Y., Hierarchical Overlapping Coordination for Large Scale Systems, *Proc. 8th Triennial World Congress/IFAC*, Kyoto, Japan, 1981.

Sung, K., *Coordination of Overlapping Hierarchical Water Resources Systems*, Ph.D. dissertation, Case Western Reserve University, Cleveland, Ohio, 1978.

Vancil, R. F., What Kind of Management Control Do you Need? in *Organization*, ed. Kotter Schlessinger, and R. D. Sathe, Irwin, Inc., Homewood, IL, 1979, originally published in *Harvard Business Review*, March–April 1973.

CHAPTER

TEN

INTEGRATION OF HIERARCHICAL OVERLAPPING COORDINATION IN MULTIOBJECTIVE OPTIMIZATION PROBLEMS

10.1 INTRODUCTION

10.1.1 Structure of the System

The use of hierarchical overlapping coordination (HOC) in multiobjective optimization problems (MOP) will be considered in this chapter. Figures 10.1*a* and *b* show a representation of the large-scale complex system that is under consideration here. The center (or the upper level) of the system that is described in the overall problem has multiple objectives. In the lower levels (or decompositions) are subsystems that are obtained by decomposing the overall system in various ways, and these subsystems also have multiple objectives.

A decisionmaker or a group of decisionmakers is necessary to the solution of a multiobjective problem. Several different decisionmaking structures are possible, for example, top-down (top-to-bottom) decisionmaking, bottom-up (bottom-to-top) decisionmaking, and hybrid decisionmaking. These terms are defined as follows:

Top-down: The decisions for the whole system are made by the decisionmaker(s) in the center, and the subsystems in the decompositions only carry out these decisions.

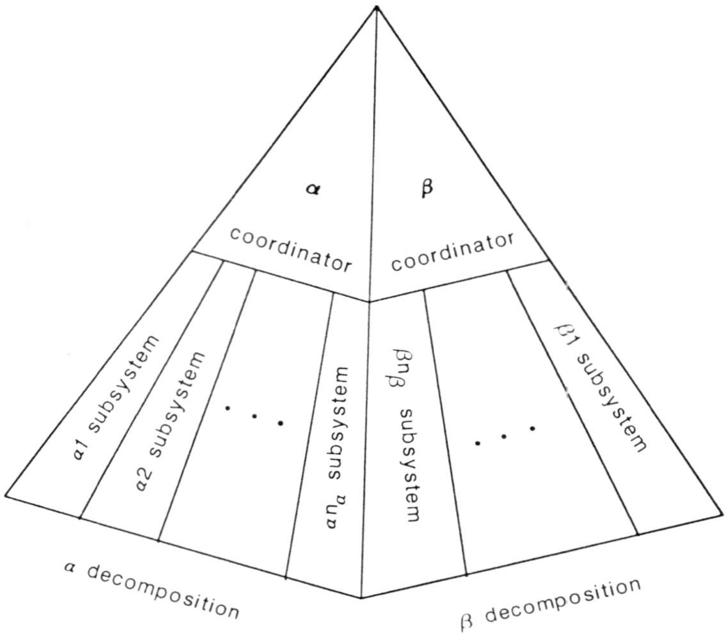

Figure 10.1 (*a*) Overlapping structure of large-scale system. Note that α and β coordinators may be identical in some cases or may be different from each other.

Bottom-up: Each subsystem in each decomposition has its decisions made by its own decisionmaker(s), and the center only coordinates these subsystems.
Hybrid: The decisionmaker(s) in the center and the decisionmaker(s) in the subsystems make the decisions.

Hybrid decisionmaking usually brings a conflict among decisionmakers, which means that they must find some way to achieve a compromise solution.

Several variations are possible within each type of decisionmaking. For example, in top-down decisionmaking, the center could consist of either a single decisionmaker or multiple decisionmakers who represent each separate decomposition. In the case of multiple decisionmakers, if all have a common preference, this becomes the equivalent of a single-decisionmaker case.

This chapter emphasizes top-down and hybrid decisionmaking.

10.1.2 Review of Necessary and Sufficient Conditions for Preferred Solution of a Multiobjective Problem

The necessary conditions for some point to be a preferred solution of a multiobjective problem were described in Sect. 5.2. We review those conditions briefly here. Suppose that the problem is represented generally as follows:

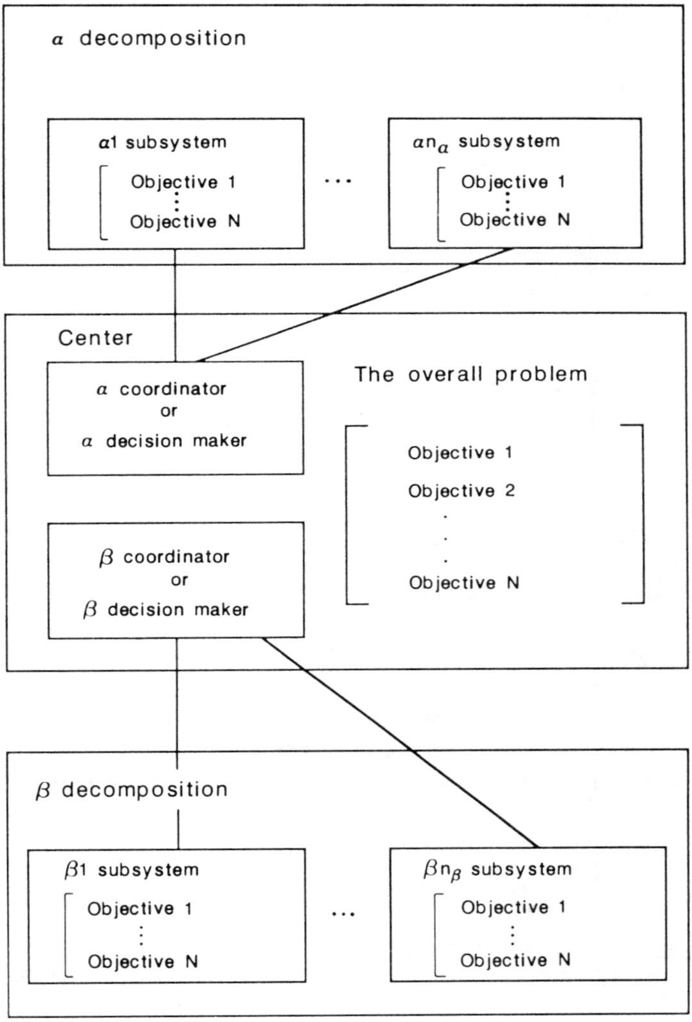

Figure 10.1 (*b*) Structure of a multiobjective hierarchical overlapping coordination problem.

$$\min\ [F_1(\mathbf{z}), F_2(\mathbf{z}), \ldots, F_N(\mathbf{z})]^T \tag{10.1}$$

subject to

$$\mathbf{w}(\mathbf{z}) \leq \mathbf{0} \tag{10.2}$$

Necessary condition Let $\mathbf{z}^0$ be a preferred solution of the multiobjective problem in Eqs. (10.1) and (10.2). There then exists a multiplier vector $\boldsymbol{\rho}$ such that

$$\boldsymbol{\lambda}^T(\mathbf{F}^0)\frac{\partial \mathbf{F}(\mathbf{z}^0)}{\partial \mathbf{z}} + \boldsymbol{\rho}^T\frac{\partial \mathbf{w}(\mathbf{z}^0)}{\partial \mathbf{z}} = \mathbf{0} \tag{10.3}$$

$$\boldsymbol{\rho}^T\mathbf{w}(\mathbf{z}^0) = 0 \qquad \boldsymbol{\rho} \geqslant \mathbf{0} \tag{10.4}$$

$$\mathbf{w}(\mathbf{z}^0) \leqslant \mathbf{0} \tag{10.5}$$

where

$$\mathbf{F} = (F_1, \ldots, F_N)^T, \mathbf{F}^0 = \mathbf{F}(\mathbf{z}^0)$$

$$\begin{aligned}\lambda(\mathbf{F}^0) &= [\lambda_1(\mathbf{F}^0), \lambda_2(\mathbf{F}^0), \ldots, \lambda_N(\mathbf{F}^0)]^T \\ &= [1, \lambda_{12}^*(\mathbf{F}^0), \ldots, \lambda_{1N}^*(\mathbf{F}^0)]^T\end{aligned}$$

$\boldsymbol{\rho}$ is a multiplier vector and $\lambda_{1k}(\mathbf{F}^0)$ is an indifference trade-off (or marginal rate of substitution, MRS) between F_1 and F_k ($k = 2, \ldots, N$) and its meaning is as follows (note that F_1 is assumed to be a reference objective function here):

$$\lambda_{1k}^* = \frac{\partial U/\partial F_k}{\partial U/\partial F_1} = -\frac{\partial F_1}{\partial F_k}\bigg|_{\mathbf{z}=\mathbf{z}^0, U:\text{constant}} \tag{10.6}$$

where U is a decisionmaker's underlying utility function.

Sufficient condition If all λ_{1k}^*'s ($k = 2, \ldots, N$) are strictly positive and the F_i's ($i = 1, \ldots, N$) are convex, then Eqs. (10.3)–(10.5) are also sufficient for $\mathbf{z}^0$ to be a locally preferred solution of the multiobjective problem represented by Eqs. (10.1) and (10.2). (For the proofs, see Tarvainen, 1981.)

In this chapter, we assume that all functions are concave (note that maximization problems will be considered) and that the λ_{1k}^*'s are strictly positive, so that Eqs. (10.3)–(10.5) are necessary and sufficient.

10.2 MULTIOBJECTIVE HIERARCHICAL OVERLAPPING COORDINATION (MHOC)

10.2.1 Problem Formulation

In this section, we formulate the mathematical model of the overall problem and decompose it into subsystems in various ways. The decomposition of the overall system in the multiobjective case is almost the same as that in the single-objective case that was described in Chap. 9. However, we will describe it carefully for intuitive reasons. For the sake of illustration, we will assume that the system has only two different hierarchical structures, namely, the α and the β decompositions. More than two decompositions are, of course, possible if the system has more than two hierarchical structures.

Suppose we have the following overall system:

$$\max_{\mathbf{x}} \begin{pmatrix} F_1(\mathbf{x}) \\ F_2(\mathbf{x}) \\ \vdots \\ F_N(\mathbf{x}) \end{pmatrix} \tag{10.7}$$

subject to

$$A\mathbf{x} = \mathbf{c} \tag{10.8}$$

$$\mathbf{x} \in S \tag{10.9}$$

where

F_i = a real-valued concave differentiable function defined on S
$\mathbf{x}$ = an $n \times 1$ decision vector
A = an $m \times n$ coefficient matrix
$\mathbf{c}$ = an $m \times 1$ fixed vector
S = an arbitrary subset of Euclidean space R^n; S may be $\{\mathbf{x}|\mathbf{x} \geqslant \mathbf{0}\}$ or $S = R^n$

Let F_i $(i = 1, \ldots, N)$ satisfy the following assumption:

$$F_i = F_i^{\alpha}(f_i^{\alpha 1}, f_i^{\alpha 2}, \ldots, f_i^{\alpha n_\alpha}) = F_i^{\beta}(f_i^{\beta 1}, f_i^{\beta 2}, \ldots, f_i^{\beta n_\beta}) \tag{10.10}$$

where the α (or β) decomposition is assumed to have n_α (or n_β) subsystems. $f_i^{\alpha_k}$ (or $f_i^{\beta_k}$) is an ith objective function of subsystem k in the α (or β) decomposition and is a real-valued concave differentiable function.

Suppose the system represented by Eq. (10.8) is equivalently represented as follows, which is accomplished by rearranging vectors $\mathbf{x}$ and $\mathbf{c}$ for each decomposition:

$$A_\alpha \mathbf{x}_\alpha = \mathbf{c}_\alpha \tag{10.11}$$

$$A_\beta \mathbf{x}_\beta = \mathbf{c}_\beta \tag{10.12}$$

where

$$\mathbf{x}_\alpha = P_\alpha \mathbf{x} \qquad \mathbf{x}_\beta = P_\beta \mathbf{x} \tag{10.13}$$

$$\mathbf{c}_\alpha = Q_\alpha \mathbf{c} \qquad \mathbf{c}_\beta = Q_\beta \mathbf{c} \tag{10.14}$$

$$A_\alpha = Q_\alpha A P_\alpha^{-1} \qquad A_\beta = Q_\beta A P_\beta^{-1} \tag{10.15}$$

The nonsingular P and Q matrices mathematically represent the rearrangement of vectors $\mathbf{x}$ and $\mathbf{c}$, as described in the previous chapter. $\mathbf{x}_\alpha$ and $\mathbf{x}_\beta$ are related by a nonsingular transformation, as was described before:

$$\mathbf{x}_\alpha = P_{\alpha\beta} \mathbf{x}_\beta \tag{10.16}$$

$$\mathbf{x}_\beta = P_{\beta\alpha} \mathbf{x}_\alpha \tag{10.17}$$

where

$$P_{\alpha\beta} = P_\alpha P_\beta^{-1} \tag{10.18}$$

$$P_{\beta\alpha} = P_\beta P_\alpha^{-1} \tag{10.19}$$

The system and its objectives, when viewed for the α decomposition, become

$$\max_{\mathbf{x}_\alpha} \begin{pmatrix} F_1^\alpha(f_1^{\alpha 1}, f_1^{\alpha 2}, \ldots, f_1^{\alpha n_\alpha}) \\ F_2^\alpha(f_2^{\alpha 1}, f_2^{\alpha 2}, \ldots, f_2^{\alpha n_\alpha}) \\ \vdots \\ F_N^\alpha(f_N^{\alpha 1}, f_N^{\alpha 2}, \ldots, f_N^{\alpha n_\alpha}) \end{pmatrix} \tag{10.20}$$

$$A_\alpha \mathbf{x}_\alpha = \mathbf{c}_\alpha \tag{10.21}$$

$$\mathbf{x}_\alpha \in S_\alpha \tag{10.22}$$

Likewise, the system and its objectives, when viewed for the β decomposition, become

$$\max_{\mathbf{x}_\beta} \begin{pmatrix} F_1^\beta(f_1^{\beta 1}, f_1^{\beta 2}, \ldots, f_1^{\beta n_\beta}) \\ F_2^\beta(f_2^{\beta 1}, f_2^{\beta 2}, \ldots, f_2^{\beta n_\beta}) \\ \vdots \\ F_N^\beta(f_N^{\beta 1}, f_N^{\beta 2}, \ldots, f_N^{\beta n_\beta}) \end{pmatrix} \tag{10.23}$$

$$A_\beta \mathbf{x}_\beta = \mathbf{c}_\beta \tag{10.24}$$

$$\mathbf{x}_\beta \in S_\beta \tag{10.25}$$

In the original overall system represented by Eqs. (10.7)–(10.9), the α and β problems are identical since both problems are obtained by rearranging the order of vectors $\mathbf{x}$ and $\mathbf{c}$.

The matrices A_α and A_β are assumed to be decomposable into the block diagonal matrices B_α, H_α, B_β, and H_β, as follows:

$$A_\alpha = B_\alpha + G_\alpha H_\alpha \tag{10.26}$$

$$A_\beta = B_\beta + G_\beta H_\beta \tag{10.27}$$

Then the system Eqs. (10.21) and (10.24) in the α and β decompositions become

$$B_\alpha \mathbf{x}_\alpha + G_\alpha \mathbf{y}_\alpha = \mathbf{c}_\alpha \tag{10.28}$$

$$H_\alpha \mathbf{x}_\alpha = \mathbf{y}_\alpha \tag{10.29}$$

$$B_\beta \mathbf{x}_\beta + G_\beta \mathbf{y}_\beta = \mathbf{c}_\beta \tag{10.30}$$

$$H_\beta \mathbf{x}_\beta = \mathbf{y}_\beta \tag{10.31}$$

where $\mathbf{y}_\alpha$ (or $\mathbf{y}_\beta$) is a vector of interaction or coupling variables, which are also partitioned into n_α (or n_β) subvectors $\mathbf{y}_{\alpha i}$ (or $\mathbf{y}_{\beta i}$) associated with the ith subsystem in the α (or β) decomposition.

It is assumed that matrices B_α, H_α, B_β, and H_β are block diagonal and are compatible with partitioning of the vector, so that the α and β problems can be rewritten as given below.

α Decomposition Problem:

$$\max_{\mathbf{x}_\alpha} \begin{pmatrix} F_1^\alpha(f_1^{\alpha 1}, \ldots, f_1^{\alpha n_\alpha}) \\ F_2^\alpha(f_2^{\alpha 1}, \ldots, f_2^{\alpha n_\alpha}) \\ \vdots \\ F_N^\alpha(f_N^{\alpha 1}, \ldots, f_N^{\alpha n_\alpha}) \end{pmatrix} \tag{10.32}$$

subject to

$$B_{\alpha i}\mathbf{x}_{\alpha i} + \sum_{j=1}^{n_\alpha} G_{\alpha ij}\mathbf{y}_{\alpha j} = \mathbf{c}_{\alpha i} \tag{10.33}$$

$$\mathbf{y}_{\alpha i} = H_{\alpha i}\mathbf{x}_{\alpha i} \tag{10.34}$$

$$\mathbf{x}_{\alpha i} \in S_{\alpha i} \tag{10.35}$$

$$i = 1, 2, \ldots, n_\alpha$$

β Decomposition Problem:

$$\max_{\mathbf{x}_\beta} \begin{pmatrix} F_1^\beta(f_1^{\beta 1}, \ldots, f_1^{\beta n_\beta}) \\ F_2^\beta(f_2^{\beta 1}, \ldots, f_2^{\beta n_\beta}) \\ \vdots \\ F_N^\beta(f_N^{\beta 1}, \ldots, f_N^{\beta n_\beta}) \end{pmatrix} \tag{10.36}$$

subject to

$$B_{\beta i}\mathbf{x}_{\beta i} + \sum_{j=1}^{n_\beta} G_{\beta ij}\mathbf{y}_{\beta j} = \mathbf{c}_{\beta i} \tag{10.37}$$

$$\mathbf{y}_{\beta i} = H_{\beta i}\mathbf{x}_{\beta i} \tag{10.38}$$

$$\mathbf{x}_{\beta i} \in S_{\beta i} \tag{10.39}$$

$$i = 1, 2, \ldots, n_\beta$$

where

$\mathbf{x}_{\alpha i}$ (or $\mathbf{x}_{\beta i}$) = a decision vector of the ith subsystem in the α (or β) decomposition

$\mathbf{x}_\alpha = (\mathbf{x}_{\alpha 1}^T, \mathbf{x}_{\alpha 2}^T, \ldots, \mathbf{x}_{\alpha n_\alpha}^T)^T$

$\mathbf{x}_\beta = (\mathbf{x}_{\beta 1}^T, \mathbf{x}_{\beta 2}^T, \ldots, \mathbf{x}_{\beta n_\beta}^T)^T$

$S_\alpha = S_{\alpha 1} \times S_{\alpha 2} \times \ldots \times S_{\alpha n_\alpha}$

$S_\beta = S_{\beta 1} \times S_{\beta 2} \times \ldots, \times S_{\beta n_\beta}$

$\mathbf{y}_{\alpha i}$ (or $\mathbf{y}_{\beta i}$) can be regarded as an "output" vector of the ith subsystem affecting other subsystems in the α (or β) decomposition, where

$$\mathbf{y}_\alpha = (\mathbf{y}_{\alpha 1}^T, \mathbf{y}_{\alpha 2}^T, \ldots, \mathbf{y}_{\alpha n_\alpha}^T)^T$$

$$\mathbf{y}_\beta = (\mathbf{y}_{\beta 1}^T, \mathbf{y}_{\beta 2}^T, \ldots, \mathbf{y}_{\beta n_\beta}^T)^T$$

For simplicity, we assume in most cases that the objective functions are additively separable and are represented as follows:

$$\begin{aligned} F_i(\mathbf{x}) &= F_i(P_\alpha^{-1}\mathbf{x}_\alpha) = F_i^\alpha(\mathbf{x}_\alpha, \mathbf{y}_\alpha) \\ &= \sum_{k=1}^{n_\alpha} f_i^{\alpha_k}(\mathbf{x}_{\alpha k}, \mathbf{y}_{\alpha 1}, \ldots, \mathbf{y}_{\alpha n_\alpha}) \qquad i = 1, \ldots, N \end{aligned} \tag{10.40}$$

and

$$\begin{aligned} F_i(\mathbf{x}) &= F_i(P_\beta^{-1}\mathbf{x}_\beta) = F_i^\beta(\mathbf{x}_\beta, \mathbf{y}_\beta) \\ &= \sum_{k=1}^{n_\beta} f_i^{\beta_k}(\mathbf{x}_{\beta k}, \mathbf{y}_{\beta 1}, \ldots, \mathbf{y}_{\beta n_\beta}) \qquad i = 1, \ldots, N \end{aligned} \tag{10.41}$$

The α and β problems are now decomposable into subproblems.

Suppose that the subproblem in the ith subsystem in each decomposition is represented as follows.

The i*th Subsystem in the α Decomposition (α*i *subsystem),* i = *1, . . .,* n_α*:*

$$\max_{\mathbf{x}_{\alpha i}} \begin{pmatrix} f_1^{\alpha i}(\mathbf{x}_{\alpha i}, \mathbf{y}_\alpha) \\ f_2^{\alpha i}(\mathbf{x}_{\alpha i}, \mathbf{y}_\alpha) \\ \vdots \\ f_N^{\alpha i}(\mathbf{x}_{\alpha i}, \mathbf{y}_\alpha) \end{pmatrix} \tag{10.42}$$

subject to

$$B_{\alpha i}\mathbf{x}_{\alpha i} + G_{\alpha i}\mathbf{y}_\alpha = \mathbf{c}_{\alpha i} \tag{10.43}$$

$$H_{\alpha i}\mathbf{x}_{\alpha i} = \mathbf{y}_{\alpha i} \tag{10.44}$$

$$\mathbf{x}_{\alpha i} \in S_{\alpha i} \tag{10.45}$$

The i*th Subsystem in the β Decomposition (β*i *subsystem),* i = *1, . . .,* n_β*:*

$$\max_{\mathbf{x}_{\beta i}} \begin{pmatrix} f_1^{\beta i}(\mathbf{x}_{\beta i}, \mathbf{y}_\beta) \\ f_2^{\beta i}(\mathbf{x}_{\beta i}, \mathbf{y}_\beta) \\ \vdots \\ f_N^{\beta i}(\mathbf{x}_{\beta i}, \mathbf{y}_\beta) \end{pmatrix} \tag{10.46}$$

subject to

$$B_{\beta i}\mathbf{x}_{\beta i} + G_{\beta i}\mathbf{y}_\beta = \mathbf{c}_{\beta i} \tag{10.47}$$

$$H_{\beta i}\mathbf{x}_{\beta i} = \mathbf{y}_{\beta i} \tag{10.48}$$

$$\mathbf{x}_{\beta i} \in S_{\beta i} \tag{10.49}$$

Note that

$$G_{\alpha i}\mathbf{y}_{\alpha} = \sum_{j=1}^{n_{\alpha}} G_{\alpha ij}\mathbf{y}_{\alpha j} \tag{10.50}$$

and

$$G_{\beta i}\mathbf{y}_{\beta} = \sum_{j=1}^{n_{\beta}} G_{\beta ij}\mathbf{y}_{\beta j} \tag{10.51}$$

The interaction vectors $\mathbf{y}_{\alpha}$ and $\mathbf{y}_{\beta}$ are given by the following relationships [see Eqs. (10.16), (10.17), (10.29), and (10.31)]:

$$\mathbf{y}_{\alpha} = H_{\alpha}\mathbf{x}_{\alpha} = H_{\alpha}P_{\alpha\beta}\mathbf{x}_{\beta} \tag{10.52}$$

$$\mathbf{y}_{\beta} = H_{\beta}\mathbf{x}_{\beta} = H_{\beta}P_{\beta\alpha}\mathbf{x}_{\alpha} \tag{10.53}$$

The overall system and the decomposed subsystems of the α and β decompositions are now represented as shown in Fig. 10.2.

10.2.2 Multiobjective Coordination Procedures

We will now consider coordination procedures for the system shown in Fig. 10.2. The multiobjective case differs from the single-objective case in the requirement of one or more decisionmakers who state preferences among objectives. As described before, several decisionmaking structures are possible. Here, we will use the following decisionmaking structures:

1. Top-down decisionmaking with a single decisionmaker in the center,
2. Hybrid decisionmaking, and
3. Top-down decisionmaking with multiple decisionmakers, each of whom represents one of the decompositions.

Bottom-up decisionmaking will not be considered here because the noninferiority of the solution of the overall problem cannot usually be satisfied if decisionmakers in the subsystems are the source for the preferences.

The basic idea of the top-down decisionmaking procedure was first proposed in Sung et al. (1981) and fully developed in subsequent papers such as Shima (1982) and Shima et al. (1982a, b).

10.2.2.1 Top-down decisionmaking: A single-decisionmaker case The term top-down decisionmaking implies that the decisionmaker(s) in the center state all the preferences for the whole system and the subsystems in the decompositions follow these preferences, as already described. In this section, we will assume that there exists a single decisionmaker in the center. The case of multiple decisionmakers in the center who share a common preference can be reduced to this case.

We will examine the case where there exists a decisionmaker who deals with the overall objectives. This decisionmaker bases his or her stated preferences on information, such as the values of objectives, that has been obtained by analyzing

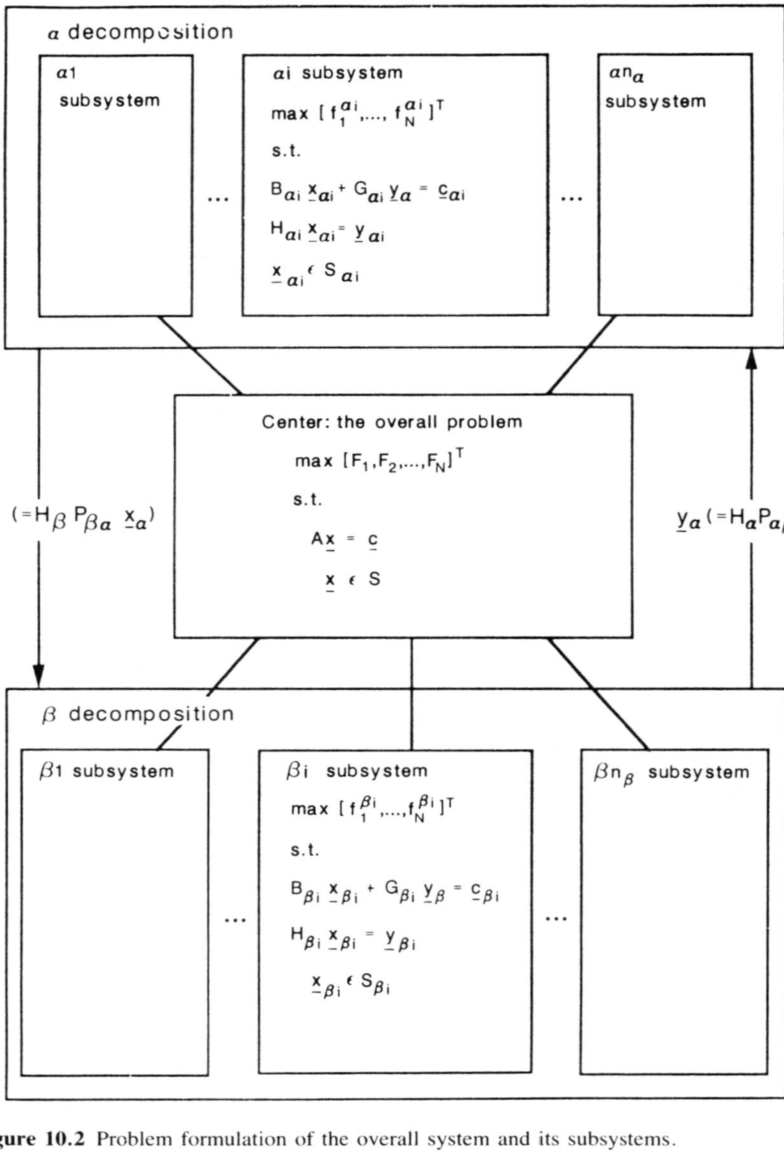

Figure 10.2 Problem formulation of the overall system and its subsystems.

another decomposition. Each decomposition is handled in this way, yielding a group of single-objective problems that are then solved in the subsystems.

Before explaining this coordination procedure, we will first consider the indifference trade-offs.

On the trade-offs Suppose $\boldsymbol{\lambda}(\mathbf{F}^0)$ is the indifference trade-off vector evaluated at $\mathbf{F}^0$ in the objective space, and it is given as follows:

$$\boldsymbol{\lambda}(\mathbf{F}^0) = [\lambda_1(\mathbf{F}^0), \lambda_2(\mathbf{F}^0), \ldots, \lambda_N(\mathbf{F}^0)]^T = \lambda_1(\mathbf{F}^0)[1, \lambda_{12}(\mathbf{F}^0), \ldots, \lambda_{1N}(\mathbf{F}^0)]^T \tag{10.54}$$

$$\boldsymbol{\lambda}(\mathbf{F})^0 = [1, \lambda_{12}(\mathbf{F}^0), \ldots, \lambda_{1N}(\mathbf{F}^0)]^T \tag{10.55}$$

where $\mathbf{F}^0 = (F_1^0, F_2^0, \ldots, F_N^0)^T$. $\lambda_1(\mathbf{F}^0) = 1$ is assumed. $\lambda_{1j}(\mathbf{F}^0)$ is the indifference trade-off between F_1 and F_j evaluated at $\mathbf{F}^0$. That is,

$$\lambda_{1j}(\mathbf{F}^0) = \frac{\partial U/\partial F_j}{\partial U/\partial F_1} = -\left.\frac{\partial F_1}{\partial F_j}\right|_{\mathbf{F}=\mathbf{F}^0}$$

Here F_1 is assumed to be the reference objective function. U is a decisionmaker's implicit utility function (see Geoffrion et al., 1972, and Sect. 5.4 concerning this indifference trade-off vector).

Since a single decisionmaker (or a group of several decisionmakers who have common preferences) makes the choices for the overall problem, he or she makes identical choices for the α and the β problems, which have earlier been defined to be identical. That is [note Eq. (10.10)],

$$\lambda_{1j}(\mathbf{F}^0) = \lambda_{1j}^{\alpha}(\mathbf{F}^{\alpha 0}) = \lambda_{1j}^{\beta}(\mathbf{F}^{\beta 0}) \tag{10.56}$$

and

$$\boldsymbol{\lambda}(\mathbf{F}^0) = \boldsymbol{\lambda}^{\alpha}(\mathbf{F}^{\alpha 0}) = \boldsymbol{\lambda}^{\beta}(\mathbf{F}^{\beta 0}) \tag{10.57}$$

which is at the preferred point $\mathbf{F}^0$ $(= \mathbf{F}^{\alpha 0} = \mathbf{F}^{\beta 0})$, where

$$\mathbf{F}^{\alpha} = (F_1^{\alpha}, \ldots, F_N^{\alpha})^T$$

$$\mathbf{F}^{\beta} = (F_1^{\beta}, \ldots, F_N^{\beta})^T$$

$\boldsymbol{\lambda}^{\alpha}(\mathbf{F}^{\alpha})$ [or $\boldsymbol{\lambda}^{\beta}(\mathbf{F}^{\beta})$] is an indifference trade-off vector for the α (or β) decomposition:

$$\begin{aligned}\boldsymbol{\lambda}^{\alpha}(\mathbf{F}^{\alpha}) &= [\lambda_1^{\alpha}(\mathbf{F}^{\alpha}), \lambda_2^{\alpha}(\mathbf{F}^{\alpha}), \ldots, \lambda_N^{\alpha}(\mathbf{F}^{\alpha})]^T \\ &= \lambda_1^{\alpha}(\mathbf{F}^{\alpha})[1, \lambda_{12}^{\alpha}(\mathbf{F}^{\alpha}), \ldots, \lambda_{1N}^{\alpha}(\mathbf{F}_1^{\alpha})]^T \\ &= [1, \lambda_{12}^{\alpha}(\mathbf{F}^{\alpha}), \ldots, \lambda_{1N}^{\alpha}(\mathbf{F}^{\alpha})]^T \\ \boldsymbol{\lambda}^{\beta}(\mathbf{F}^{\beta}) &= [\lambda_1^{\beta}(\mathbf{F}^{\beta}), \lambda_2^{\beta}(\mathbf{F}^{\beta}), \ldots, \lambda_N^{\beta}(\mathbf{F}^{\beta})]^T \\ &= \lambda_1^{\beta}(\mathbf{F}^{\beta})[1, \lambda_{12}^{\beta}(\mathbf{F}^{\beta}), \ldots, \lambda_{1N}^{\beta}(\mathbf{F}^{\beta})]^T \\ &= [1, \lambda_{12}^{\beta}(\mathbf{F}^{\beta}), \ldots, \lambda_{1N}^{\beta}(\mathbf{F}^{\beta})]^T\end{aligned}$$

where $\lambda_1\ (\mathbf{F}^{\alpha}) = 1$ and $\lambda_1\ (\mathbf{F}^{\beta}) = 1$ are assumed. Then, the internal trade-off vector of each subsystem in each decomposition is given as follows (see Sect. 5.4; Shima, 1982; Shima et al., 1982a, b):

$$\boldsymbol{\lambda}^{\alpha i}(\mathbf{f}^{\alpha}) = \left[\sum_{k=1}^{N} \lambda_k^{\alpha}(\mathbf{F}^{\alpha}) \frac{\partial F_k^{\alpha}(\mathbf{f}^{\alpha})}{\partial \mathbf{f}^{\alpha i}}\right]^T \tag{10.58}$$

$$\boldsymbol{\lambda}^{\beta i}(\mathbf{f}^{\beta}) = \left[\sum_{k=1}^{N} \lambda_k^{\beta}(\mathbf{F}^{\beta}) \frac{\partial F_k^{\beta}(\mathbf{f}^{\beta})}{\partial \mathbf{f}^{\beta i}}\right]^T \tag{10.59}$$

where

$\boldsymbol{\lambda}^{\alpha i}(\mathbf{f}^{\alpha})$ [or $\boldsymbol{\lambda}^{\beta i}(\mathbf{f}^{\beta})$] = an internal indifference trade-off vector of the ith subsystem in α (or β) decomposition

$\mathbf{f}^{\alpha} = (\mathbf{f}^{\alpha 1}, \mathbf{f}^{\alpha 2}, \ldots, \mathbf{f}^{\alpha n_\alpha})^T$

$\mathbf{f}^{\beta} = (\mathbf{f}^{\beta 1}, \mathbf{f}^{\beta 2}, \ldots, \mathbf{f}^{\beta n_\beta})^T$

$\mathbf{f}^{\alpha i} = (f_1^{\alpha i}, f_2^{\alpha i}, \ldots, f_N^{\alpha i})^T, \; i = 1, \ldots, n_\alpha$

$\mathbf{f}^{\beta i} = (f_1^{\beta i}, f_2^{\beta i}, \ldots, f_N^{\beta i})^T, \; i = 1, \ldots, n_\beta$

$\mathbf{f}^{\alpha i}$ (or $\mathbf{f}^{\beta i}$) = an objective vector of the ith subsystem in the α (or β) decomposition defined in $S_{\alpha i}$ (or $S_{\beta i}$)

Once the indifference trade-off vector $\boldsymbol{\lambda}(\mathbf{F})$ is given by the central decisionmaker, then the trade-off vectors $\boldsymbol{\lambda}^{\alpha i}(\mathbf{f}^{\alpha})$ ($i = 1, \ldots, n_\alpha$) and $\boldsymbol{\lambda}^{\beta i}(\mathbf{f}^{\beta})$ ($i = 1, \ldots, n_\beta$) of the subsystems in the α and β decompositions are determined uniquely by $\boldsymbol{\lambda}(\mathbf{F})$.

If the objective functions are of the general forms shown in Eq. (10.10), then the scaled internal trade-off vectors are defined as follows:

$$\boldsymbol{\lambda}_S^{\alpha i} = \frac{1}{\lambda_1^{\alpha i}(\mathbf{f}^{\alpha})} \boldsymbol{\lambda}^{\alpha i}(\mathbf{f}^{\alpha}) \qquad i = 1, \ldots, n_\alpha \tag{10.60}$$

$$\boldsymbol{\lambda}_S^{\beta i} = \frac{1}{\lambda_1^{\beta i}(\mathbf{f}^{\beta})} \boldsymbol{\lambda}^{\beta i}(\mathbf{f}^{\beta}) \qquad i = 1, \ldots, n_\beta \tag{10.61}$$

(The subscript S denotes the scaling.)

The trade-offs between the subsystems can also be defined as follows:

$$\lambda^{\alpha k/\alpha l}(\mathbf{f}^{\alpha}) = \frac{\lambda_1^{\alpha k}(\mathbf{f}^{\alpha})}{\lambda_1^{\alpha l}(\mathbf{f}^{\alpha})} \qquad k, l = 1, \ldots, n_\alpha \tag{10.62}$$

$$\lambda^{\beta k/\beta l}(\mathbf{f}^{\beta}) = \frac{\lambda_1^{\beta k}(\mathbf{f}^{\beta})}{\lambda_1^{\beta l}(\mathbf{f}^{\beta})} \qquad k, l = 1, \ldots, n_\beta \tag{10.63}$$

The necessary and sufficient conditions for the preferred solution have been developed in Tarvainen (1981) along with other theoretical work (e.g., on the convergence theorem). These were briefly explained in Sect. 10.1.2.

Overlapping coordination in top-down decisionmaking In the case where the decisionmaker deals with the overall objectives, the multiobjective HOC can be explained as follows.

1. First, the decisionmaker decides his indifference trade-off vector $\boldsymbol{\lambda}^{\alpha}(\mathbf{F}^{\alpha})$, basing it on the point obtained by the initial feasible solution, which is estimated arbitrarily. Set n equal to 1.

2. The decisionmaker gives his indifference trade-off vector $\boldsymbol{\lambda}^{\alpha}(\mathbf{F}^{\alpha})^{(n)}$ to the α subsystems. The trade-off vectors in the α subsystems are completely fixed by Eq. (10.58). The α manager in each α subsystem finds a solution $\mathbf{x}_{\alpha}^{(n)}$ by optimizing his subproblem, using the preferences given by the decisionmaker and the value $\mathbf{y}_{\alpha}^{(n)}$ of the interaction vector ($\mathbf{y}_{\alpha}^{(n)} = H_{\alpha}P_{\alpha\beta}\mathbf{x}_{\beta}^{(n-1)}$ for $n \geqslant 2$, $\mathbf{y}_{\alpha}^{(1)}$ is the feasible value that is initially estimated). The α managers send their proposal (such as the solution and the values of its objectives) to the decisionmaker. The α managers also send their proposal, which is the value of the interaction vector $\mathbf{y}_{\beta}^{(n)}$ of the β decomposition, to the β managers, using Eq. (10.53).
3. The decisionmaker decides now his indifference trade-off vector based on the point that has been obtained by solving the α decomposition problem.

   In order to find the indifference trade-off vector $\boldsymbol{\lambda}^{\beta}(\mathbf{F}^{\beta})^{(n)}$, the decisionmaker is asked: "How much in F_j exactly compensates for a change Δ_1 in F_1, with all other objectives remaining at their current values?" That is, we approximate the indifference trade-off $\lambda_{1j}^{\beta}(\mathbf{F}^{\beta})^{(n)}$ on the basis of the point ($F_1^{\alpha(n)}$, . . ., $F_N^{\alpha(n)}$) in the objective space. If the decisionmaker chooses Δ_j, then the indifference trade-off is approximated by

$$\lambda_{1j}^{\beta}(\mathbf{F}^{\beta})^{(n)} = -\frac{\Delta_1}{\Delta_j} \tag{10.64}$$

   [See Geoffrion et al., 1972, and Dyer, 1973, about how to find the indifference trade-off; also see Eq. (5.2).]
4. The decisionmaker gives his indifference trade-off vector $\boldsymbol{\lambda}^{\beta}(\mathbf{F}^{\beta})^{(n)}$ to the β subsystems. The trade-off vectors in the β subsystems are completely fixed by Eq. (10.59). The β manager in each β subsystem optimizes his subproblem, using the preference given by the decisionmaker and the value $\mathbf{y}_{\beta}^{(n)}$ of the interaction vector ($\mathbf{y}_{\beta}^{(n)} = H_{\beta}P_{\beta\alpha}\mathbf{x}_{\alpha}^{(n)}$), and finds a solution $\mathbf{x}_{\beta}^{(n)}$. The β managers send their proposals to the decisionmaker and the α managers.
5. The decisionmaker evaluates the indifference trade-off vector based on the information given by the β managers. Set $n = n + 1$. Then go to step 2. These procedures are repeated until the terminal conditions are satisfied.

The process terminates when the improvement from one stage to the next is less than a prespecified small number ϵ. That is,

$$|F_i^{\alpha}(\mathbf{x}_{\alpha}^{(n+1)}, \mathbf{y}_{\alpha}^{(n+1)}) - F_i^{\alpha}(\mathbf{x}_{\alpha}^{(n)}, \mathbf{y}_{\alpha}^{(n)})| < \epsilon \qquad \text{for all } i \qquad i = 1, \ldots, N \tag{10.65}$$

and

$$|F_i^{\beta}(\mathbf{x}_{\beta}^{(n+1)}, \mathbf{y}_{\beta}^{(n+1)}) - F_i^{\beta}(\mathbf{x}_{\beta}^{(n)}, \mathbf{y}_{\beta}^{(n)})| < \epsilon \qquad \text{for all } i \qquad i = 1, \ldots, N \tag{10.66}$$

$\mathbf{x}^{(n)}$ means the value of the vector at iteration n. This also implies that

$$|\lambda_{1j}^{\alpha(n+1)} - \lambda_{1j}^{\beta(n+1)}| < \epsilon' \qquad \text{for } j = 2, \ldots, N \tag{10.67}$$

where ϵ' is a prespecified small number.

The coordination process is summarized in Fig. 10.3. The structure of the coordination procedure is also described in Figs. 10.4*a* and *b*.

After reaching convergence, the preferred solution is obtained. The value of

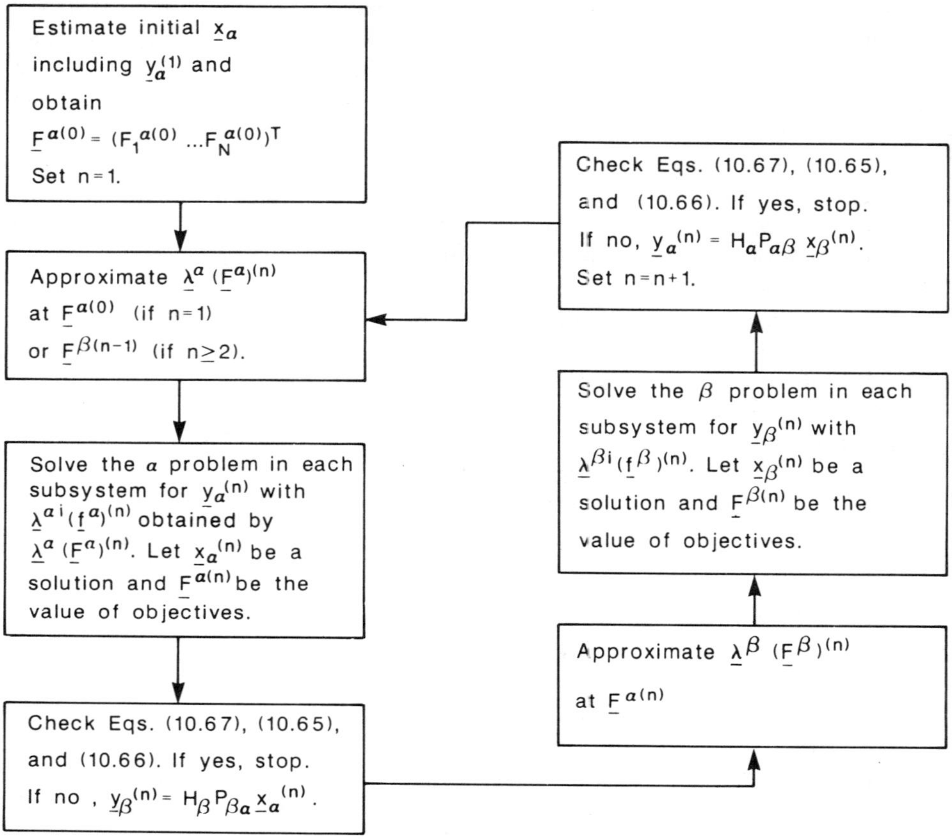

Figure 10.3 Coordination process in top-down decisionmaking.

the indifference trade-off vector $\boldsymbol{\lambda}(\mathbf{F})$ [$= \boldsymbol{\lambda}^\alpha(\mathbf{F}^\alpha) = \boldsymbol{\lambda}^\beta(\mathbf{F}^\beta)$] and the values of the trade-off vectors $\boldsymbol{\lambda}^{\alpha i}(\mathbf{f}^\alpha)$ $(i = 1, \ldots, n_\alpha)$ and $\boldsymbol{\lambda}^{\beta i}(\mathbf{f}^\beta)$ $(i = 1, \ldots, n_\beta)$ are also obtained. The values of the objective function $F_i = F_i^\alpha = F_i^\beta$ $(i = 1, \ldots, N)$ and the values of the vector objective functions $\mathbf{f}^{\alpha i}$ $(i = 1, \ldots, n_\alpha)$ and $\mathbf{f}^{\beta i}$ $(i = 1, \ldots, n_\beta)$ of the subsystems in the α and β decompositions are also obtained.

This approach is computationally very attractive, but there are restrictions that the objective functions must be concave and at least one objective function (e.g., a reference objective function F_1) must be strictly concave. In the subsystems, for example, the reference objective functions $f_1^{\alpha i}$ and $f_1^{\beta j}$ $(i = 1, \ldots, n_\alpha;\ j = 1, \ldots, n_\beta)$ must also be strictly concave.

Some notes on real decisionmaking In the top-down decisionmaking discussed in the previous section, only one noninferior solution (which is a preferred solution) was generated. In real-life decisionmaking, however, decisionmakers might feel uncomfortable about basing their decision on only one noninferior solution, even if it is obtained through several interactions with the analyst, so that it could be called a preferred solution. The decisionmakers may want to have more information in the neighborhood of this solution, and they might want to know what results would be obtained if they would change their preferences slightly. This is the sort of situation that might occur in real decisionmaking. To resolve such a problem, it is better to have several other noninferior solutions in the neighborhood of the obtained solution.

One method of generating several noninferior solutions is for the decisionmaker to change her preferences slightly and solve the problem each time she

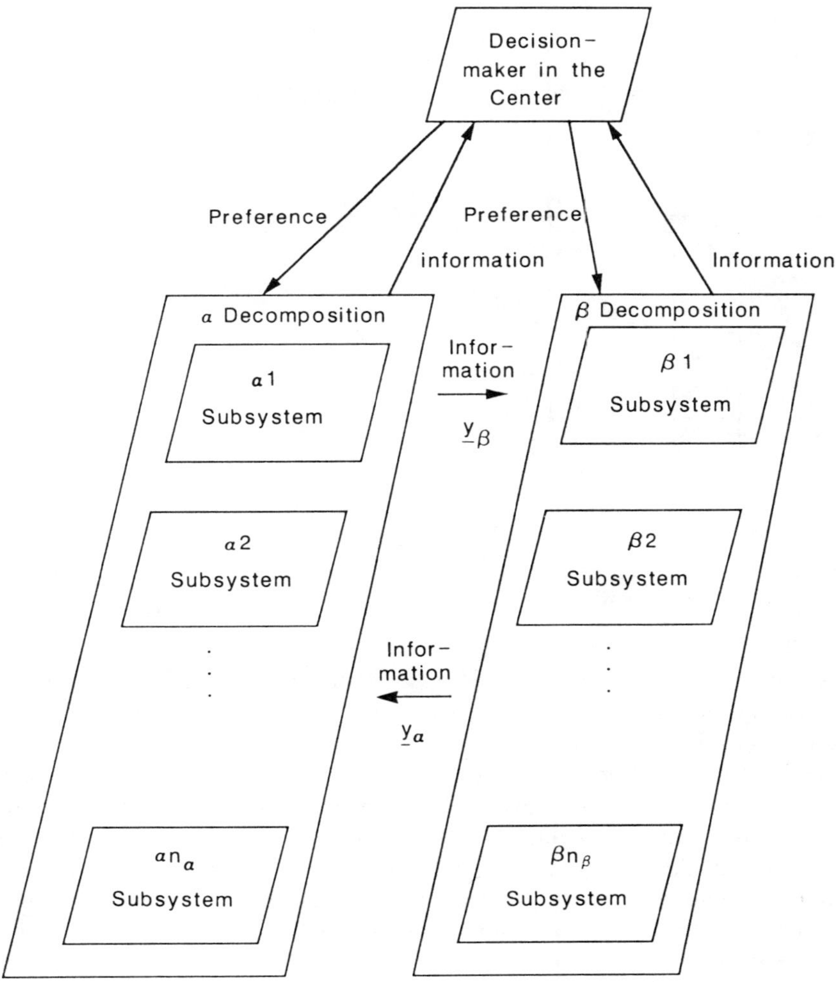

Figure 10.4 (*a*) Information structure of multiobjective hierarchical overlapping coordination.

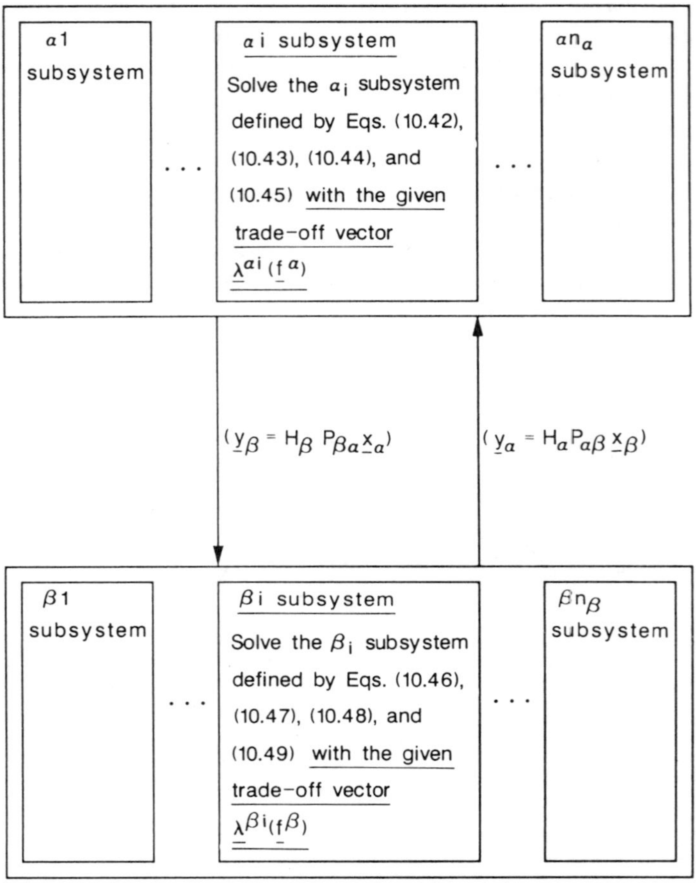

Figure 10.4 (*b*) Structure of the coordination procedure in top-down decisionmaking in the case where a decisionmaker deals with the overall objectives.

makes a change. If several noninferior solutions are obtained in the region of interest to the decisionmaker, she will feel more comfortable with her decision, because these solutions are in the neighborhood of her preferred solution. In order to reach her final decision (that is, her final preferred solution), the SWT method (for details, see Chap. 2; Haimes et al., 1975; Haimes, 1977) and/or the interactive SWT (Chankong and Haimes, 1978) can be applied. The indifference band of the decisionmaker's trade-offs is obtained, and the decisionmaker can choose her final indifference trade-off vector $\boldsymbol{\lambda}^*$, based on this information. If the decisionmaker can feel comfortable about the final indifference trade-off vector $\boldsymbol{\lambda}^*$, she will also be happier with her final preferred solution, which is based on it. Real-life decisionmaking demands an approach that generates several options (as discussed in this section) rather than a single preferred solution.

10.2.2.2 Hybrid decisionmaking In the previous section, the decisionmaker made choices based on the values of the overall objectives that were obtained by the managers in the subsystems. That is, the central decisionmakers make the decisions on the trade-offs.

In this section, we consider the case where both central decisionmakers and the managers (subdecisionmakers) in the subsystems of the α and β decompositions together state their preferences. As previously defined, this decisionmaking structure is called hybrid decisionmaking.

Conflict among decisionmakers may occur. The central decisionmaker is concerned with the overall problem and the managers in the α and β decompositions are concerned with their own subproblems. If the managers (subdecisionmakers) in the subsystems state preferences independently, the preferred solutions for the subproblems may be obtained. However, it is rarely possible to maintain the noninferiority of the overall problem. The coordination between the α and β decompositions may not be achieved, either. In order to maintain the noninferiority for both overall the problem and subproblems, some relationships (which we call the coordination equation) works as the coordinator. The central decisionmaker must make a final decision if the preferences among the subdecisionmakers do not converge, that is, the relationships are not satisfied. In this sense, the central decisionmaker is a leader and the subdecisionmakers are followers.

Here, we first derive these relationships (called the coordination equation) and then explain the hybrid overlapping coordination procedure.

Derivation of the coordination equation Let the overall problem be represented as follows:

$$\max\ [F_1(\mathbf{x}), F_2(\mathbf{x}), \ldots, F_N(\mathbf{x})]^T$$

subject to $A\mathbf{x} = \mathbf{c}$.

Note that we have equality constraints here and we ignore Eq. (10.9) ($\mathbf{x} \in S$) for simplicity, because it is not necessary to derive the coordination relationships. As described in the previous sections, we have the following assumptions:

Assumption 1:

$$F_k = F_k^{\alpha}(f_k^{\alpha 1}, \ldots, f_k^{\alpha n_\alpha}) = F_k^{\beta}\,(f_k^{\beta 1}, \ldots, f_k^{\beta n_\beta})$$

for $\forall k = 1, 2, \ldots, N$. That is, F_k is a function of $f_k^{\alpha i}$ or $f_k^{\beta i}$ only and not the function of $f_j^{\alpha i}$ or $f_j^{\beta i}$ $(j \neq k)$.

Assumption 2:

$$F_k = F_k^{\alpha} = F_k^{\beta}$$

$f_k^{\alpha i}$ and $f_k^{\beta j}$ are differentiable concave functions for $k = 1, \ldots, N$, $i = 1, \ldots, n_\alpha$, and $j = 1, \ldots, n_\beta$.

Likewise, the α decomposition problem is represented as follows:

$$\max_{\mathbf{x}_\alpha} [F_1^\alpha(f_1^{\alpha 1}, \ldots, f_1^{\alpha n_\alpha}), \ldots, F_N^\alpha(f_N^{\alpha 1}, \ldots, f_N^{\alpha n_\alpha})]^T \tag{10.68}$$

subject to

$$B_{\alpha i}\mathbf{x}_{\alpha i} + G_{\alpha i}\mathbf{y}_\alpha = \mathbf{C}_{\alpha i} \tag{10.69}$$

$$H_{\alpha i}\mathbf{x}_{\alpha i} = \mathbf{y}_{\alpha i} \tag{10.70}$$

for $i = 1, \ldots, n_\alpha$.

Then, from the conditions described in Sect. 10.1.2 (see also Chap. 5 or Tarvainen, 1981), we have the following property.

Property 10.1 Let every element of $\boldsymbol{\lambda}^\alpha(\mathbf{F}^{\alpha 0})$ be strictly positive. $\mathbf{z}_\alpha^0$ is a preferred solution of the α decomposition problem given in Eqs. (10.68)–(10.70) for the proposed (feasible and fixed) value of $\mathbf{y}_\alpha$ if and only if there exist vectors $\boldsymbol{\mu}_\alpha$ and $\boldsymbol{\sigma}_\alpha$ which, together with $\mathbf{z}_\alpha^0$, satisfy Eqs. (10.71)–(10.73). Note that in these equations, $\mathbf{S}(\mathbf{z}_\alpha^0, \mathbf{y}_\alpha) = \mathbf{0}$, $\mathbf{T}(\mathbf{z}_\alpha^0, \mathbf{y}_\alpha) = \mathbf{0}$, $\mathbf{S}_{\alpha i}(\mathbf{z}_{\alpha i}^0, \mathbf{y}_\alpha) = \mathbf{0}$, and $\mathbf{T}_{\alpha i}(\mathbf{z}_{\alpha i}^0, \mathbf{y}_\alpha) = \mathbf{0}$ correspond to Eqs. (10.28), (10.29), (10.69), and (10.70), respectively; $\mathbf{S}_\alpha = (\mathbf{S}_{\alpha 1}^T, \ldots, \mathbf{S}_{\alpha n_\alpha}^T)^T$, $\mathbf{T} = (\mathbf{T}_{\alpha 1}^T, \ldots, \mathbf{T}_{\alpha n_\alpha}^T)^T$; $\boldsymbol{\mu}_\alpha$ and $\boldsymbol{\sigma}_\alpha$ are the Lagrange multiplier vectors, each of which corresponds to each condition, respectively; $\boldsymbol{\mu}_\alpha = (\boldsymbol{\mu}_{\alpha 1}^T, \ldots, \boldsymbol{\mu}_{\alpha n_\alpha}^T)^T$; and $\boldsymbol{\sigma}_\alpha = (\boldsymbol{\sigma}_{\alpha 1}^T, \ldots, \boldsymbol{\sigma}_{\alpha n_\alpha}^T)^T$.

$$\boldsymbol{\lambda}^\alpha(\mathbf{F}^{\alpha 0})^T \frac{\partial \mathbf{F}^\alpha(\mathbf{z}_\alpha^0, \mathbf{y}_\alpha)}{\partial \mathbf{z}_\alpha} + \boldsymbol{\mu}_\alpha^T \frac{\partial \mathbf{S}(\mathbf{z}_\alpha^0, \mathbf{y}_\alpha)}{\partial \mathbf{z}_\alpha} + \boldsymbol{\sigma}_\alpha^T \frac{\partial \mathbf{T}(\mathbf{z}_\alpha^0, \mathbf{y}_\alpha)}{\partial \mathbf{z}_\alpha} = \mathbf{0} \tag{10.71}$$

$$\mathbf{S}_{\alpha i}(\mathbf{z}_{\alpha i}^0, \mathbf{y}_\alpha) = \mathbf{0} \tag{10.72}$$

$$\mathbf{T}_{\alpha i}(\mathbf{z}_{\alpha i}^0, \mathbf{y}_\alpha) = \mathbf{0} \tag{10.73}$$

for $i = 1, \ldots, n_\alpha$.

Here,

$$\{\mathbf{x}_\alpha\} = \{\mathbf{z}_\alpha, \mathbf{y}_\alpha\} = \{\mathbf{z}_\alpha, H_\alpha P_\alpha \mathbf{x}\} = \{\mathbf{z}_\alpha, H_\alpha P_{\alpha\beta}\mathbf{x}_\beta\}$$

where $\{\cdot\}$ implies a set.

$$\mathbf{F}^{\alpha 0} = \mathbf{F}^\alpha(\mathbf{z}_\alpha^0, \mathbf{y}_\alpha)$$

$$\boldsymbol{\lambda}^\alpha(\mathbf{F}^\alpha) = [\lambda_1^\alpha(\mathbf{F}^\alpha), \ldots, \lambda_N^\alpha(\mathbf{F}^\alpha)]^T = \lambda_1^\alpha(1, \lambda_{12}^\alpha, \ldots, \lambda_{1N}^\alpha)^T = (1, \lambda_{12}, \ldots, \lambda_{1N})^T$$

$$\mathbf{z}_\alpha = (\mathbf{z}_{\alpha 1}, \ldots, \mathbf{z}_{\alpha n_\alpha})^T$$

Here, $\lambda_1^\alpha = 1$ is assumed for simplicity, and $\mathbf{z}_\alpha$ corresponds to the decision vectors except those corresponding to $\mathbf{y}_\alpha$; that is, $\mathbf{z}_\alpha$ is a decision vector that is not fixed in the α decomposition.

PROOF The proof holds from the conditions in 10.1.2. Note that the only difference here is that we have equality constraints and some fixed (feasible) variables.

Now,

$$(\boldsymbol{\lambda}^{\alpha})^T \frac{\partial \mathbf{F}^{\alpha}}{\partial \mathbf{z}_{\alpha}} = \sum_{k=1}^{N} \lambda_k^{\alpha} \frac{\partial F_k^{\alpha}}{\partial \mathbf{z}_{\alpha}} = \sum_{i=1}^{n_\alpha} \sum_{k=1}^{N} \lambda_k^{\alpha} \frac{\partial F_k^{\alpha}}{\partial f_k^{\alpha i}} \frac{\partial f_k^{\alpha i}}{\partial \mathbf{z}_{\alpha i}}$$

$$\boldsymbol{\mu}_{\alpha}^T \frac{\partial \mathbf{S}}{\partial \mathbf{z}_{\alpha}} = \sum_{i=1}^{n_\alpha} \boldsymbol{\mu}_{\alpha i}^T \frac{\partial \mathbf{S}_{\alpha i}}{\partial \mathbf{z}_{\alpha i}}, \quad \boldsymbol{\sigma}_{\alpha}^T \frac{\partial \mathbf{T}}{\partial \mathbf{z}_{\alpha}} = \sum_{i=1}^{n_\alpha} \boldsymbol{\sigma}_{\alpha i}^T \frac{\partial \mathbf{T}_{\alpha i}}{\partial \mathbf{z}_{\alpha i}}$$

Hence, Eq. (10.71) is

$$\sum_{i=1}^{n_\alpha} \left(\sum_{k=1}^{N} \lambda_k^{\alpha} \frac{\partial F_k^{\alpha}}{\partial f_k^{\alpha i}} \frac{\partial f_k^{\alpha i}}{\partial \mathbf{z}_{\alpha i}} + \boldsymbol{\mu}_{\alpha i}^T \frac{\partial \mathbf{S}_{\alpha i}}{\partial \mathbf{z}_{\alpha i}} + \boldsymbol{\sigma}_{\alpha i}^T \frac{\partial \mathbf{T}_{\alpha i}}{\partial \mathbf{z}_{\alpha i}} \right)$$

$$= \sum_{i=1}^{n_\alpha} \left(\lambda_1^{\alpha} \frac{\partial F_1^{\alpha}}{\partial f_1^{\alpha i}} \frac{\partial f_1^{\alpha i}}{\partial \mathbf{z}_{\alpha i}} + \lambda_2^{\alpha} \frac{\partial F_2^{\alpha}}{\partial f_2^{\alpha i}} \frac{\partial f_2^{\alpha i}}{\partial \mathbf{z}_{\alpha i}} + \cdots + \lambda_N^{\alpha} \frac{\partial F_N^{\alpha}}{\partial f_N^{\alpha i}} \frac{\partial f_N^{\alpha i}}{\partial \mathbf{z}_{\alpha i}} \right.$$

$$\left. + \boldsymbol{\mu}_{\alpha i}^T \frac{\partial \mathbf{S}_{\alpha i}}{\partial \mathbf{z}_{\alpha i}} + \boldsymbol{\sigma}_{\alpha i}^T \frac{\partial \mathbf{T}_{\alpha i}}{\partial \mathbf{z}_{\alpha i}} \right) = \sum_{i=1}^{n_\alpha} \left(\frac{\partial F_1^{\alpha}}{\partial f_1^{\alpha i}} \frac{\partial f_1^{\alpha i}}{\partial \mathbf{z}_{\alpha i}} + \lambda_{12}^{\alpha} \frac{\partial F_2^{\alpha}}{\partial f_2^{\alpha i}} \frac{\partial f_2^{\alpha i}}{\partial \mathbf{z}_{\alpha i}} + \cdots \right.$$

$$\left. + \lambda_{1N}^{\alpha} \frac{\partial F_N^{\alpha}}{\partial f_N^{\alpha i}} \frac{\partial f_N^{\alpha i}}{\partial \mathbf{z}_{\alpha i}} + \boldsymbol{\mu}_{\alpha i}^T \frac{\partial \mathbf{S}_{\alpha i}}{\partial \mathbf{z}_{\alpha i}} + \boldsymbol{\sigma}_{\alpha i}^T \frac{\partial \mathbf{T}_{\alpha i}}{\partial \mathbf{z}_{\alpha i}} \right)$$

$$= \sum_{i=1}^{n_\alpha} \frac{\partial F_1^{\alpha}}{\partial f_1^{\alpha i}} \left[\frac{\partial f_1^{\alpha i}}{\partial \mathbf{z}_{\alpha i}} + \lambda_{12}^{\alpha} \frac{\partial F_2^{\alpha}}{\partial f_2^{\alpha i}} \left(\frac{\partial F_1^{\alpha}}{\partial f_1^{\alpha i}} \right)^{-1} \frac{\partial f_2^{\alpha i}}{\partial \mathbf{z}_{\alpha i}} + \cdots \right.$$

$$\left. + \lambda_{1N}^{\alpha} \frac{\partial F_N^{\alpha}}{\partial f_N^{\alpha i}} \left(\frac{\partial F_1^{\alpha}}{\partial f_1^{\alpha i}} \right)^{-1} \frac{\partial f_N^{\alpha i}}{\partial \mathbf{z}_{\alpha i}} + \hat{\boldsymbol{\mu}}_{\alpha i}^T \frac{\partial \mathbf{S}_{\alpha i}}{\partial \mathbf{z}_{\alpha i}} + \hat{\boldsymbol{\sigma}}_{\alpha i}^T \frac{\partial \mathbf{T}_{\alpha i}}{\partial \mathbf{z}_{\alpha i}} \right]$$

$$= \sum_{i=1}^{n_\alpha} \frac{\partial F_1^{\alpha}}{\partial f_1^{\alpha i}} \left(\frac{\partial f_1^{\alpha i}}{\partial \mathbf{z}_{\alpha i}} + \sum_{k=2}^{N} \lambda_{1k}^{\alpha i} \frac{\partial f_k^{\alpha i}}{\partial \mathbf{z}_{\alpha i}} + \hat{\boldsymbol{\mu}}_{\alpha i}^T \frac{\partial \mathbf{S}_{\alpha i}}{\partial \mathbf{z}_{\alpha i}} + \hat{\boldsymbol{\sigma}}_{\alpha i}^T \frac{\partial \mathbf{T}_{\alpha i}}{\partial \mathbf{z}_{\alpha i}} \right) = \mathbf{0}$$

where

$$\lambda_{1k}^{\alpha i} = \lambda_{1k}^{\alpha} \frac{\partial F_k^{\alpha}}{\partial f_k^{\alpha i}} \left(\frac{\partial F_1^{\alpha}}{\partial f_1^{\alpha i}} \right)^{-1}$$

$$\hat{\boldsymbol{\sigma}}_{\alpha i} = \boldsymbol{\sigma}_{\alpha i} \left(\frac{\partial F_1^{\alpha}}{\partial f_1^{\alpha i}} \right)^{-1}$$

$$\hat{\boldsymbol{\mu}}_{\alpha i} = \boldsymbol{\mu}_{\alpha i} \left(\frac{\partial F_1^{\alpha}}{\partial f_1^{\alpha i}} \right)^{-1}$$

$$\mathbf{z}_{\alpha} = (\mathbf{z}_{\alpha 1}^T, \mathbf{z}_{\alpha 2}^T, \ldots, \mathbf{z}_{\alpha n_\alpha}^T)^T$$

We assume that the inverse always exists.

We will summarize the above discussion in Property 10.2.

Property 10.2 Let $\mathbf{z}^0_{\alpha i}$ satisfy the following conditions:

$$\frac{\partial f_1^{\alpha i}(\mathbf{z}^0_{\alpha_i}, \mathbf{y}_\alpha)}{\partial \mathbf{z}_{\alpha i}} + \sum_{k=2}^{N} \lambda_{1k}^{\alpha i} \frac{\partial f_k^{\alpha i}(z^0_{\alpha i}, y_\alpha)}{\partial \mathbf{z}_{\alpha i}} + \hat{\boldsymbol{\mu}}_{\alpha i} \frac{\partial \mathbf{S}_{\alpha i}(\mathbf{z}^0_{\alpha_i}, \mathbf{y}_\alpha)}{\partial \mathbf{z}_{\alpha_i}} + \hat{\boldsymbol{\sigma}}_{\alpha i} \frac{\partial \mathbf{T}_{\alpha i}(\mathbf{z}^0_{\alpha_i}, \mathbf{y}_\alpha)}{\partial \mathbf{z}_{\alpha_i}} = \mathbf{0} \quad (10.74)$$

$$\mathbf{S}_{\alpha i}(\mathbf{z}^0_{\alpha i}, \mathbf{y}_\alpha) = \mathbf{0} \quad (10.75)$$

$$\mathbf{T}_{\alpha i}(\mathbf{z}^0_{\alpha i}, \mathbf{y}_\alpha) = \mathbf{0} \quad (10.76)$$

for $i = 1, \ldots, n_\alpha$. Then, $\mathbf{z}^0_\alpha = (\mathbf{z}^0_{\alpha 1}, \ldots, \mathbf{z}^0_{\alpha n_\alpha})^T$ satisfies the conditions in Eqs. (10.71)–(10.73). The converse is also true.

Proof The proof follows from the above discussion. Equation (10.71) is clearly satisfied if Eq. (10.74) is satisfied.

Equations (10.75) and (10.76) are exactly the same as Eqs. (10.72) and (10.73). And $\{\mathbf{z}_{\alpha i}\} \cap \{\mathbf{z}_{\alpha j}\} = \phi$ for $i \neq j$; hence, the converse is also true.

Now, the conditions in (10.74), (10.75), and (10.76) imply that $\mathbf{z}^0_{\alpha i}$ satisfies Property 10.3.

Property 10.3 Let $\lambda_{1k}^{\alpha i}$ $(k = 2, \ldots, N)$ be strictly positive. $\{\mathbf{z}^0_{\alpha i}, \mathbf{y}_{\alpha i}\} = \{\mathbf{x}^0_{\alpha i}\}$ is a preferred solution of the following αi subproblem for the proposed (fixed) value of $\mathbf{y}_\alpha$ if and only if there exist vectors $\hat{\boldsymbol{\mu}}_{\alpha i}$ and $\hat{\boldsymbol{\sigma}}_{\alpha i}$ which, together with $\mathbf{z}^0_{\alpha i}$, satisfy the conditions in Eqs. (10.74)–(10.76).

$$\max\ (f_1^{\alpha i}, f_2^{\alpha i}, \ldots, f_N^{\alpha i})^T$$

subject to

$$B_{\alpha i}\mathbf{x}_{\alpha i} + G_{\alpha i}\mathbf{y}_\alpha = \mathbf{c}_{\alpha i} \qquad H_{\alpha i}\mathbf{x}_{\alpha i} = \mathbf{y}_{\alpha i} \qquad i = 1, \ldots, n_\alpha$$

Proof The proof is clear from the conditions in 10.1.2, Property 10.2, and the above discussion.

Finally, we have Property 10.4 as a summary of the above discussion.

Property 10.4 If $\{\mathbf{z}^0_{\alpha i}, \mathbf{y}_{\alpha i}\} = \{\mathbf{x}^0_{\alpha i}\}$ is a preferred solution of the αi subproblem $(i = 1, \ldots, n_\alpha)$ for the proposed (fixed and feasible) value of $\mathbf{y}_\alpha$, then $\{\mathbf{z}^0_\alpha, \mathbf{y}_\alpha\} = \{\mathbf{x}^0_\alpha\}$ is a preferred solution of the problem in Eqs. (10.68)–(10.70) for the same proposed value of $\mathbf{y}_\alpha$. The converse is also true, where

$$\lambda_{1k}^{\alpha i} = \lambda_{1k}^{\alpha} \frac{\partial F_k^\alpha}{\partial f_k^{\alpha i}} \left(\frac{\partial F_1^\alpha}{\partial f_1^{\alpha i}}\right)^{-1} \qquad \begin{array}{l} i = 1, \ldots, n_\alpha \\ k = 2, \ldots, N \end{array} \quad (10.77)$$

λ_{1k}^α and $\lambda_{1k}^{\alpha i}$ are strictly positive; $\{\mathbf{x}_\alpha\} = \{\mathbf{x}_{\alpha i};\ i = 1, \ldots, n_\alpha\}$, $\{\mathbf{z}_\alpha\} = \{\mathbf{z}_{\alpha i};\ i = 1, \ldots, n_\alpha\}$.

PROOF The proof is also clear from the above properties.

From Eq. (10.77),

$$\lambda_{1k}^{\alpha} = \lambda_{1k}^{\alpha 1} \frac{\partial F_1^{\alpha}}{\partial f_1^{\alpha 1}} \left(\frac{\partial F_k^{\alpha}}{\partial f_k^{\alpha 1}} \right)^{-1} = \cdots = \lambda_{1k}^{\alpha n_\alpha} \frac{\partial F_1^{\alpha}}{\partial f_1^{\alpha n_\alpha}} \left(\frac{\partial F_k^{\alpha}}{\partial f_k^{\alpha n_\alpha}} \right)^{-1}$$

$$\text{for } k = 2, \ldots, N \tag{10.78a}$$

This is the α part of the coordination relationships. The coordination equation of the β decomposition can be derived in the same way. The β part of the coordination equation is

$$\lambda_{1k}^{\beta} = \lambda_{1k}^{\beta 1} \frac{\partial F_1^{\beta}}{\partial f_1^{\beta 1}} \left(\frac{\partial F_k^{\beta}}{\partial f_k^{\beta 1}} \right)^{-1} = \cdots = \lambda_{1k}^{\beta n_\beta} \frac{\partial F_1^{\beta}}{\partial f_1^{\beta n_\beta}} \left(\frac{\partial F_k^{\beta}}{\partial f_k^{\beta n_\beta}} \right)^{-1}$$

$$\text{for } k = 2, \ldots, N \tag{10.78b}$$

The α and β problems are essentially the same as the original overall problem, so the convergence of the preferences can be obtained if $\lambda_{1k} = \lambda_{1k}^{\alpha} = \lambda_{1k}^{\beta}$ holds, that is, if the next coordination relationships that follow are satisfied:

$$\begin{aligned} \lambda_{1k} &= \lambda_{1k}^{\alpha 1} \frac{\partial F_1^{\alpha}}{\partial f_1^{\alpha 1}} \left(\frac{\partial F_k^{\alpha}}{\partial f_k^{\alpha 1}} \right)^{-1} = \cdots = \lambda_{1k}^{\alpha n_\alpha} \frac{\partial F_1^{\alpha}}{\partial f_1^{\alpha n_\alpha}} \left(\frac{\partial F_k^{\alpha}}{\partial f_k^{\alpha n_\alpha}} \right)^{-1} \\ &= \lambda_{1k}^{\beta 1} \frac{\partial F_1^{\beta}}{\partial f_1^{\beta 1}} \left(\frac{\partial F_k^{\beta}}{\partial f_k^{\beta 1}} \right)^{-1} = \cdots = \lambda_{1k}^{\beta n_\beta} \frac{\partial F_1^{\beta}}{\partial f_1^{\beta n_\beta}} \left(\frac{\partial F_k^{\beta}}{\partial f_k^{\beta n_\beta}} \right)^{-1} \end{aligned}$$

$$\text{for } k = 2, \ldots, N \tag{10.78c}$$

We call Eq. (10.78c) the coordination equation. These relationships are depicted in Fig. 10.5 in the case where we have only two objectives, F_1 and F_2.

If the values of the trade-offs given by the decisionmaker and the subdecisionmakers of each decomposition satisfy Eq. (10.78c) and the interaction vectors $\mathbf{y}_\alpha$ and $\mathbf{y}_\beta$ converge to the solutions in the noninferior set through information exchange, then the preferred solution for the overall problem in the center and the subsystems of each decomposition can be obtained in noninferior sets for each corresponding problem.

Coordination procedure From the discussion in the previous section, what we have to achieve by the coordination of the center and the subsystems in each decomposition are the convergence of the preferences among the decisionmakers in the center and subsystems and the convergence of the exchanged information $\mathbf{y}_\alpha$ and $\mathbf{y}_\beta$ to the solutions. This section considers the coordination procedures that are designed to accomplish these aims.

Mainly, two possible approaches can be considered. The first is to apply the conventional overlapping coordination. That is, the problem is first solved in each subsystem of the α (or β) decomposition, based on the feasible value of the fixed information $\mathbf{y}_\alpha$ ($\mathbf{y}_\beta$). The preferences of the central decisionmaker and subdeci-

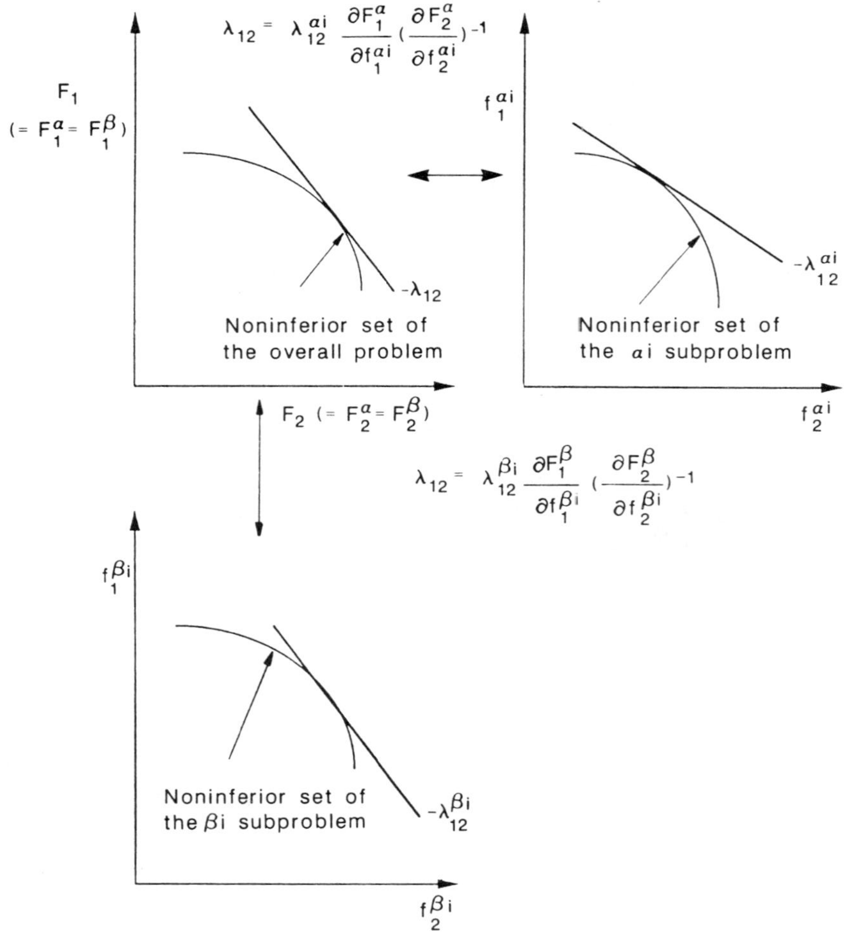

Figure 10.5 Relationships represented by the coordination equation.

sionmakers in the α (or β) decomposition are coordinated so that the relationships given in Eq. (10.78a) [or Eq. (10.78b)] are satisfied, and then the solution of the α (or β) decomposition at this step is obtained, based on the obtained preference. The information $\mathbf{y}_\beta$ (or $\mathbf{y}_\alpha$) is sent from the α (or β) decomposition to the β (or α) decomposition, and the problem is solved again in the β (or α) decomposition. The same procedures are repeated until convergence of the preferences [Eq. (10.78c)] and convergence of the exchanged information $\mathbf{y}_\alpha$ and $\mathbf{y}_\beta$ are obtained.

The second approach does not use the conventional overlapping coordination. The problem is solved in the center, so we do not have to worry about convergence of $\mathbf{y}_\alpha$ and $\mathbf{y}_\beta$, because the information ($\mathbf{y}_\alpha$ and $\mathbf{y}_\beta$) that they provide is in the noninferior set and is always given from the center to the subsystems in each decomposition. Information is exchanged between the center and the α decomposition and between the center and the β decomposition, but it is not exchanged between the α and the β decompositions. The set of the noninferior solutions, the corresponding trade-offs, and the values of the objectives make up the information that is sent from the center to the subsystems in each decomposition, for example, through the decision support system (DSS) networks, as will be described in a later section. The decisionmaker in the center and subdecisionmakers of each decomposition negotiate in order for the relationships in Eq. (10.78c) to be satisfied so that the compromise solution can be found, based on the exchanged information.

We are considering the second method in this section because the first method, though it is very interesting, may have the same convergence problem as the overlapping coordination. It may be applicable to weakly coupled problems (see Macko and Haimes, 1978; Shima and Haimes 1984). It may be especially difficult to obtain convergence of the preferences and the interaction vectors $\mathbf{y}_\alpha$ and $\mathbf{y}_\beta$ by information exchange between the α and β decomposition, and it may be too much of a burden for the decisionmakers. In the second method, we have to worry only about the convergence of the preferences among the decisionmakers. The weak point of the second method may be that it requires a large computer if the overall problem is large and if an appropriate decomposition method for computational purposes is not available. However, the coordination may be easily achieved and decisionmakers may acquire deep insight into the problems and conflicts that are part of the problem that is under consideration.

Next, we will describe steps of the procedure.

1. The following ϵ-constraint (Haimes et al., 1971) overall problem is solved in the center:

$$\max F_1(\mathbf{x})$$

subject to

$$F_2(\mathbf{x}) \geq \epsilon_2^i$$

$$\vdots$$

$$F_N(\mathbf{x}) \geq \epsilon_N^i$$

$$A\mathbf{x} = \mathbf{c} \qquad \mathbf{x} \in S$$

where

$$i = 1, \ldots, I \ (i \text{ is the iteration number})$$
$$\boldsymbol{\epsilon}^i = (\epsilon_2^i, \ldots, \epsilon_N^i)^T$$

The necessary information (the set of noninferior solutions, etc.) is obtained by changing the value of $\boldsymbol{\epsilon}^i$ slightly. Here, we assume the trade-offs are obtained as the Kuhn-Tucker multipliers of the ϵ constraints (for details, see Haimes and Chankong, 1979; Chankong and Haimes, 1983).

2. The center sends the obtained information to the subsystems in the α and β decompositions by using the following relationships:

$$\mathbf{x}_\alpha = P_\alpha \mathbf{x} \qquad \mathbf{x}_\beta = P_\beta \mathbf{x} \qquad \mathbf{y}_\alpha = H_\alpha P_\alpha \mathbf{x} \qquad \mathbf{y}_\beta = H_\beta P_\beta \mathbf{x}$$

$$\lambda_{1k}^{\alpha i} = \lambda_{1k} \frac{\partial F_k^\alpha}{\partial f_k^{\alpha i}} \left(\frac{\partial F_1^\alpha}{\partial f_1^{\alpha i}} \right)^{-1} \qquad \lambda_{1k}^{\beta_i} = \lambda_{1k} \frac{\partial F_k^\beta}{\partial f_k^{\beta i}} \left(\frac{\partial F_1^\beta}{\partial f_1^{\beta i}} \right)^{-1}$$

The information given to each subsystem is the noninferior solution for each corresponding subproblem, and the information necessary for decisionmaking (the set of noninferior solutions of the subproblems, the corresponding trade-offs, the values of the objectives, etc.) can be given to each subdecisionmaker in each subsystem. This information is exchanged through the decision support system network (described in the next section).

3. The central decisionmaker and each subdecisionmaker find the indifference band of trade-offs for their own problems. (This step is the same as the one in the SWT method; see Haimes, 1977.)
4. The obtained indifference band of each subsystem is changed to the level of the indifference band of the center by using the following relationships:

$$\hat{\lambda}_{1k}^{\alpha_i} = \lambda_{1k}^{\alpha i} \frac{\partial F_1^\alpha}{\partial f_1^{\alpha i}} \left(\frac{\partial F_k^\alpha}{\partial f_k^{\alpha i}} \right)^{-1} \qquad \hat{\lambda}_{1k}^{\beta j} = \lambda_{1k}^{\beta j} \frac{\partial F_1^\beta}{\partial f_1^{\beta j}} \left(\frac{\partial F_k^\beta}{\partial f_k^{\beta j}} \right)^{-1} \qquad \begin{aligned} i &= 1, \ldots, n_\alpha \\ j &= 1, \ldots, n_\beta \\ k &= 2, \ldots, N \end{aligned}$$

The indifference bands of the trade-offs of the subsystems are sent to the center, and indifference trade-offs are obtained from the common band of trade-offs by negotiation (see Fig. 10.6, for example). This step is the same as the one in the SWT method in the multiple decisionmakers case (see Hall and Haimes 1976). Finally, the compromise solution is obtained by solving the problem with these indifference trade-offs (see Haimes, 1977). Of course, there may exist a case where no common band of preferences is obtained (see Fig. 10.6*b*). Even in that case, we may apply the methods described in Hall and Haimes (1976) or other methods such as minimizing each decisionmaker's dissatisfaction. In any case, decisionmakers must negotiate and the central decisionmaker is responsible for the final agreement on the trade-offs.

Comments on the hybrid decisionmaking Computer technology has advanced rapidly and it is now possible to construct the large decision support systems on the large-scale computer networks. On the other hand, the organizations have also become larger and larger and now many organizations are widely distributed around the world. Typical examples of such organizations are multinational corporations.

Decisionmaking in such large-scale systems is very complex and many aspects must be considered to make an acceptable decision. (Even in rather small systems, more effective decisions can be made if many aspects are taken into account.)

As an example, let us think of the very simple structure of an industrial corporation which has a production system and a sales system. Sales information,

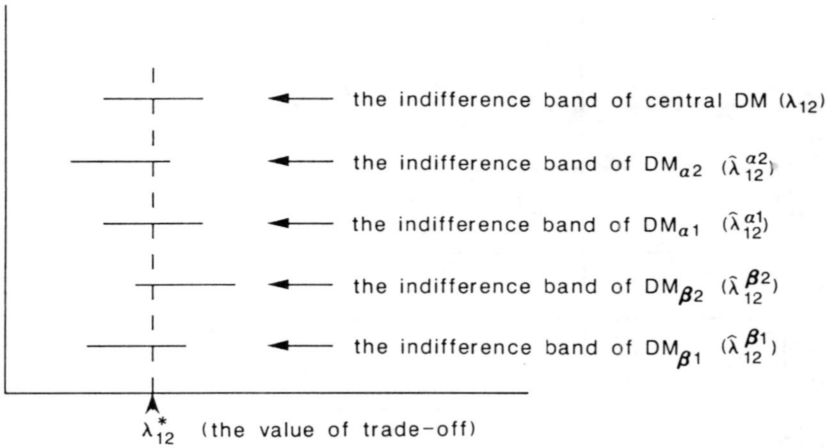

(*a*)

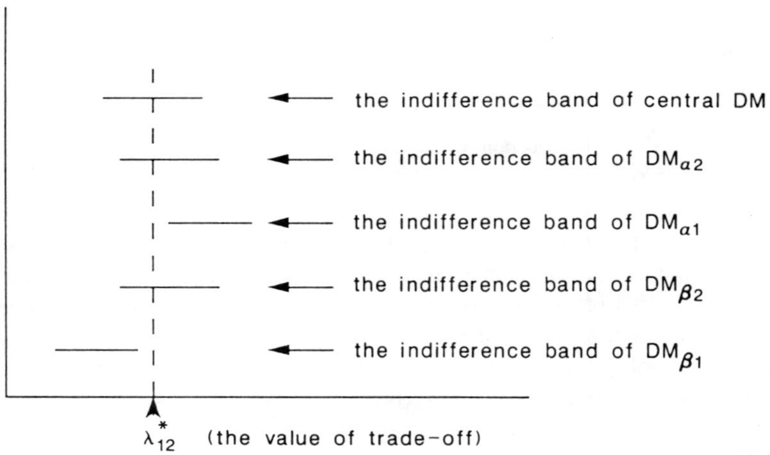

(*b*)

Figure 10.6 (*a*) Ideal case in the simple example, where $n_\alpha = 2$, $n_\beta = 2$, $k = 2$. (*b*) Probable case.

such as consumers' demands and expectations, is obtained by the sales system, and manufacturing information, such as prices of materials and production costs, is obtained by the production systems. The coordination of these different aspects of the system can bring better performance to the systems.

The integrated decisionmaking structure is depicted in matrix form in Fig. 10.7. In large-scale corporations, several subsystems exist in each aspect of the system. The physical structure of large industrial corporations can be depicted as shown in Fig. 10.8.

Now, it is possible to connect the center (headquarters) with the subsystems (plants of the production system and local branches of the sales system) by com-

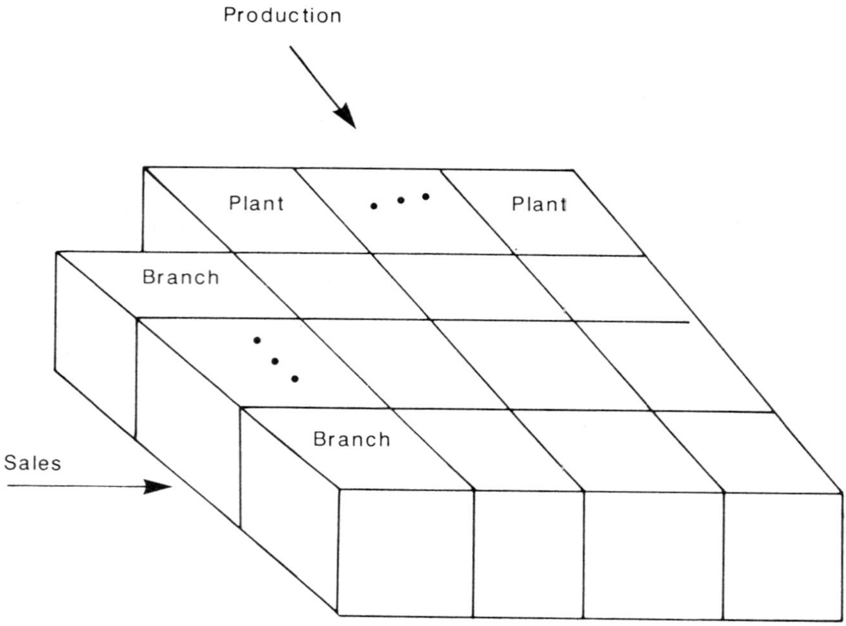

Figure 10.7 Matrix structure of the production division.

puter network. A decision support system can be constructed on these large computer networks and the conceptual structure of such large-scale decision support systems is also shown in Fig. 10.8. To make efficient and effective production plans, for example, many aspects must be considered as described before. These large-scale decision support systems are really helpful for such purposes. Of course, not only the production plans but also many tactical or operational plans can be created effectively and efficiently on these large-scale decision networks.

In world-wide distributed systems, such as multinational corporations, managers cannot often meet one another, so communication through these decision support networks is very helpful because the basic plans must be created, modified, and changed many times until the final plan is reached.

A very important issue is that the different hierarchies (decompositions) of such large systems must be coordinated to create efficient and effective decision-making processes through these decision support networks. The hybrid decision-making coordination procedure that was described in the previous section can be implemented on these computer networks for some planning problems. The overall problem may be formulated in the company headquarters with the help of the analyst, and subproblems are sent to the subsystems of each decomposition through the computer network. Then the coordination procedure can be applied to get agreement. Efficient and effective plans for the central office and the subsystems may be created through the computer network. The above example considered a multinational industrial corporation. However, public (governmental) systems can also be considered in the same framework.

The idea of implementing overlapping coordination on large decision support networks is now in its infancy stage and direct implementation may be ideal at this stage. Further research is necessary and should be an interesting area of research. We hope that our approach will also stimulate other work in the area of this issue.

10.2.2.3 Test for the preferred solution This section considers the convergence properties of the multiobjective hierarchical overlapping coordination (MHOC) for the top-down procedure. The convergence properties of a single-objective HOC for problems with equality constraints were studied first by Macko and Haimes

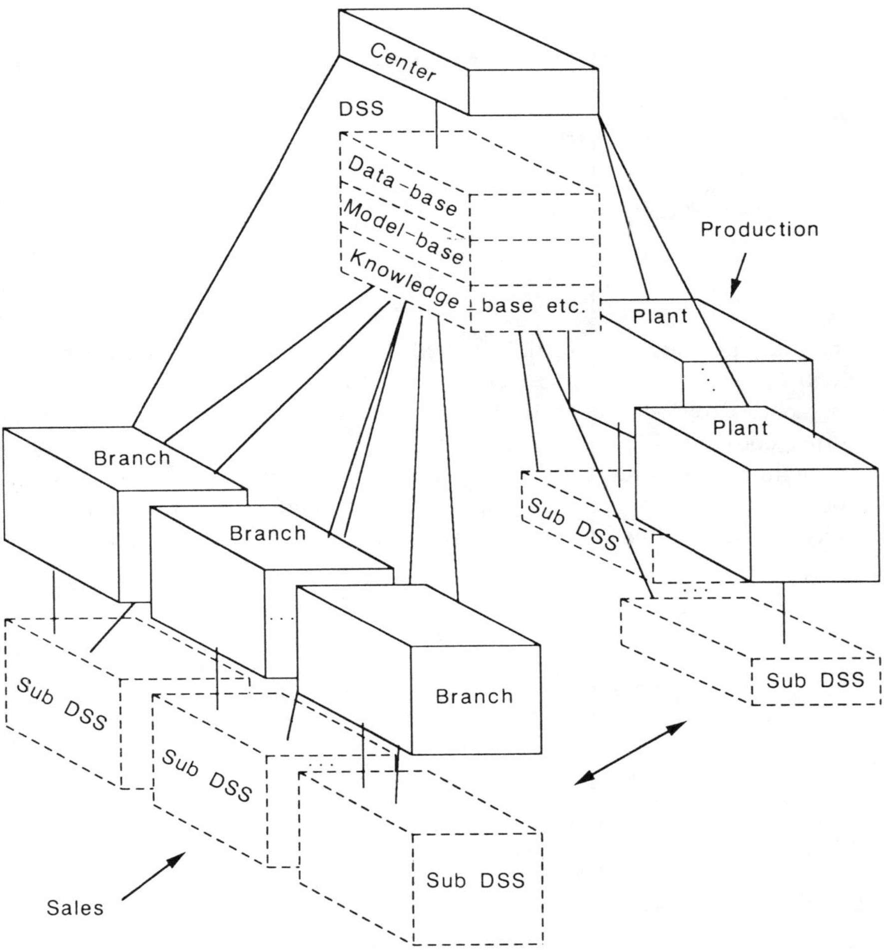

Figure 10.8 Physical structure of system and decision support system (DSS).

(1978), and later by Shima and Haimes (1984), not only for problems with equality constraints but also for problems with linear inequality constraints.

The convergence properties of the single-objective HOC are all applicable to the MHOC, because they depend on the structure of the constraints. The convergence properties of HOC (see details in Macko and Haimes, 1978; Shima and Haimes, 1984; and Chap. 9 of this book) will not be discussed here, but we will briefly discuss the test for optimality (a test for the preferred solution, in this case) of the top-down procedure of the HMOC.

The necessary conditions for a preferred solution to the multiobjective problem were considered in Tarvainen (1981) and Tarvainen and Haimes (1980, 1982), and were briefly reviewed in Sect. 10.1.2. The necessary conditions for $\mathbf{x}^0$, which is a regular point, to be a preferred point for the problem represented by Eqs. (10.7)–(10.9) are that there exists a vector of multipliers $\boldsymbol{\mu}$ such that

$$\boldsymbol{\lambda}^T(\mathbf{F}^0)\frac{\partial \mathbf{F}(\mathbf{x}^0)}{\partial \mathbf{x}} + \boldsymbol{\mu}^T A \leqslant 0 \tag{10.79}$$

$$\left[\boldsymbol{\lambda}^T(\mathbf{F}^0)\frac{\partial \mathbf{F}(\mathbf{x}^0)}{\partial \mathbf{x}} + \boldsymbol{\mu}^T A\right]\mathbf{x} = 0 \tag{10.80}$$

where $\mathbf{x} \geqslant 0$, $A\mathbf{x} = \mathbf{c}$, and we assume that

$$S = \{\mathbf{x} | \mathbf{x} \geqslant 0\}$$

$$\mathbf{F} = (F_1, F_2, \ldots, F_N)^T$$

These conditions are also sufficient if all components of $\boldsymbol{\lambda}(\mathbf{F}^0)$ are strictly positive and F_i is concave, as described in Sect. 10.1.2.

Let $\mathbf{x}_{\alpha y}$ be a decision vector corresponding to $\mathbf{y}_\alpha$, and let $\mathbf{x}_{\beta y}$ be a decision vector corresponding to $\mathbf{y}_\beta$. Since, in the α decomposition $\mathbf{y}_\beta$ is fixed, the α problem is not optimized with respect to $\mathbf{x}_{\alpha y}$. Since $\mathbf{y}_\alpha$ is fixed in the β decomposition, the β problem is not optimized with respect to $\mathbf{x}_{\beta y}$. Therefore, the following conditions may not be satisfied in the α and β decompositions, respectively:

$$\begin{gathered}\left(\boldsymbol{\lambda}^{\alpha^T}(\mathbf{F}^\alpha)\frac{\partial \mathbf{F}^\alpha(\mathbf{x}_\alpha^0)}{\partial \mathbf{x}_\alpha} + \boldsymbol{\mu}^{\alpha^T} A_\alpha\right)_{\alpha y} \leqslant \mathbf{0} \\ \left(\boldsymbol{\lambda}^{\alpha^T}(\mathbf{F}^\alpha)\frac{\partial \mathbf{F}^\alpha(\mathbf{x}_\alpha^0)}{\partial \mathbf{x}_\alpha} + \boldsymbol{\mu}^{\alpha^T} A_\alpha\right)_{\alpha y} \mathbf{x}_{\alpha y} = \mathbf{0}\end{gathered} \tag{10.81}$$

and

$$\begin{gathered}\left(\boldsymbol{\lambda}^{\beta^T}(\mathbf{F}^\beta)\frac{\partial \mathbf{F}^\beta(\mathbf{x}_\beta^0)}{\partial \mathbf{x}_\beta} + \boldsymbol{\mu}^{\beta^T} A_\beta\right)_{\beta y} \leqslant \mathbf{0} \\ \left(\boldsymbol{\lambda}^{\beta^T}(\mathbf{F}^\beta)\frac{\partial \mathbf{F}^\beta(\mathbf{x}_\beta^0)}{\partial \mathbf{x}_\beta} + \boldsymbol{\mu}^{\beta^T} A_\beta\right)_{\beta y} \mathbf{x}_{\beta y} = \mathbf{0}\end{gathered} \tag{10.82}$$

where $(\cdot)_{\alpha y}$ and $(\cdot)_{\beta y}$ imply conditions or coefficients corresponding to $\mathbf{x}_{\alpha y}$ and $\mathbf{x}_{\beta y}$.

Recall that $\mathbf{F} = \mathbf{F}^\alpha = \mathbf{F}^\beta$ and $\boldsymbol{\lambda}(\mathbf{F}) = \boldsymbol{\lambda}^\alpha(\mathbf{F}^\alpha) = \boldsymbol{\lambda}^\beta(\mathbf{F}^\beta)$, and also that the α, the β, and the original problems are essentially the same, so that these have essentially the same conditions for a preferred solution.

We can use these conditions as a test for the preferred solution. After the coordination process is terminated, we will check if these conditions are satisfied. The reason we must check is that a point that is not preferred is sometimes obtained because of the cycling problem of HOC. Convergence to a preferred point depends on the structure of the coefficient matrix A, and we recommend using HOC for cases where the subsystems are "weakly" coupled (see details in Shima and Haimes, 1984).

10.2.2.4 Example 10.1 Example 10.1 and its continuation in a later section as Example 10.2 constitute the only numerical example given in this chapter. The application problems of MHOC are considered in the next chapter. Example 10.1 will be solved using the top-down decisionmaking procedure.

Suppose the overall problem is represented as follows:

$$\max \begin{pmatrix} F_1 = 132 - (x_1 - 8)^2 - x_2^2 - (x_3 - 6)^2 - x_4^2 - x_5^2 \\ F_2 = 2x_1 + 5x_2 + x_3 + 4x_4 + 6x_5 \end{pmatrix} \qquad \begin{matrix} (10.83) \\ (10.84) \end{matrix}$$

subject to

$$x_1 + 2x_2 + 4x_5 \leqslant 8 \qquad (10.85)$$

$$x_2 \leqslant 8 \qquad (10.86)$$

$$x_3 + 3x_4 \leqslant 12 \qquad (10.87)$$

$$x_3 + x_4 \leqslant 6 \qquad (10.88)$$

$$x_3 + 2x_5 \leqslant 6 \qquad (10.89)$$

$$x_i \geqslant 0 \qquad i = 1, \ldots, 5$$

Suppose the following subsystems of the α and β decompositions exist.

α Decomposition

$\alpha 1$ Subsystem:

$$\max \begin{pmatrix} f_1^{\alpha 1} = 80 - (x_1 - 8)^2 - x_2^2 - x_5^2 \\ f_2^{\alpha 1} = 2x_1 + 5x_2 + 6x_5 \end{pmatrix} \qquad \begin{matrix} (10.90) \\ (10.91) \end{matrix}$$

subject to

$$x_1 + 2x_2 + 4x_5 \leqslant 8 \qquad (10.92)$$

$$x_2 \leqslant 8 \qquad (10.93)$$

$$2x_5 \leqslant 6 - y_\alpha \qquad (10.94)$$

$$x_i \geqslant 0 \qquad i = 1, 2, 5$$

α2 Subsystem:

$$\max\begin{pmatrix} f_1^{\alpha 2} = \frac{1}{2}(52 - (x_3 - 6)^2 - x_4^2) \\ f_2^{\alpha 2} = x_3 + 4x_4 \end{pmatrix} \qquad (10.95),\ (10.96)$$

subject to

$$x_3 + 3x_4 \leq 12 \qquad (10.97)$$

$$x_3 + x_4 \leq 6 \qquad (10.98)$$

$$x_3 = y_\alpha \qquad (10.99)$$

$$x_i \geq 0 \quad i = 3, 4$$

β Decomposition

β1 Subsystem:

$$\max\begin{pmatrix} f_1^{\beta 1} = \frac{1}{2}(80 - (\bar{x}_1 - 8)^2 - \bar{x}_2^2) \\ f_2^{\beta 1} = 2\bar{x}_1 + 5\bar{x}_2 \end{pmatrix} \qquad (10.100),\ (10.101)$$

subject to

$$\bar{x}_1 + 2\bar{x}_2 \leq 8 - 4y_\beta \qquad (10.102)$$

$$\bar{x}_2 \leq 8 \qquad (10.103)$$

$$\bar{x}_i \geq 0 \qquad i = 1, 2$$

β2 Subsystem:

$$\max\begin{pmatrix} f_1^{\beta 2} = 52 - (\bar{x}_3 - 6)^2 - \bar{x}_4^2 - \bar{x}_5^2 \\ f_2^{\beta 2} = \frac{1}{2}(\bar{x}_3 + 4\bar{x}_4 + 6\bar{x}_5) \end{pmatrix} \qquad (10.104),\ (10.105)$$

subject to

$$\bar{x}_3 + 3\bar{x}_4 \leq 12 \qquad (10.106)$$

$$\bar{x}_3 + \bar{x}_4 \leq 6 \qquad (10.107)$$

$$\bar{x}_3 + 2\bar{x}_5 \leq 6 \qquad (10.108)$$

$$\bar{x}_5 = y_\beta \qquad (10.109)$$

$$\bar{x}_i \geq 0 \qquad i = 3, 4, 5$$

where $\bar{x}_i$ means the variable in the β decomposition. Note that

$$F_1 = f_1^{\alpha 1} + 2f_1^{\alpha 2} = 2f_1^{\beta 1} + f_1^{\beta 2} \qquad (10.110)$$

$$F_2 = f_2^{\alpha 1} + f_2^{\alpha 2} = f_2^{\beta 1} + 2f_2^{\beta 2} \qquad (10.111)$$

During the computation, y_α and y_β are given by

$$y_\alpha = H_\alpha P_{\alpha\beta}\mathbf{x}_\beta = \bar{x}_3 \qquad (10.112)$$

$$y_\beta = H_\beta P_{\beta\alpha}\mathbf{x}_\alpha = x_5 \tag{10.113}$$

The entire structure of this problem is shown in Fig. 10.9.

For illustrative purposes, let us assume that the decisionmaker has the following utility function:

$$U(F_1, F_2) = \mathbf{R}_1\mathbf{F} + \mathbf{F}^T R_2 \mathbf{F} = F_1 + aF_2 + \frac{b}{2}F_2^2 \tag{10.114}$$

where

$$\mathbf{R}_1 = [1, a]$$

$$R_2 = \begin{bmatrix} 0 & 0 \\ 0 & \frac{b}{2} \end{bmatrix}$$

$$\mathbf{F} = [F_1, F_2]^T$$

Then the indifference trade-off vector can be represented as follows:

$$\boldsymbol{\lambda}(\mathbf{F}) = [\lambda_1(\mathbf{F}), \lambda_2(\mathbf{F})]^T = [1, \lambda_{12}(\mathbf{F})]^T = [1, a + bF_2]^T \tag{10.115}$$

For the purpose of computation, we assume the existence of a mathematical decisionmaker who makes the above choice. However, we should note that in real decisionmaking, the human decisionmaker interacts with the systems analyst and makes his or her choices known after each iteration, when an α or β decomposition problem has been solved. This was described in the previous section. The mathematical decisionmaker is assumed here only to demonstrate the decisionmaking process.

According to the methodologies given under "On the trade-off" in Sect. 10.2.2.1,

$$\begin{aligned} \boldsymbol{\lambda}^\alpha(\mathbf{F}^\alpha) &= [\lambda_1^\alpha(\mathbf{F}^\alpha), \lambda_2^\alpha(\mathbf{F}^\alpha)]^T \\ &= [1, \lambda_{12}^\alpha(\mathbf{F}^\alpha)]^T \\ &= [1, a + bF_2^\alpha]^T \end{aligned} \tag{10.116}$$

$$\begin{aligned} \boldsymbol{\lambda}^\beta(\mathbf{F}^\beta) &= [\lambda_1^\beta(\mathbf{F}^\beta), \lambda_2^\beta(\mathbf{F}^\beta)]^T \\ &= [1, \lambda_{12}^\beta(\mathbf{F}^\beta)]^T \\ &= [1, a + bF_2^\beta]^T \end{aligned} \tag{10.117}$$

where $F_1 = F_1^\alpha = F_1^\beta$ and $F_2 = F_2^\alpha = F_2^\beta$.

According to Eqs. (10.58) and (10.59), the trade-off vector in each subsystem is given as follows:

$$\boldsymbol{\lambda}^{\alpha 1} = \begin{pmatrix} \lambda_1^{\alpha 1} \\ \lambda_2^{\alpha 1} \end{pmatrix} = \lambda_1^{\alpha 1}\begin{pmatrix} 1 \\ \lambda_{12}^{\alpha 1} \end{pmatrix} = \begin{pmatrix} 1 \\ a + bF_2^\alpha \end{pmatrix} \tag{10.118}$$

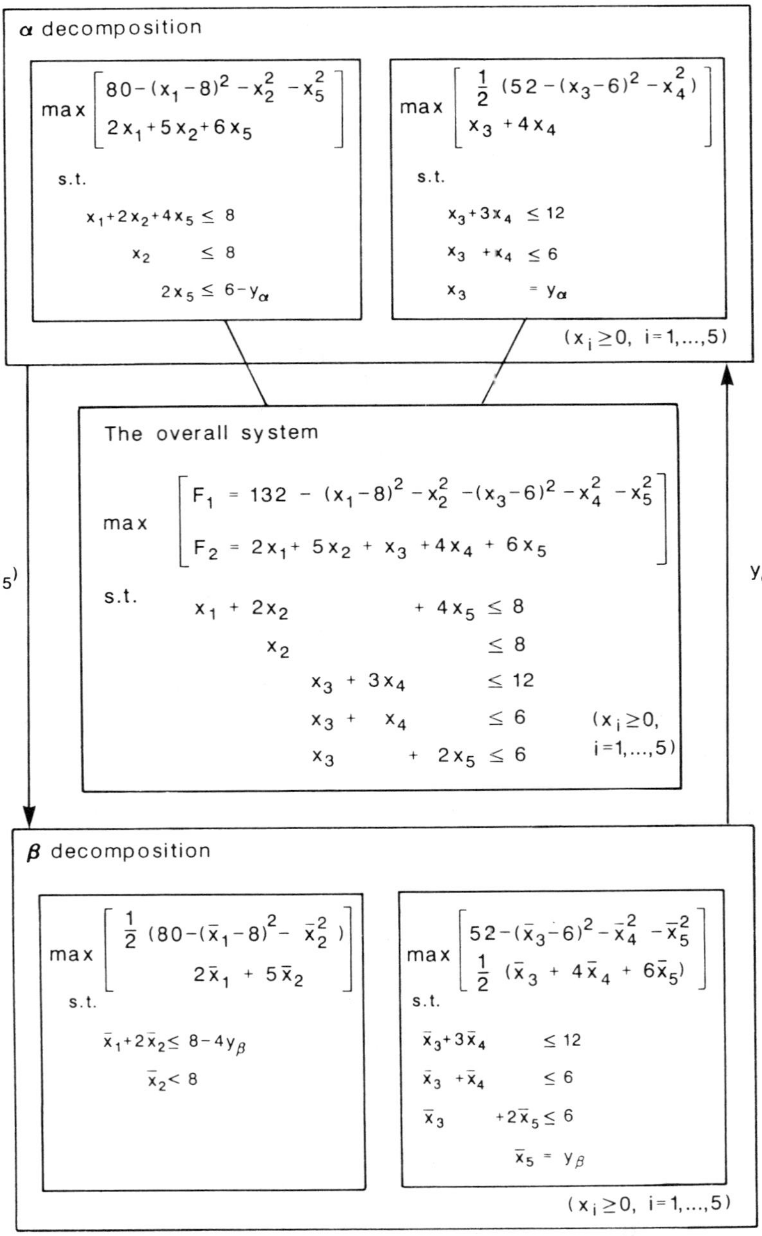

Figure 10.9 Structure of the system in Example 10.1.

$$\boldsymbol{\lambda}^{\alpha 2} = \begin{pmatrix} \lambda_1^{\alpha 2} \\ \lambda_2^{\alpha 2} \end{pmatrix} = \lambda_1^{\alpha 2} \begin{pmatrix} 1 \\ \lambda_{12}^{\alpha 2} \end{pmatrix} = \begin{pmatrix} 2 \\ a + bF_2^{\alpha} \end{pmatrix} = 2 \begin{pmatrix} 1 \\ \dfrac{a}{b} + \dfrac{b}{2} F_2^{\alpha} \end{pmatrix} \tag{10.119}$$

$$\boldsymbol{\lambda}^{\beta 1} = \begin{pmatrix} \lambda_1^{\beta 1} \\ \lambda_2^{\beta 1} \end{pmatrix} = \lambda_1^{\beta 1} \begin{pmatrix} 1 \\ \lambda_{12}^{\beta 1} \end{pmatrix} = \begin{pmatrix} 2 \\ a + bF_2^{\beta} \end{pmatrix} = 2 \begin{pmatrix} 1 \\ \dfrac{a}{2} + \dfrac{b}{2} F_2^{\beta} \end{pmatrix} \tag{10.120}$$

$$\boldsymbol{\lambda}^{\beta 2} = \begin{pmatrix} \lambda_1^{\beta 2} \\ \lambda_2^{\beta 2} \end{pmatrix} = \lambda_1^{\beta 2} \begin{pmatrix} 1 \\ \lambda_{12}^{\beta 2} \end{pmatrix} = \begin{pmatrix} 1 \\ 2a + 2bF_2^{\beta} \end{pmatrix} \tag{10.121}$$

(Note that $\lambda^{\alpha 2/\alpha 1} = \lambda_1^{\alpha 2}/\lambda_1^{\alpha 1} = 2$, $\lambda^{\beta 2/\beta 1} = \lambda_1^{\beta 2}/\lambda_1^{\beta 1} = \frac{1}{2}$.) $\boldsymbol{\lambda}^{\alpha i}$ and $\boldsymbol{\lambda}^{\beta j}$ at the nth iteration are obtained by using the approximation approach of trade-offs that was described before:

$$\boldsymbol{\lambda}^{\alpha 1(n)} = [1,\, a + b(F_2^{\beta(n-1)})]^T \tag{10.122}$$

$$\boldsymbol{\lambda}^{\alpha 2(n)} = \left[1,\, \frac{a}{2} + \frac{b}{2}(F_2^{\beta(n-1)})\right]^T \tag{10.123}$$

where $n \geqslant 1$ and $F_2^{\beta(0)} = F_2^{\alpha(0)}$ and

$$\boldsymbol{\lambda}^{\beta 1(n)} = \left[1,\, \frac{a}{2} + \frac{b}{2}(F_2^{\alpha(n)})\right]^T \tag{10.124}$$

$$\boldsymbol{\lambda}^{\beta 2(n)} = [1,\, 2a + 2b(F_2^{\alpha(n)})]^T \tag{10.125}$$

where $n \geqslant 1$.

Suppose that $a = 4.00$ and $b = -0.02$. We will solve the problem.

First, we estimate $\mathbf{x}_{\alpha}^{(0)}$. Let $\mathbf{x}_{\alpha}^{(0)} = [0, 0, 0, 0, 0]^T$; that is, $y_{\alpha}^{(0)} = x_3 = 0$; $F_1^{\alpha(0)} = 32$; $F_2^{\alpha(0)} = 0$. At this point, suppose the decisionmaker makes the following choices by using Eqs. (10.122) and (10.123) (where $F_1^{\alpha(0)} = F_1^{\beta(0)}$ and $F_2^{\alpha(0)} = F_2^{\beta(0)}$):

$$\begin{aligned} [1, \lambda_{12}^{\alpha 1(1)}]^T &= [1, 4.00 - 0.02\,(F_2^{\beta(0)})]^T \\ &= [1, 4.00]^T \\ [1, \lambda_{12}^{\alpha 2(1)}]^T &= \left[1, 2.00 - \frac{0.02}{2}(F_2^{\beta(0)})\right]^T \\ &= [1, 2.00]^T \end{aligned}$$

We solve the $\alpha 1$ and $\alpha 2$ subsystems with these trade-offs and $y_{\alpha} = 0$.

$$\mathbf{x}_{\alpha 1}^{(1)} = [x_1, x_2, x_5]^T = [7.2, 0.4, 0]^T$$

$$\mathbf{x}_{\alpha 2}^{(1)} = [x_3, x_4]^T = [3.0, 3.0]^T$$

$$F_1^{\alpha(1)} = 110.20$$

$$F_2^{\alpha(1)} = 31.40$$

Second, at the above point, the trade-offs are obtained as follows:

$$[1, \lambda_{12}^{\beta1(1)}]^T = \left[1, 2.00 - \frac{0.02}{2}(F_2^{\alpha(1)})\right]^T$$

$$= [1, 1.6860]^T$$

$$[1, \lambda_{12}^{\beta2(1)}]^T = [1, 8.00 - 0.04(F_2^{\alpha(1)})]^T$$

$$= [1, 6.7440]^T$$

$$y_\beta = x_5 = 0$$

We solve the $\beta1$ and $\beta2$ subsystems with these trade-offs and $y_\beta = 0$ and find the solution:

$$\mathbf{x}_{\beta1}^{(1)} = (\bar{x}_1, \bar{x}_2)^T = (7.3256, 0.3372)^T$$

$$\mathbf{x}_{\beta2}^{(1)} = (\bar{x}_3, \bar{x}_4, \bar{x}_5)^T = (3.4710, 2.5290, 0)^T$$

$$F_1^{\beta(1)} = 116.11$$

$$F_2^{\beta(1)} = 29.92$$

If we continue to solve the problem, the solution converges to the following value:

$$\mathbf{x} = (x_1, x_2, x_3, x_4, x_5)^T$$

$$= (7.3200, 0.3400, 3.4498, 2.5502, 0)^T \tag{10.126}$$

$$F_1 = F_1^\alpha = F_1^\beta = 115.86$$

$$F_2 = F_2^\alpha = F_2^\beta = 29.99 \tag{10.127}$$

$$\lambda = \lambda^\alpha = \lambda^\beta = (1, \lambda_{12})^T = (1, 3.4002)^T \tag{10.128}$$

The following information is obtained in the subsystems.

α1 Subsystem:

$$f_1^{\alpha1} = 79.42 \qquad f_2^{\alpha1} = 16.34 \qquad \lambda_{12}^{\alpha1} = 3.4002$$

α2 Subsystem:

$$f_1^{\alpha2} = 18.22 \qquad f_1^{\alpha2} = 13.65 \qquad \lambda_{12}^{\alpha2} = 1.7001$$

β1 Subsystem:

$$f_1^{\beta1} = 39.71 \qquad f_2^{\beta1} = 16.34 \qquad \lambda_{12}^{\beta1} = 1.7001$$

β2 Subsystem:

$$f_1^{\beta 2} = 36.44 \qquad f_2^{\beta 2} = 6.82 \qquad \lambda_{12}^{\beta 2} = 6.8004$$

If we substitute the solutions and the indifference trade-offs to the conditions of the preferred solution with $\mu_1 = 8.1606$, $\mu_2 = \mu_3 = 0$, $\mu_4 = 8.5007$, and $\mu_5 = 0$, then we can find that these conditions are all satisfied. (Note that μ_1 is a multiplier corresponding to the constraint of Eq. (10.85) and μ_2 is a multiplier corresponding to the constraint of Eq. (10.86), and so on. Also note that $\mu_i = \mu_i^\alpha = \mu_i^\beta$ ($i = 1, \ldots, 5$). μ_i^α is a multiplier of the α decomposition problem corresponding to the same constraint as μ_i (as is μ_i^β).

10.2.2.5 Top-down decisionmaking: Multiple (central) decisionmaker case It was assumed in the previous section using top-down decisionmaking that in the central office there exists a single decisionmaker (or a group of common preferences). In this section, we will consider the case where the central office has multiple decisionmakers, each of whom may represent a particular decomposition.

First, we will consider how each decisionmaker identifies a preferred solution. Second, we will consider how the decisionmakers find a compromise solution. Two cases will be considered. One is the case where the noninferior solutions that are near the preferred solutions are found and are given to the decisionmakers, with majority rule being applied.

The second case is one where a compromise solution, based on indifference trade-offs at the preferred solutions, is suggested by the analyst, and that solution is taken as a starting point for negotiations that are based on trade-off information provided by the analyst. This procedure is not restricted to the overlapping case. However, we will consider this in the framework of the hierarchical overlapping coordination.

Preferred solutions Suppose, for the sake of illustration, that there exist two decisionmakers in the center. Each decisionmaker is assumed to be a representative of one of the decompositions. A group of decisionmakers from a particular decomposition may be treated as a single decisionmaker because they are concerned with the same aspects of the system and will usually have a common preference. Let us call any decisionmaker from the α decomposition an α decisionmaker and any decisionmaker from the β decomposition a β decisionmaker.

The preferred solutions can be found by any available method. We can apply the HMO coordination procedures described in Chap. 5 to the α and the β decompositions. However, let us assume here that each decisionmaker finds his preferred solution by the MHOC procedure described in the previous section. There exists no difficulty in finding the decisionmakers' preferred solutions. It is easily seen that this discussion can be generalized to the case where there exist more than two decisionmakers in the center.

Compromise solution For the sake of illustration, we will assume that the system has only two objectives, F_1 and F_2. Suppose that $\lambda_{12}^{\alpha *}$ and $\lambda_{12}^{\beta *}$ are the α and β decisionmakers' indifference trade-offs. When a preferred solution lies in the non-inferior set, the indifference trade-offs are equal to the actual trade-offs, which are directly available when using, for example, the SWT or interactive SWT (ISWT) methods. Let $F_1^{\alpha*}$, $F_2^{\alpha*}$, $F_1^{\beta*}$, and $F_2^{\beta*}$ be the decisionmakers' preferred values of the objectives.

Decisionmakers usually have their indifference bands in the neighborhood of their preferred solutions. If the points in the neighborhood of decisionmakers' preferred solutions, which are also noninferior solutions, are available, then it may be possible to find a compromise point that lies in both the α decisionmaker's and the β decisionmaker's indifference bands. Or, the two decisionmakers can negotiate with each other to find a compromise solution, even if they cannot find a common point in their indifference bands.

How can we find such information? Suppose F_1 is strictly concave and $\lambda_{12}^{\alpha*} = \max(\lambda_{12}^{\alpha*}, \lambda_{12}^{\beta*}) > \min(\lambda_{12}^{\alpha*}, \lambda_{12}^{\beta*}) = \lambda_{12}^{\beta*}$. Then the indifference trade-offs $\lambda_{12}^{\alpha*}$ and $\lambda_{12}^{\beta*}$ are equal to the actual trade-offs. We can find values λ_{12}^{i*}'s ($i = 1, \ldots, k$) that satisfy the following inequality:

$$\lambda_{12}^{\alpha*} > \lambda_{12}^{1*} > \lambda_{12}^{2*} > \cdots > \lambda_{12}^{k*} > \lambda_{12}^{\beta*} \tag{10.129}$$

Then, by solving the following problem,

$$\left.\begin{aligned} &\max_{\mathbf{x}} F_1(\mathbf{x}) + \lambda_{12}^{i*} F_2(\mathbf{x}) \\ &\text{subject to } A\mathbf{x} = \mathbf{c} \\ &\mathbf{x} \in S \end{aligned}\right\} \quad \text{for } i = 1, \ldots, k \tag{10.130}$$

we can find the points that are also noninferior solutions, and that interest the decisionmakers the most, in order to reach a compromise solution. The above problem can be solved by any method, and can be solved efficiently by HOC. Once these points are available, a compromise solution can be found, as described above. If there are more than two decisionmakers, the majority rule can be applied (see Hall and Haimes, 1976, or Chap. 6).

If F_1 and F_2 are linear and $F_2^{\alpha*} = \max(F_2^{\alpha*}, F_2^{\beta*}) > \min(F_2^{\alpha*}, F_2^{\beta*}) = F_2^{\beta*}$, then the following problem should be solved instead of the above problem:

$$\left.\begin{aligned} &\max F_1(\mathbf{x}) \\ &\text{subject to } F_2(\mathbf{x}) \geq F_2^{i*} \\ &A\mathbf{x} = \mathbf{c} \\ &\mathbf{x} \in S \end{aligned}\right\} \tag{10.131}$$

where

$$i = 1, \ldots, k$$

$$F_2^{\alpha*} > F_2^{1*} > F_2^{2*} > \cdots > F_2^{k*} > F_2^{\beta*}$$

There is another way of finding a compromise solution: the analyst can calculate a compromise solution to be used as a starting point for negotiations. The compromise solution, denoted by (F_1^0, F_2^0), is determined as an objective vector that minimizes the sum of the loss that occurs to the decisionmakers in order to get it. That is, the following sum is minimized with respect to F_1 and F_2:

$$L(F_1, F_2) \stackrel{\Delta}{=} U^\alpha(F_1^{\alpha*}, F_2^{\alpha*}) - U^\alpha(F_1, F_2) + w^{\alpha\beta}[U^\beta(F_1^{\beta*}, F_2^{\beta*}) - U^\beta(F_1, F_2)] \quad (10.132)$$

where $U^\alpha(F_1, F_2)$ and $U^\beta(F_1, F_2)$ are the α and β decisionmakers' underlying utility functions, respectively. $w^{\alpha\beta}$ is a relative weight between the α and β decisionmakers.

By using the Taylor expansion, the loss function given by Eq. (10.132) is approximated as follows:

$$\begin{aligned} L(F_1, F_2) \approx & -\frac{\partial U^\alpha}{\partial F_1}(F_1^{\alpha*}, F_2^{\alpha*})(F_1 - F_1^{\alpha*}) \\ & -\frac{\partial U^\alpha}{\partial F_2}(F_1^{\alpha*}, F_2^{\alpha*})(F_2 - F_2^{\alpha*}) \\ & + w^{\alpha\beta}\left[-\frac{\partial U^\beta}{\partial F_1}(F_1^{\beta*}, F_2^{\beta*})(F_1 - F_1^{\beta*}) \right. \\ & \left. -\frac{\partial U^\beta}{\partial F_2}(F_1^{\beta*}, F_2^{\beta*})(F_2 - F_2^{\beta*})\right] \\ = & -\frac{\partial U^\alpha}{\partial F_1}(F_1^{\alpha*}, F_2^{\alpha*})[F_1 - F_1^{\alpha*} + \lambda_{12}^{\alpha*}(F_2 - F_2^{\alpha*}) \\ & + \lambda^{\alpha\beta*}(F_1 - F_1^{\beta*} + \lambda_{12}^{\beta*}(F_2 - F_2^{\beta*}))] \end{aligned} \quad (10.133)$$

where the relationships between the utility function and the indifference trade-offs have been used, and we define

$$\lambda^{\alpha\beta*} = w^{\alpha\beta}\frac{\partial U^\beta(F_1^{\beta*}, F_2^{\beta*})/\partial F_1}{\partial U^\alpha(F_1^{\alpha*}, F_2^{\alpha*})/\partial F_1} \quad (10.134)$$

Assuming the usual case, where $\partial U^\alpha/\partial F_1 > 0$, a compromise solution can be found by minimizing the loss function, that is, by solving the following problem:

$$\max \{F_1(\mathbf{x}) - F_1^{\alpha*} + \lambda_{12}^{\alpha*}(F_2(\mathbf{x}) - F_2^{\alpha*}) + \lambda^{\alpha\beta*}[(F_1(\mathbf{x}) - F_1^{\beta*}) + \lambda_{12}^{\beta*}(F_2(\mathbf{x}) - F_2^{\beta*})]\} \quad (10.135)$$

subject to

$$A\mathbf{x} = \mathbf{c}$$

$$\mathbf{x} \in S$$

$\lambda_{12}^{\alpha*}$ and $\lambda_{12}^{\beta*}$, which are the α and β decisionmakers' indifference tradeoffs, respectively, are found as described in the previous section. The decisionmakers must agree on the value of $\lambda^{\alpha\beta*}$. If the evaluation of $\lambda^{\alpha\beta*}$ causes difficulties, the analyst can take $\lambda^{\alpha\beta*} = 1$.

Even if the decisionmakers can agree on $\lambda^{\alpha\beta*} = 1$, the compromise solution may not be satisfactory to both decisionmakers because of the simple way the compromise solution was determined. Hence negotiations around the compromise solution may be necessary. In this case, the points around the compromise solution that are necessary for the negotiation can be found by solving the problem given in Eq. (10.130) or (10.131). In this case λ_{12}^{i*} or F_2^{i*} must be given from the neighborhood of the compromise solution. Finally, we should note that these discussions can be extended to the general case where there exist more than two decisionmakers and more than two objectives.

10.2.2.6 Example 10.2 (a continuation of Example 10.1) Suppose that we have two decisionmakers. Let the α decisionmaker be a representative of the α decomposition and the β decisionmaker be a representative of the β decomposition.

Step 1: The two decisionmakers can find their preferred solutions by using the MHOC procedures described earlier. Suppose $\lambda_{12}^{\alpha*} = 5.58$ and $\lambda_{12}^{\beta*} = 11.61$ are the α and β decisionmakers' indifference trade-offs. There are indifference bands of trade-offs around these given values. However, we are only interested in the values between $\lambda_{12}^{\alpha*}$ and $\lambda_{12}^{\beta*}$ so that we can arrive at a compromise solution.

Step 2: Let

$$\lambda_{12}^{(1)} = 6.00 \qquad \lambda_{12}^{(2)} = 7.00 \qquad \lambda_{12}^{(3)} = 7.59$$

$$\lambda_{12}^{(4)} = 8.00 \qquad \lambda_{12}^{(5)} = 9.00 \qquad \lambda_{12}^{(6)} = 9.60$$

$$\lambda_{12}^{(7)} = 10.00$$

We will solve the problem given in Eq. (10.130) with these $\lambda_{12}^{(i)}$'s ($i = 1, \ldots, 7$). Note that HOC solves this problem efficiently. Then we will find the corresponding noninferior solutions and values of objectives that are given to decisionmakers as the information on which to base the compromise solution. Those values are shown in Table 10.1.

Table 10.1 Information given to the decisionmakers

k	$\lambda_{12}^{(k)}$	x_1	x_2	x_3	x_4	x_5	F_1	F_2
α	5.58	6.88	0.56	3	3	0	112.43	31.56
1	6.00	6.80	0.60	3	3	0	112.20	31.60
2	7.00	6.60	0.70	3	3	0	111.55	31.70
3	7.59	6.48	0.76	3	3	0	111.12	31.76
4	8.00	6.40	0.80	3	3	0	110.80	31.80
5	9.00	6.20	0.90	3	3	0	109.95	31.90
6	9.60	6.08	0.96	3	3	0	109.39	31.96
7	10.00	6.00	1.00	3	3	0	109.00	32.00
β	11.61	5.68	1.16	3	3	0	107.26	32.16

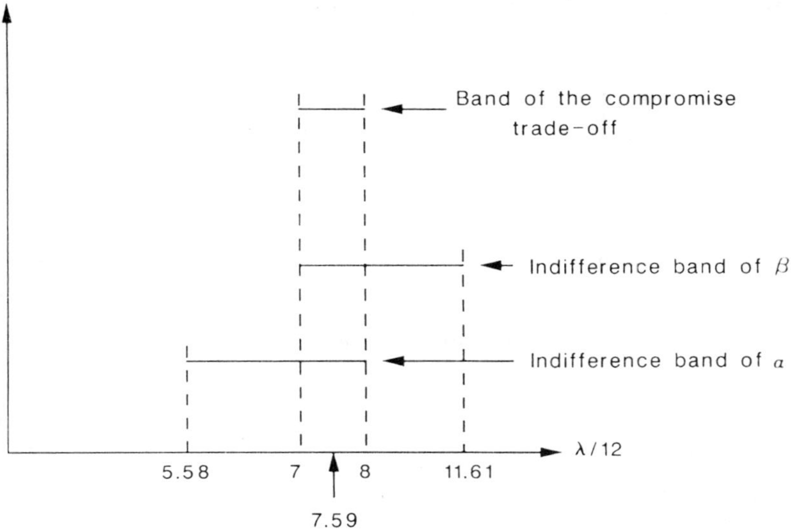

Figure 10.10 Compromise trade-offs of Example 10.2.

From this table, suppose the α decisionmaker finds that his indifference band is (5.58, 8.00) and the β decisionmaker finds that her indifference band is (7.00, 11.61). Suppose that they agree that $\lambda_{12}^* = 7.59$ is a compromise value of trade-off, as shown in Fig. 10.10. Then the compromise solution can be obtained by solving the problem in Eq. (10.130) again, using $\lambda_{12}^* = 7.59$. We should note that, once λ_{12}^* is given, the problem in Eq. (10.130) can be solved by HOC very efficiently. The compromise solution, which has already been given in Table 10.1, will be

$$\mathbf{x} = [6.08, 0.76, 3, 3, 0]^T$$

and

$$F_1 = 111.12 \qquad F_2 = 31.76$$

If the decisionmakers cannot find a common compromise value of trade-off, that is, if they cannot find the band of compromise trade-offs, they will have to negotiate with each other until they agree on some value. If the number of decisionmakers is more than two, majority rule can be applied (see details in Hall and Haimes, 1976).

APPENDIX: DIRECT DERIVATION OF COORDINATION EQUATION FROM THE TRADE-OFF RELATIONSHIPS

The coordination equation (10.78) can be derived directly from the trade-off equations (10.58) and (10.59). We consider only the α decomposition; the β decomposition can be dealt with in the same way.

Suppose we have the same assumptions represented in this chapter. Equation (10.58) is

$$\boldsymbol{\lambda}^{\alpha i}(\mathbf{f}^{\alpha}) = \left[\sum_{k=1}^{n} \lambda_k^{\alpha}(\mathbf{F}^{\alpha}) \frac{\partial F_k^{\alpha}(\mathbf{f}^{\alpha})}{\partial \mathbf{f}^{\alpha i}}\right]^T$$

Here

$$\boldsymbol{\lambda}^{\alpha} = (\lambda_1^{\alpha}, \lambda_2^{\alpha}, \ldots, \lambda_N^{\alpha})^T = \lambda_1^{\alpha}(1, \lambda_{12}^{\alpha}, \ldots, \lambda_{1N}^{\alpha})^T$$

We assume that $\lambda_1^{\alpha} = 1$, so that $\boldsymbol{\lambda}^{\alpha} = (1, \lambda_{12}^{\alpha}, \ldots, \lambda_{1N}^{\alpha})^T$. Hence, $\lambda_k^{\alpha} = \lambda_{1k}^{\alpha}$ in this case, and

$$\frac{\partial F_k^{\alpha}(\mathbf{f}^{\alpha})}{\partial \mathbf{f}^{\alpha i}} = \left(\frac{\partial F_k^{\alpha}}{\partial f_1^{\alpha i}}, \frac{\partial F_k^{\alpha}}{\partial f_2^{\alpha i}}, \ldots, \frac{\partial F_k^{\alpha}}{\partial f_N^{\alpha i}}\right)^T$$

Hence,

$$\lambda_k^{\alpha} \frac{\partial F_k^{\alpha}}{\partial \mathbf{f}^{\alpha i}} = \left(\lambda_k^{\alpha} \frac{\partial F_k^{\alpha}}{\partial f_1^{\alpha i}}, \lambda_k^{\alpha} \frac{\partial F_k^{\alpha}}{\partial f_2^{\alpha i}}, \ldots, \lambda_k^{\alpha} \frac{\partial F_k^{\alpha}}{\partial f_N^{\alpha i}}\right)^T$$

So,

$$\boldsymbol{\lambda}^{\alpha i} = (\lambda_1^{\alpha i}, \lambda_2^{\alpha i}, \ldots, \lambda_N^{\alpha i})^T = \lambda_1^{\alpha i}(1, \lambda_{12}^{\alpha i}, \ldots, \lambda_{1N}^{\alpha i})^T$$

where $\lambda_{1j}^{\alpha i} = \lambda_j^{\alpha i}/\lambda_1^{\alpha i}$, that is, $\lambda_j^{\alpha i} = \lambda_1^{\alpha i} \cdot \lambda_{1j}^{\alpha i}$. From the above equations,

$$\boldsymbol{\lambda}^{\alpha i} = \left(\sum_{k=1}^{N} \lambda_k^{\alpha} \frac{\partial F_k^{\alpha}}{\partial f_1^{\alpha i}}, \sum_{k=1}^{N} \lambda_k^{\alpha} \frac{\partial F_k^{\alpha}}{\partial f_2^{\alpha i}}, \ldots, \sum_{k=1}^{N} \lambda_k^{\alpha} \frac{\partial F_k^{\alpha}}{\partial f_N^{\alpha i}}\right)^T$$

Hence,

$$\lambda_j^{\alpha i} = \lambda_1^{\alpha i} \cdot \lambda_{1j}^{\alpha i} = \sum_{k=1}^{N} \lambda_k^{\alpha} \frac{\partial F_k^{\alpha}}{\partial f_j^{\alpha i}} = \lambda_1^{\alpha} \frac{\partial F_1^{\alpha}}{\partial f_j^{\alpha i}} + \cdots + \lambda_N^{\alpha} \frac{\partial F_N^{\alpha}}{\partial f_j^{\alpha i}}$$

We assumed that $F_j^{\alpha} = F_j^{\alpha}(f_j^{\alpha 1}, f_j^{\alpha 2}, \ldots, f_j^{\alpha n_{\alpha}})$, that is, F_j^{α} is the function of $f_j^{\alpha i}$ $(i = 1, \ldots, n_{\alpha})$ (and not that of $f_k^{\alpha_i}$, $k \neq j$). Hence,

$$\lambda_j^{\alpha i} = \lambda_1^{\alpha i} \cdot \lambda_{1j}^{\alpha i} = \lambda_j^{\alpha} \frac{\partial F_j^{\alpha}}{\partial f_j^{\alpha i}}$$

and

$$\lambda_1^{\alpha i} = \lambda_1^{\alpha} \frac{\partial F_1^{\alpha}}{\partial f_1^{\alpha i}} = \frac{\partial F_1^{\alpha}}{\partial f_1^{\alpha i}} \qquad (\lambda_1^{\alpha} = 1)$$

Then from $\lambda_j^{\alpha} = \lambda_{1j}^{\alpha}$,

$$\lambda_j^{\alpha i} = \lambda_1^{\alpha i} \cdot \lambda_{1j}^{\alpha i} = \frac{\partial F_1^{\alpha}}{\partial f_1^{\alpha i}} \cdot \lambda_{1j}^{\alpha i} = \lambda_j^{\alpha} \frac{\partial F_j^{\alpha}}{\partial f_j^{\alpha i}} = \lambda_{1j}^{\alpha} \frac{\partial F_j^{\alpha}}{\partial f_j^{\alpha i}}$$

That is,

$$\lambda_{1j}^{\alpha i} \frac{\partial F_1^\alpha}{\partial f_1^{\alpha i}} = \lambda_{1j}^\alpha \frac{\partial F_j^\alpha}{\partial f_j^{\alpha i}}$$

Then

$$\lambda_{1j}^\alpha = \lambda_{1j}^\alpha \frac{\partial F_1^\alpha}{\partial f_1^\alpha} \left(\frac{\partial F_j^\alpha}{\partial f_j^\alpha} \right)^{-1}$$

for $i = 1, \ldots, n_\alpha$. That is,

$$\lambda_{1j}^\alpha = \lambda_{1j}^{\alpha 1} \frac{\partial F_1^\alpha}{\partial f_1^{\alpha 1}} \frac{(\partial F_j^\alpha)^{-1}}{\partial f_j^{\alpha 1}} = \cdots = \lambda_{1j}^{\alpha n_\alpha} \frac{\partial F_1^\alpha}{\partial f_1^{\alpha n_\alpha}} \left(\frac{\partial F_j^\alpha}{\partial f_j^{\alpha n_\alpha}} \right)^{-1}$$

Since $\lambda_{1j} = \lambda_{1j}^\alpha$ and for $j = k$, then

$$\lambda_{1k} = \lambda_{1k}^\alpha = \lambda_{1k}^{\alpha 1} \frac{\partial F_1^\alpha}{\partial f_1^{\alpha 1}} \left(\frac{\partial F_k^\alpha}{\partial f_k^{\alpha 1}} \right)^{-1} = \cdots = \lambda_{1k}^{\alpha n_\alpha} \frac{\partial F_1^\alpha}{\partial f_1^{\alpha n_\alpha}} \left(\frac{\partial F_k^\alpha}{\partial f_k^{\alpha n_\alpha}} \right)^{-1}$$

With this, the α part of the coordination equation is derived.

The assumption $\lambda_1^\alpha = 1$ is not necessary for the derivation of the coordination equation. If $\lambda_1^\alpha \neq 1$, then

$$\lambda_1^\alpha \frac{\partial F_1^\alpha}{\partial f_1^{\alpha i}} \lambda_{1j}^{\alpha i} = \lambda_j^\alpha \frac{\partial F_j^\alpha}{\partial f_j^{\alpha i}}$$

Hence,

$$\frac{\lambda_j^\alpha}{\lambda_1^\alpha} = \lambda_{1j}^\alpha = \lambda_{1j}^{\alpha i} \left(\frac{\partial F_1^\alpha}{\partial f_1^{\alpha i}} \right) \left(\frac{\partial F_j}{\partial f_j^{\alpha i}} \right)^{-1}$$

That is, the same equation can be derived. However, in this case, note that $\lambda_1 = \lambda_1^\alpha$ $(= \lambda_1^\beta)$ holds. The β part of the coordination equation can be derived in the same way.

REFERENCES

Chankong, V., and Haimes, Y. Y., The Interactive Surrogate Worth Trade-off (ISWT) Method for Multiobjective Decision-Making, in *Multicriteria Problem Solving*, Vol. 155, ed. S. Zoints, Springer-Verlag, New York, 1978.

Chankong, V., and Haimes, Y. Y., *Multiobjective Decision Making: Theory and Methodology*, North-Holland, Amsterdam, 1983.

Dyer, J. S., An Empirical Investigation of a Man-Machine Interactive Approach to the Solution of Multiple Criteria Problem, in *Multiple Criteria Decision Making*, ed. J. L. Cochrane and M. Zeleny, University of South Carolina Press, Columbia, S.C., 1973.

Geoffrion, A. M., Dyer, J., and Feinberg, A., An Interactive Approach for Multicriterion Optimization with an Application to the Operation of an Academic Department, *Management Sci.*, vol. 19, pp. 257–368, 1972.

Haimes, Y. Y. *Hierarchical Analyses of Water Resources Systems: Modeling and Optimization of Large-Scale Systems*, McGraw-Hill, New York, 1977.

Haimes, Y. Y., and Chankong, V., Kuhn-Tucker Multipliers as Trade-offs in Multiobjective Decision-Making Analysis, *Automatica*, vol. 15, no. 1, pp. 59–72, 1979.

Haimes, Y. Y., Hall, W. A., and Freedman, H. T., *Multiobjective Optimization in Water Resource Systems: The Surrogate Worth Trade-off Method*, Elsevier, Amsterdam, 1975.

Haimes, Y. Y., Lasdon, L. S., and Wismer, D. A., On the Bicriterion Formulation of the Integrated System Identification and Systems Optimization, *IEEE Trans. Syst. Man Cybern.*, vol. SMC-1, pp. 286–287, 1971.

Haimes, Y. Y., and Shima, T., Overlapping Coordination, in *Systems & Control Encyclopedia: Theory, Technology, Applications*, ed. M. G. Singh, Pergamon Press, Oxford, 1987, pp. 3564–3568.

Haimes, Y. Y., and Tarvainen, K., Hierarchical-Multiobjective Framework for Large Scale Systems, in *Multicriteria Analysis in Practice*, ed. P. Nijkamp and J. Spronk, Gower Press, London, 1980.

Hall, W. A., and Haimes, Y. Y., The Surrogate Worth Trade-Off Method with Multiple Decision Makers, in *Multiple Criteria Decision Making, Kyoto, 1975*, ed. M. Zeleny, Springer-Verlag, New York, 1976.

Lasdon, L., *Optimization Theory for Large Systems*, Macmillan, London, 1970.

Macko, D., and Haimes, Y. Y., Overlapping Coordination of Hierarchical Structures, *IEEE Trans. Syst. Man Cybern.*, vol. SMC-8, no. 10, pp. 745–751, October 1978.

Shima, T., *Hierarchical Overlapping Coordination of Large Scale Systems with Multiple Objectives*, Ph.D. dissertation, Case Western Reserve University, Cleveland, Ohio, 1982.

Shima, T., Haimes, Y. Y., Tarvainen, K., and Sung, K., Applications of Hierarchical Overlapping Coordination with Multiple Objectives, *Inform. Decision Technol.*, vol. 15, no. 1, pp. 33–57, 1989.

Shima, T., and Haimes, Y. Y., Hierarchical Overlapping Coordination with Multiple Objectives, Working paper, Case Western Reserve University, Cleveland, Ohio, Oct. 1986.

Shima, T., and Haimes, Y. Y., The Convergence Properties of Hierarchical Overlapping Coordination, *IEEE Trans. Syst. Man Cybern.*, vol. SMC-14, no. 1, Jan./Feb., 1984.

Shima, T., Tarvainen, K., and Haimes, Y. Y., Hierarchical Overlapping Coordination for Large Scale Systems, *Proc. 8th Triennial World Congress/IFAC, Kyoto, Japan*, 1982a.

Shima, T., Haimes, Y. Y., and Tarvainen, K., Multiobjective Hierarchical Coordination, presented at the Multiple Criteria Decision Making Conference, Belgium, August 1982b.

Sung, K., Haimes, Y. Y., and Tarvainen, K., Some Extensions of Overlapping Coordination, Working Paper, Systems Engineering Department, Case Western Reserve University, April 1981.

Tarvainen, K., *Hierarchical Multiobjective Optimization*, Ph.D. dissertation, Case Western Reserve University, Cleveland, Ohio, May 1981.

Tarvainen, K., and Haimes, Y. Y., Hierarchical-Multiobjective Framework for Energy Storage Systems, *Proc. Fourth Multiple Criteria Decision Making Conference in Newark, Delaware, August 10–15, 1980*, Springer-Verlag, New York, pp. 424–446, 1981.

Tarvainen, K., and Haimes, Y. Y., Coordination of Hierarchical Multiobjective Systems: Theory and Methodology, *IEEE Trans. Syst., Man, and Cybernetics*, Vol. SMC-12, No. 6, pp. 751–764, Nov./Dec. 1982.

CHAPTER

ELEVEN

EXAMPLES OF INTEGRATION OF HIERARCHICAL OVERLAPPING COORDINATION WITH MULTIOBJECTIVE OPTIMIZATION

This chapter considers examples of the approach developed in Chap. 10, which combines the features of hierarchical overlapping coordination (HOC) and multiobjective optimization into one method: multiobjective hierarchical overlapping coordination (MHOC).

11.1 WATER RESOURCES MANAGEMENT PROBLEM

A representative example of the HOC (see Chap. 9) was first discussed in Haimes and Macko (1973) and subsequently in Mendu et al. (1980). This example was developed in terms of the management of a water resources system. Some of the ways a typical river basin planning problem might be decomposed (Macko and Haimes, 1978) are

- Planning subareas (geographic decomposition),
- Watersheds (hydrological decomposition),
- Planning periods (temporal decomposition), and
- Objectives (functional decomposition).

In addition to having a number of different decompositions, a water resources management and planning problem can generally be characterized by several objectives, such as

- Minimizing the cost,
- Protecting the quality of the environment,
- Reducing flood damage,
- Supplying water needs, and
- Maximizing agricultural production.

Water resource systems constitute a promising area for applications of the MHOC, which was outlined in Chap. 10.

A single-objective HOC as applied to water resource problems was considered in Sung (1978). We will modify the problem considered there and reformulate it in a multiobjective framework. Then, to solve it, we will apply the MHOC scheme.

11.1.1 Problem Statement

Consider a land management problem with three major types of areas classified according to activity: an agricultural area, a floodplain (which also has agriculture), and an area we will call area 3. The crops produced are corn and beans. Suppose that this land is also used to supply the needs for land-based outdoor recreation, such as picnicking, camping, playfields, and golf courses. Our objectives are to minimize the cost of development and the reduction in environmental quality (e.g., increased sedimentation rate) while satisfying such things as the need for recreation. Figure 11.1 shows the structure of a problem with overlapping usage of the land.

The following crop rotation practices are used:

C_1 = corn—corn—corn (no conservation practice)
C_2 = corn—corn—corn (with contouring)
B_1 = beans—beans—beans (no conservation practice)
B_2 = beans—beans—beans (with contouring)

Conventional tillage is assumed to be the management practice in use.

The decision variables are defined as follows:

x_1 = C_1 acreage in the agricultural area
x_2 = C_2 acreage in the agricultural area
x_3 = B_1 acreage in the agricultural area
x_4 = B_2 acreage in the agricultural area
x_5 = C_1 acreage in the floodplain
x_6 = C_2 acreage in the floodplain
x_7 = B_1 acreage in the floodplain
x_8 = B_2 acreage in the floodplain
x_9 = picnicking acreage in the floodplain
x_{10} = camping acreage in the floodplain
x_{11} = playfield acreage in the floodplain
x_{12} = golf course acreage in the floodplain
x_{13} = picnicking acreage in area 3

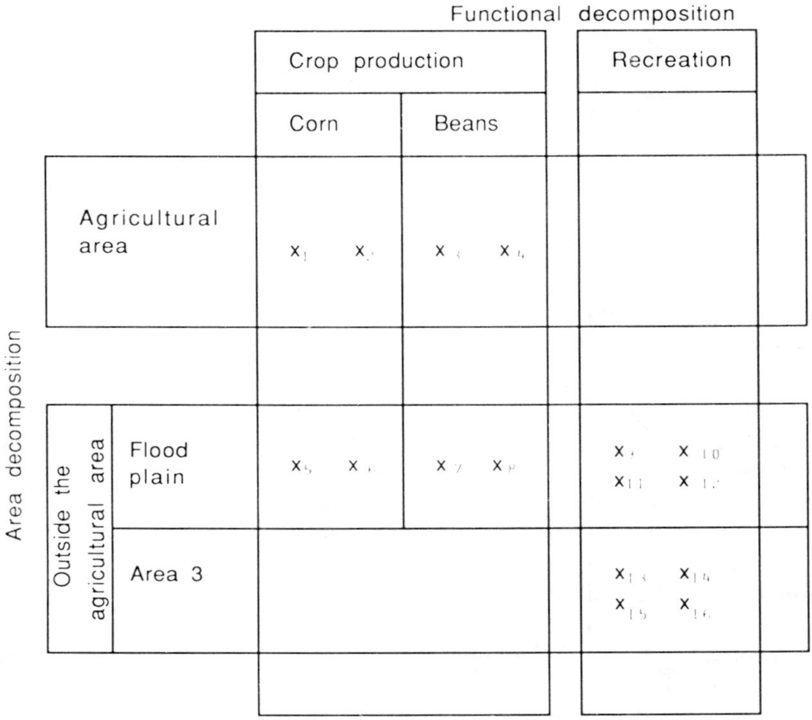

Figure 11.1 Land use problem with two overlapping decompositions.

x_{14} = camping acreage in area 3
x_{15} = playfield acreage in area 3
x_{16} = golf course acreage in area 3

The overall problem can now be formulated as follows:

$$\min \begin{bmatrix} F_1 = \sum_{i=1}^{16} (b_{1i}x_i + b_{2i}x_i^2) \\ F_2 = \sum_{i=1}^{16} d_i x_i \end{bmatrix} \tag{11.1}$$

subject to

$$a_1x_1 + a_2x_2 + a_3x_5 + a_4x_6 \geqslant c_1 \tag{11.2}$$

$$a_5x_3 + a_6x_4 + a_7x_7 + a_8x_8 \geqslant c_2 \tag{11.3}$$

$$x_1 + x_2 + x_3 + x_4 \leqslant c_3 \tag{11.4}$$

$$x_5 + x_6 + x_7 + x_8 + x_9 + x_{10} + x_{11} + x_{12} \leqslant c_4 \tag{11.5}$$

$$a_9x_9 + a_{10}x_{13} \geqslant c_5 \tag{11.6}$$

$$a_{11}x_{10} + a_{12}x_{14} \geqslant c_6 \tag{11.7}$$

$$a_{13}x_{11} + a_{14}x_{15} \geqslant c_7 \tag{11.8}$$

$$a_{15}x_{12} + a_{16}x_{16} \geqslant c_8 \tag{11.9}$$

$$x_{13} + x_{14} + x_{15} + x_{16} \leqslant c_9 \tag{11.10}$$

$$x_i \geqslant 0 \qquad i = 1, \ldots, 16$$

where

F_1 = approximated cost function (dollars)
F_2 = approximated amount of sediment (tons)
[F_1 and F_2 are assumed for simplicity to be of the forms as shown in Eq. (11.1).]
Equation (11.2) is a constraint on the production of corn
Equation (11.3) is a constraint on the production of beans
Equation (11.4) is a land constraint for the agricultural area
Equation (11.5) is a land constraint for the floodplain
Equations (11.6)–(11.9) are constraints on recreational activities
Equation (11.10) is a land constraint for area 3
c_1 is a production requirement of corn (in bushels)
c_2 is a production requirement of beans (in bushels)
c_3 is a land availability in the agricultural area
c_4 is land availability in the flood plain
c_5 is a recreational need for picnicking acreage
c_6 is a recreational need for camping acreage
c_7 is a recreational need for playfield acreage
c_8 is a recreational need for golf course acreage
c_9 is land availability in area 3
$a_1 \ldots a_4$ = corn production rate (bushel/acre)
$a_5 \ldots a_8$ = bean production rate (bushel/acre)
a_9, a_{10} = recreation days/picnic acreage
a_{11}, a_{12} = recreation days/camping acreage
a_{13}, a_{14} = recreation days/playfield acreage
a_{15}, a_{16} = recreation days/golf course acreage

Let the α decomposition be an area decomposition and let the β decomposition be a function decomposition. The subproblems of each decomposition are represented as follows:

α Decomposition

α_l Subsystem (agricultural area):

$$\min \begin{bmatrix} f_1^{\alpha 1} = \sum_{i=1}^{4} (b_{1i}x_i + b_{2i}x_i^2) \\ f_2^{\alpha 1} = \sum_{i=1}^{4} d_i x_i \end{bmatrix}$$

subject to

$$a_1x_1 + a_2x_2 + a_3y_{\alpha1} + a_4y_{\alpha2} \geq c_1$$

$$a_5x_3 + a_6x_4 + a_7y_{\alpha3} + a_8y_{\alpha4} \geq c_2$$

$$x_1 + x_2 + x_3 + x_4 \leq c_3$$

$$x_i \geq 0 \qquad (i = 1, \ldots, 4)$$

α_2 Subsystem (outside the agricultural area):

$$\min \begin{bmatrix} f_1^{\alpha2} = \sum_{i=5}^{16} (b_{1i}x_i + b_{2i}x_i^2) \\ f_2^{\alpha2} = \sum_{i=5}^{16} d_i x_i \end{bmatrix}$$

subject to

$$x_5 + x_6 + x_7 + x_8 + x_9 + x_{10} + x_{11} + x_{12} \leq c_4$$

$$a_9x_9 + a_{10}x_{13} \geq c_5$$

$$a_{11}x_{10} + a_{12}x_{14} \geq c_6$$

$$a_{13}x_{11} + a_{14}x_{15} \geq c_7$$

$$a_{15}x_{12} + a_{16}x_{16} \geq c_8$$

$$x_{13} + x_{14} + x_{15} + x_{16} \leq c_9$$

$$x_5 = y_{\alpha1}$$

$$x_6 = y_{\alpha2}$$

$$x_7 = y_{\alpha3}$$

$$x_8 = y_{\alpha4}$$

$$x_i \geq 0 \qquad i = 5, \ldots, 16$$

β Decomposition

β_1 Subsystem (crops):

$$\min \begin{bmatrix} f_1^{\beta1} = \sum_{i=1}^{8} (b_{1i}x_i + b_{2i}x_i^2) \\ f_2^{\beta1} = \sum_{i=1}^{8} d_i x_i \end{bmatrix}$$

subject to

$$a_1x_1 + a_2x_2 + a_3x_5 + a_4x_6 \geq c_1$$

$$a_5x_3 + a_6x_4 + a_7x_7 + a_8x_8 \geq c_2$$

$$x_1 + x_2 + x_3 + x_4 \leq c_3$$

$$x_5 + x_6 + x_7 + x_8 + y_{\beta 1} + y_{\beta 2} + y_{\beta 3} + y_{\beta 4} \leq c_4$$

$$x_i \geq 0 \qquad i = 1, \ldots, 8$$

β_2 Subsystem (recreation):

$$\min \begin{bmatrix} f_1^{\beta 2} = \sum_{i=9}^{16} (b_{1i} x_i + b_{2i} x_i^2) \\ f_2^{\beta 2} = \sum_{i=9}^{16} d_i x_i \end{bmatrix}$$

subject to

$$a_9 x_9 + a_{10} x_{13} \geq c_5$$

$$a_{11} x_{10} + a_{12} x_{14} \geq c_6$$

$$a_{13} x_{11} + a_{14} x_{15} \geq c_7$$

$$a_{15} x_{12} + a_{16} x_{16} \geq c_8$$

$$x_{13} + x_{14} + x_{15} + x_{16} \leq c_9$$

$$x_9 = y_{\beta 1}$$

$$x_{10} = y_{\beta 2}$$

$$x_{11} = y_{\beta 3}$$

$$x_{12} = y_{\beta 4}$$

$$x_i \geq 0 \qquad i = 9, \ldots, 16$$

11.1.2 Solution Procedure

We solve the problem using a top-down decisionmaking procedure. For illustrative purposes, we will assume the existence of a single decisionmaker who has the following preference implicitly in mind:

$$U(F_1, F_2) = p_0 \mathbf{F} + 0.5\, \mathbf{F}^T Q_0 \mathbf{F}$$

where $\mathbf{F} = [F_1, F_2]^T$.

For simplicity, $\mathbf{p}_0$ and Q_0 are assumed as follows:

$$\mathbf{p}_0 = [p_1, p_2]$$

$$Q_0 = \begin{bmatrix} 0 & 0 \\ 0 & q \end{bmatrix}$$

The term *single decisionmaker* may refer either to an individual or to a group of decisionmakers who by consensus act as a single decisionmaker. The α managers are concerned with the development of the land and the β decision managers are concerned with both crop production and with recreational development. We should note that a "mathematical decisionmaker" is assumed for illustrative purposes. In actual decisionmaking in a systems analysis context, a decisionmaker would interact with a systems analyst and would state his or her preferences each time the α or β decomposition was to be solved.

As in Chap. 10, the trade-offs are

$$\boldsymbol{\lambda}(\mathbf{F}) = [\lambda_1(\mathbf{F}), \lambda_2(\mathbf{F})]^T = [1, \lambda_{12}(\mathbf{F})]^T$$

$$\lambda_{12}(\mathbf{F}) = \frac{\partial U/\partial F_2}{\partial U/\partial F_1} = (p_2 + qF_2)/p_1 \qquad \text{(dollars/ton)}$$

$$\boldsymbol{\lambda}(\mathbf{F}) = \boldsymbol{\lambda}^\alpha(\mathbf{F}^\alpha) = \boldsymbol{\lambda}^\beta(\mathbf{F}^\beta)$$

Suppose that the coefficients have the values given in Table 11.1. Also assume the following values: $p_1 = 1$; $p_2 = 12$; $q = -4 \times 10^{-5}$.

We will now solve the problem.

0. First, the initial feasible values are arbitrarily given:

$$x_1 = 0$$
$$x_2 = 58.3896 \times 10^3$$
$$x_3 = 0$$
$$x_4 = 58.5714 \times 10^3$$
$$x_5 = 0$$
$$x_6 = 76.8365 \times 10^3$$
$$x_7 = 0$$
$$x_8 = 0$$

Table 11.1 Values of the coefficients for the land use problem

i	b_{1i}	b_{2i} ($\times 10^{-6}$)	d_i	a_i	c_i ($\times 10^3$)
1	56.2	1	0.5	93	1450
2	58	1.1	0.55	97	2050
3	45.4	1	1.1	32	132.95
4	47	1.1	1.3	35	78
5	57	1	0.5	108	680
6	59	1.1	0.6	115	650
7	46	1	1.1	34	5000
8	48	1.1	1.4	41	489
9	3370	100		1915	4
10	5000	110		1910	
11	1650	105		2355	
12	1480	105		2350	
13	3364	100		1970	
14	4991	115		1960	
15	1607	105		255	
16	1429	100		250	

$x_9 = 0.3550 \times 10^3$
$x_{10} = 0.2760 \times 10^3$
$x_{11} = 0$
$x_{12} = 0.5323 \times 10^3$
$x_{13} = 0$
$x_{14} = 0$
$x_{15} = 2.5510 \times 10^3$
$x_{16} = 1.4489 \times 10^3$

$F_1^{(0)} = 0.14982 \times 10^{11}$
$F_2^{(0)} = 0.15436 \times 10^6$
$\lambda_{12}^{(0)} = 0.58256 \times 10^1$

1. Iteration 1
   a. α Decomposition: The following are given from step 0:

$$\lambda_{12}^{\alpha(1)} = \lambda_{12}^{(0)} = (p_2 + qF_2^{(0)})/p_1 = 0.58256 \times 10^1 \qquad \text{(dollars/ton)}$$

and

$$\begin{aligned}\mathbf{y}_\alpha &= [y_{\alpha 1}, y_{\alpha 2}, y_{\alpha 3}, y_{\alpha 4}]^T \\ &= [x_5, x_6, x_7, x_8]^T \\ &= [0.0, 76.8365 \times 10^3, 0.0, 0.0]^T \qquad \text{(acres)}\end{aligned}$$

Given the above trade-off values, the trade-offs in the αi ($i = 1, 2$) subsystems are obtained by Eq. (10.58). In this case, $\lambda_{12}^{\alpha} = \lambda_{12}^{\alpha 1} = \lambda_{12}^{\alpha 2}$ by Eq. (10.58).

We solve the $\alpha 1$ and the $\alpha 2$ subsystems and obtain the solutions and the following values of objectives:

$\alpha 1$ Subsystem:

$x_1 = 0$
$x_2 = 0.$
$x_3 = 0.93469 \times 10^4$
$x_4 = 0.90482 \times 10^4$
$f_1^{\alpha 1} = 0.17827 \times 10^9$
$f_2^{\alpha 1} = 0.22044 \times 10^5$

$\alpha 2$ Subsystem:

$x_5\ (= y_{\alpha 1}) = 0$
$x_6\ (= y_{\alpha 2}) = 0.76836 \times 10^5$
$x_7\ (= y_{\alpha 3}) = 0$
$x_8\ (= y_{\alpha 4}) = 0$
$x_9 = 0$
$x_{10} = 0$

$x_{11} = 0.74485 \times 10^3$
$x_{12} = 0.41865 \times 10^3$
$x_{13} = 0.35602 \times 10^3$
$x_{14} = 0.27660 \times 10^3$
$x_{15} = 0.18024 \times 10^4$
$x_{16} = 0.15290 \times 10^4$
$f_1^{\alpha 2} = 0.71813 \times 10^{10}$
$f_2^{\alpha 2} = 0.46102 \times 10^5$

$F_1^{\alpha(1)} = 0.73595 \times 10^{10}$ (dollars)
$F_2^{\alpha(1)} = 0.68146 \times 10^5$ (tons)
$\lambda_{12}^{\alpha(1)} = 0.46102 \times 10^5$ (dollars/ton)

b. β Decomposition: The following are given from the α decomposition:

$$\lambda_{12}^{\beta(1)} = (p_2 + qF_2^{\alpha(1)})/p_1 = 9.92743 \times 10^1 \qquad \text{(ton)}$$

$$\begin{aligned} y_\beta &= (y_{\beta 1}, y_{\beta 2}, y_{\beta 3}, y_{\beta 4})^T \\ &= (x_9, x_{10}, x_{11}, x_{12})^T \\ &= (0.0, 0.0, 0.74485 \times 10^3, 0.41865 \times 10^3)^T \qquad \text{(acres)} \end{aligned}$$

Given the above trade-off values, the trade-offs in the βi ($i = 1, 2$) subsystems are obtained by Eq. (10.59). (In this case, $\lambda_{12}^{\beta} = \lambda_{12}^{\beta 1} = \lambda_{12}^{\beta 2}$ and $\lambda_{12}^{\beta 2}$ is not considered because $f_2^{\beta 2}$ is always zero.)

We solve the β1 and the β2 subsystems to obtain the following values of objectives:

β1 Subsystem:

$x_1 = 0.69788 \times 10^4$
$x_2 = 0.66173 \times 10^4$
$x_3 = 0.84014 \times 10^4$
$x_4 = 0.81326 \times 10^4$
$x_5 = 0.23810 \times 10^4$
$x_6 = 0.23687 \times 10^4$
$x_7 = 0.26255 \times 10^4$
$x_8 = 0.28814 \times 10^4$
$f_1^{\beta 1} = 0.27015 \times 10^9$ (dollars)
$f_2^{\beta 1} = 0.36476 \times 10^5$ (tons)

β2 Subsystem:

$x_9\ (= y_{\beta 1}) = 0.0$
$x_{10}\ (= y_{\beta 2}) = 0.0$
$x_{11}\ (= y_{\beta 3}) = 0.74485 \times 10^3$
$x_{12}\ (= y_{\beta 4}) = 0.41865 \times 10^5$

$x_{14} = 0.27660 \times 10^3$
$x_{15} = 0.18024 \times 10^4$
$x_{16} = 0.15290 \times 10^4$
$f_1^{\beta 2} = 0.68251 \times 10^9$ (dollars)
$f_2^{\beta 2} = 0.0$ (tons)

$F_1^{\beta(1)} = 0.95266 \times 10^9$ (dollars)
$F_2^{\beta(1)} = 0.36476 \times 10^5$ (tons)
$\lambda_{12}^{\beta(1)} = 0.92742 \times 10^1$ (dollars/ton)

2. Iteration 2
   a. α Decomposition: The following are given from the β decomposition:

$$\lambda_{12}^{\alpha(2)} = 0.10541 \times 10^2 \qquad \text{(dollars/ton)}$$

and

$$\mathbf{y}_\alpha = (0.23810 \times 10^4, 0.23687 \times 10^4, 0.26255 \times 10^4, 0.28814 \times 10^4)^T \qquad \text{(acres)}$$

We solve the $\alpha 1$ and $\alpha 2$ subsystems again, to obtain the solutions and the following values of objectives:

$\alpha 1$ Subsystem:

$x_1 = 0.69787 \times 10^4$
$x_2 = 0.66173 \times 10^4$
$x_3 = 0.84013 \times 10^4$
$x_4 = 0.81326 \times 10^4$
$f_1^{\alpha 1} = 0.24175 \times 10^9$ (dollars)
$f_2^{\alpha 1} = 0.26943 \times 10^5$ (tons)

$\alpha 2$ Subsystem:
$x_5\ (= y_{\alpha 1}) = 0.23810 \times 10^4$
$x_6\ (= y_{\alpha 2}) = 0.23687 \times 10^4$
$x_7\ (= y_{\alpha 3}) = 0.26255 \times 10^1$
$x_8\ (= y_{\alpha 4}) = 0.28814 \times 10^4$
$x_9 = 0.17802 \times 10^3$
$x_{10} = 0.14137 \times 10^3$
$x_{11} = 0.12754 \times 10^4$
$x_{12} = 0.95437 \times 10^3$
$x_{13} = 0.17754 \times 10^3$
$x_{14} = 0.13493 \times 10^3$
$x_{15} = 0.12691 \times 10^4$

$x_{16} = 0.98255 \times 10^3$
$f_1^{\alpha 2} = 0.58064 \times 10^9$ (dollars)
$f_2^{\alpha 2} = 0.95337 \times 10^4$ (tons)

$F_1^{\alpha(2)} = 0.82239 \times 10^9$ (dollars)
$F_2^{\alpha(2)} = 0.36476 \times 10^5$ (tons)
$\lambda_{12}^{\alpha(2)} = 0.10541 \times 10^2$ (dollars/tons)

b. β Decomposition: The following is given from the α decomposition:

$$\lambda_{12}^{\beta(2)} = 0.10541 \times 10^2 \qquad \text{(dollars/ton)}$$

This $\lambda_{12}^{\beta(2)}$ is exactly the same as $\lambda_{12}^{\alpha(2)}$. We can now stop the calculation because convergence of the preference has been obtained. In real decisionmaking, this implies that a decisionmaker has found the indifference trade-off. If we continue to solve the β1 and the β2 subsystems with this value of $\lambda_{12}^{\beta(2)}$ and the given $\mathbf{y}_\beta = (0.17802 \times 10^3, 0.14137 \times 10^3, 0.12754 \times 10^4, 0.95437 \times 10^3)^T$ in order to check whether the same solution is obtained, then we will obtain the solutions and the values of objectives as follows:

β1 Subsystem:

$x_1 = 0.69788 \times 10^4$
$x_2 = 0.66174 \times 10^4$
$x_3 = 0.84015 \times 10^4$
$x_4 = 0.81328 \times 10^4$
$x_5 = 0.23808 \times 10^4$
$x_6 = 0.23685 \times 10^4$
$x_7 = 0.26251 \times 10^4$
$x_8 = 0.28808 \times 10^4$
$f_1^{\beta 1} = 0.27015 \times 10^9$ (dollars)
$f_2^{\beta 1} = 0.36475 \times 10^5$ (tons)

β2 Subsystem:

$x_9\ (= y_{\beta 1}) = 0.17802 \times 10^3$
$x_{10}\ (= y_{\beta 2}) = 0.14137 \times 10^3$
$x_{11}\ (= y_{\beta 3}) = 0.12754 \times 10^4$
$x_{12}\ (= y_{\beta 4}) = 0.95437 \times 10^3$
$x_{13} = 0.17754 \times 10^3$
$x_{14} = 0.13493 \times 10^3$
$x_{15} = 0.12691 \times 10^4$
$x_{16} = 0.98255 \times 10^3$
$f_1^{\beta 2} = 0.55224 \times 10^9$ (dollars)
$f_2^{\beta 2} = 0.0$ (tons)

$F_1^{\beta(2)} = 0.82239 \times 10^9$ (dollars)
$F_2^{\beta(2)} = 0.36475 \times 10^5$ (tons)
$\lambda_{12}^{\beta(2)} = 0.10541 \times 10^2$ (dollars/ton)

Of course, these values are the same as those that were obtained in the α decomposition at iteration 2. The final result is summarized as follows:

$x_1 = 0.69788 \times 10^4$ (acres)
$x_2 = 0.66174 \times 10^4$
$x_3 = 0.84015 \times 10^4$
$x_4 = 0.81328 \times 10^4$
$x_5 = 0.23808 \times 10^4$
$x_6 = 0.23685 \times 10^4$
$x_7 = 0.26251 \times 10^4$
$x_8 = 0.28808 \times 10^4$
$x_9 = 0.17802 \times 10^3$
$x_{10} = 0.14137 \times 10^3$
$x_{11} = 1.2754 \times 10^3$
$x_{12} = 0.95437 \times 10^3$
$x_{13} = 0.17754 \times 10^3$
$x_{14} = 0.13493 \times 10^3$
$x_{15} = 1.2691 \times 10^3$
$x_{16} = 0.98255 \times 10^3$

$F_1\ (= F_1^{\alpha} = F_1^{\beta}) = 0.82239 \times 10^9$ (dollars)
$F_2\ (= F_2^{\alpha} = F_2^{\beta}) = 0.36475 \times 10^5$ (tons)
$\lambda_{12}\ (= \lambda_{12}^{\alpha} = \lambda_{12}^{\beta}) = 0.10541 \times 10^2$ (dollars/ton)

11.1.3 Checking for Noninferiority

In Sect. 11.1.2, the solution was obtained by the MHOC procedure. In this section, we check whether the solution obtained satisfies the noninferiority of the overall problem. That is, we will check whether the MHOC procedure gives a noninferior solution or not.

The noninferiority of the solution can be checked by Eq. (10.81) or (10.82). However, we will check it by solving the overall problem again, without using decomposition, and by then comparing the solution with the one obtained by the MHOC procedure.

Given $\lambda_{12}^* = 0.10541 \times 10^2$, we will solve the following problem without decomposition:

$$\min F_1(\mathbf{x}) + \lambda_{12}^* F_2(\mathbf{x}) \tag{11.11}$$

subject to the constraints given in Eqs. (11.2)–(11.10) and $\mathbf{x} \geqslant \mathbf{0}$.

We then obtain the solution shown in Table 11.2. These solutions are the same as those obtained by the MHOC procedure in the previous section, except for the last digits of the solutions. Noninferiority is satisfied; the MHOC procedure has worked very well in this example.

Table 11.2 Solutions without decomposition

Acreage parameter	Solution (acres × 10^4)	Acreage parameter	Solution (acres × 10^3)
x_1	0.69789	x_9	0.17802
x_2	0.66174	x_{10}	0.14137
x_3	0.84018	x_{11}	1.2754
x_4	0.81331	x_{12}	0.95437
x_5	0.23806	x_{13}	0.17754
x_6	0.23685	x_{14}	0.13493
x_7	0.26235	x_{15}	1.2691
x_8	0.28804	x_{16}	0.98254

In this example, λ_{12} is defined only as a function of F_2 for the sake of simplicity. In such a case, it may still be possible to improve F_1 in some problems, even if λ_{12} is convergent to a certain value. In such cases, the stopping criteria—Eqs. (10.65) and (10.66), that is, $|F_i^{\alpha(n+1)} - F_i^{\alpha(n)}| < \epsilon$ or $|F_i^{\beta(n+1)} - F_i^{\beta(n)}| < \epsilon$ or, more simply, $|F_i^{\beta(n)} - F_i^{\alpha(n)}| < \epsilon$ or $|F_i^{\alpha(n+1)} - F_i^{\beta(n)}| < \epsilon$ ($i = 1, 2$ for this example)—must be used together with the stopping criteria of trade-off λ_{12}, that is, Eq. (10.67).

In real decisionmaking, in which a decisionmaker would be interested in the values of both F_1 and F_2, this problem could not arise because the values of both F_1 and F_2 would influence the decisionmaker's preference λ_{12}. The stopping criteria of λ_{12} would be sufficient in actual decisionmaking. However, it would always be safer to check the stopping criteria of the objective functions together with the stopping criteria for λ_{12}. For a more detailed discussion on the convergence properties of HOC, the reader may refer to Shima and Haimes (1984) and Shima et al. (1982).

11.2 MANAGEMENT PLANNING FOR AN INDUSTRIAL ORGANIZATION

11.2.1 Introduction

A large industrial corporation is a typical example of a large-scale system. Such a system has many aspects and can be analyzed from several points of view. For example, the following activities are usually part of the functioning of a large industrial corporation: production, marketing, finance and accounting, logistics, information processing, and research and development. There will also be several planning levels, such as strategic, tactical, and operational planning.

Some planning problems are well structured and others are unstructured. The management of a large industrial corporation is very complex, and it is necessary to integrate many functions to achieve better performance of the corporation.

For decisionmaking for management planning problems (such as production and inventory planning) at the upper levels of large industrial corporations (such as a multinational corporation), the so-called matrix organization may be the most efficient and effective structural format.

A typical matrix organization at the higher decisionmaking level of an electric company is shown in Fig. 11.2. In this matrix organization, one side of the production system happens to be a product type of hierarchical structure (product decomposition) and another side happens to be a geographic type of hierarchical structure (geographic or market decomposition), where the market characteristics are different from each other. The overlapping hierarchical structures are shown in Figs. 11.3 and 11.4 in two different ways.

A more general type of matrix organization having more than two sides (aspects of the system) is of course possible, too. An example that has three sides (aspects) is shown in Fig. 11.5.

A matrix organization at the higher decisionmaking level provides more efficient and effective management planning because many aspects of the complex system can be considered simultaneously. For example, the production of each product is well considered and planned by means of the product type of decomposition, and the distribution of them is well considered and planned by the geographic (market) type of decomposition. These different decompositions help facilitate both decisionmaking and information exchange within the corporation. (See more details in Davis, 1976, 1979; Lawrence et al., 1977; Vancil, 1973.)

The management planning problem of this matrix type of organization seems well suited to applications of MHOC procedures, which we investigate in the next simplified numerical example. Note that the matrix structure is assumed for use only at the higher decisionmaking level of management planning. In our example,

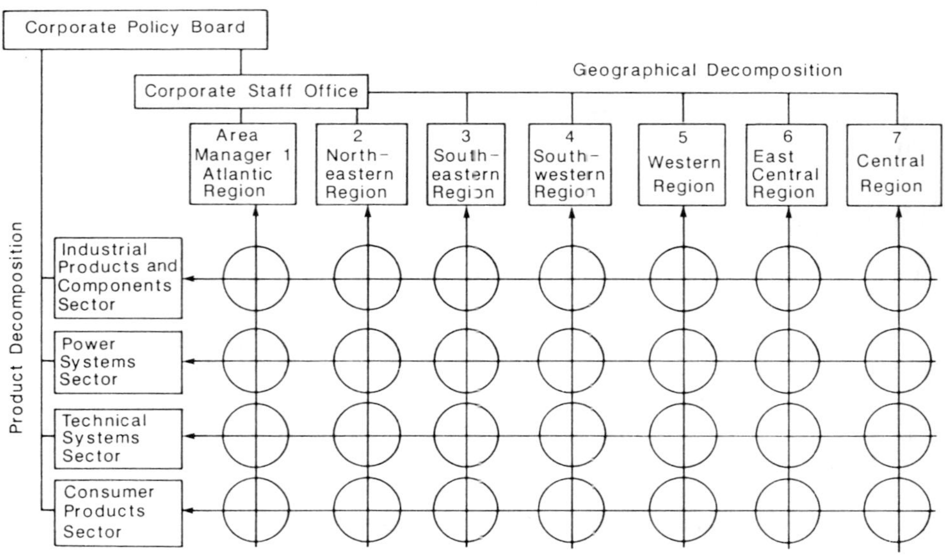

Figure 11.2 Matrix organization of products division of an electric company.

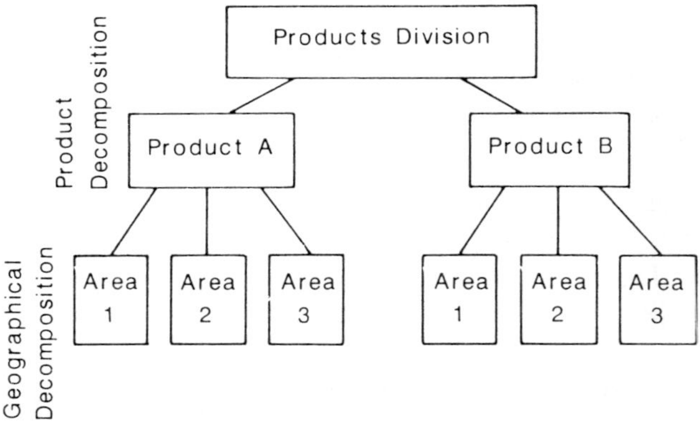

Figure 11.3 Product-geographic decomposition of matrix organization depicted in Fig. 11.2.

each aspect of the matrix organization has a different (physical) hierarchical structure, and different people work in different hierarchical structures (that is, a person belongs to only one hierarchical structure), so that the difficult problems of the human factors (such as the confusion of orders from the upper levels to the lower level, which is usual in matrix organization) cannot occur (see Lawrence et al., 1977, for a discussion of the problems of the human factor).

11.2.2 Industrial Problem: Example

We will consider the following hypothetical problem (see Shima et al., 1989). Suppose that a Canadian-based industrial company markets its product in Canada and in the United States in both the east and the south. Suppose its head office is in Toronto, it has branches in several cities in Canada and the United States, and it has four plants in or near Montreal (plant A), Toronto (plant B), Milwaukee (plant C), and Atlanta (plant D). The branches are assumed to have inventories and to supply their product to their local markets. The head office and the branches

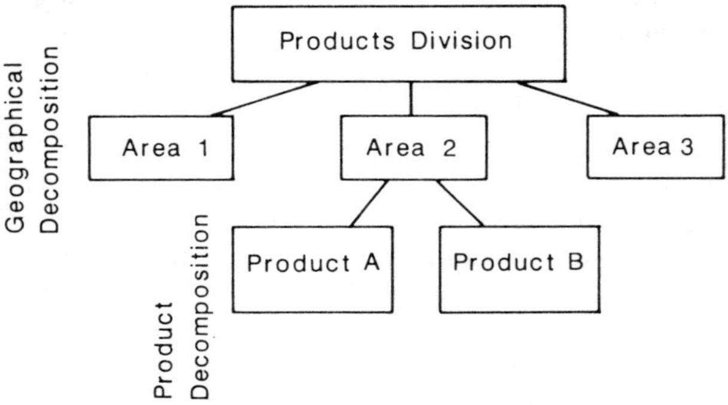

Figure 11.4 Geographic-product decomposition of matrix organization depicted in Fig. 11.2.

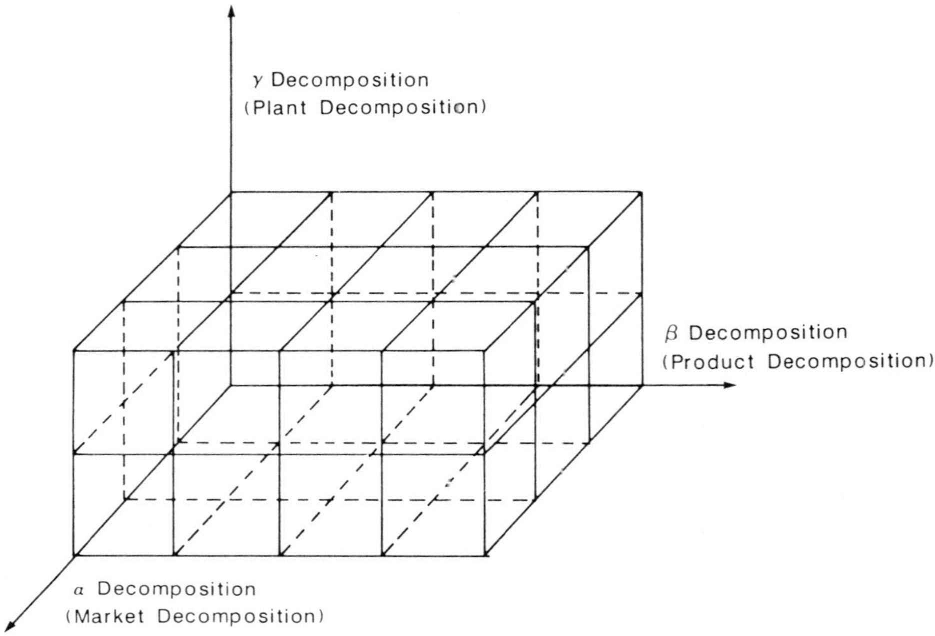

Figure 11.5 Generalized matrix organization, $n = 3$.

are connected by computer networks, and information (such as local market demand and the amount of product in each inventory) is collected in the data base of the main computer in the head office. This information is managed by several managers, who are assumed (in this example) to be of two types: plant managers and area (market) managers. Suppose there are two plant managers, one of whom takes care of the plants in Canada (plants A and B) and the other of whom takes care of the plants in the United States (plants C and D). Suppose also that there are two area (market) managers, one of whom manages the local markets in Canada and northern and eastern United States and the other of whom manages the local markets in the southern United States. This distribution arises from the consumption patterns of the two groups of local markets, which are quite different from each other, especially in the winter. This management structure is outlined in Fig. 11.6.

This company must make a production plan for a certain period and this plan must satisfy the demands of the local branches. For simplicity, it is assumed that lead time has already been taken into account in these demands. The company's distribution network (consisting of 21 local markets) is shown in Fig. 11.7. Figure 11.8 shows the conceptual form of this network, which is a simplified version of Fig. 11.7. When the managers make a production plan, they must take into account distribution costs in addition to production costs. The production plan is influenced also by the distribution plan for the product. The company must choose the least expensive plan. The company also must take into account many other aspects, such as minimizing the consumption of energy and minimizing any impact on environmental quality.

The plant managers are concerned with the production part of the problem, and the area managers are concerned with the distribution part, and this company can be operated very well by the cooperation of these managers. The plant decomposition (α decomposition) and the area or market decomposition (β decomposition) are shown in Fig. 11.9. The decisionmaking and information structure of this company has already been shown in Fig. 11.6.

11.2.2.1 Problem formulation Suppose for the sake of illustration that the company has two objectives: (1) minimizing the cost (the production cost plus distribution and other costs), and (2) minimizing the total energy consumption. Its constraints are (1) demand constraints and (2) production constraints. The overall problem is formulated as follows:

$$\min_{\mathbf{p},\mathbf{x}} \begin{bmatrix} F_1(\mathbf{x}, \mathbf{p}) = \sum_{i=1}^{4} (a_i p_i^2 + b_i p_i + c_i) + \sum_{i=1}^{37} (d_i x_i^2 + e_i) \\ F_2(\mathbf{x}, \mathbf{p}) = \sum_{i=1}^{4} (f_i p_i + g_i) \end{bmatrix} \qquad \begin{matrix} (11.12) \\ (11.13) \end{matrix}$$

subject to

$$x_1 - x_2 = s_a \tag{11.14}$$

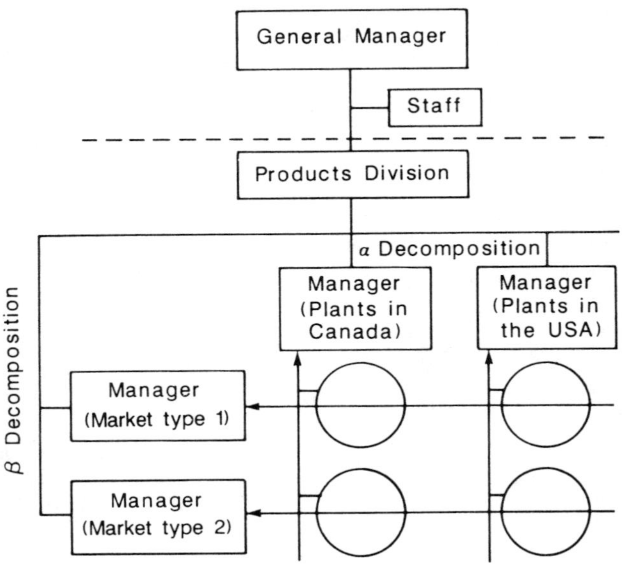

Figure 11.6 Management structure of the example problem.

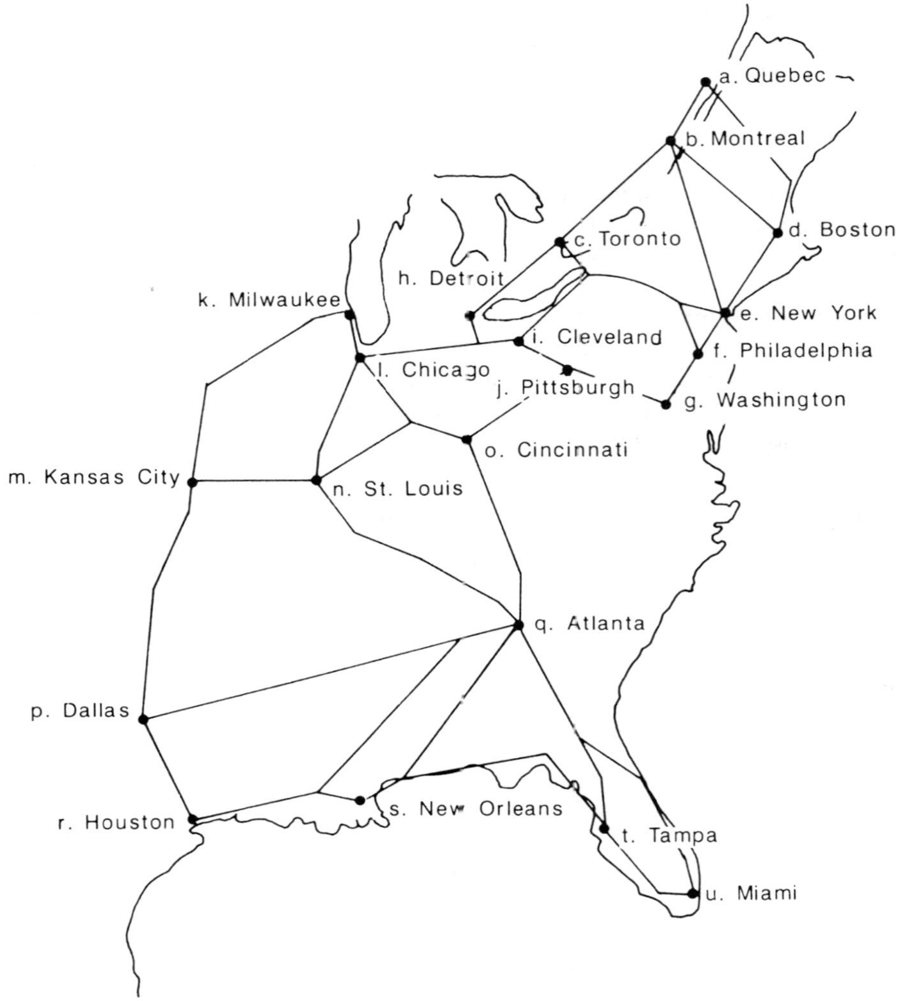

Figure 11.7 Distribution network of the company.

$$-x_1 - x_3 - x_5 - x_6 + p_1 = s_b \tag{11.15}$$

$$x_6 - x_7 - x_8 - x_{11} - x_{12} - x_{13} + p_2 = s_c \tag{11.16}$$

$$x_2 + x_3 - x_4 = s_d \tag{11.17}$$

$$x_4 + x_5 + x_7 - x_9 = s_e \tag{11.18}$$

$$x_8 + x_9 - x_{10} = s_f \tag{11.19}$$

$$x_{10} + x_{14} = s_g \tag{11.20}$$

$$x_{11} - x_{15} - x_{16} = s_h \qquad (11.21)$$

$$x_{12} + x_{16} - x_{17} + x_{18} - x_{21} = s_i \qquad (11.22)$$

$$x_{13} - x_{14} + x_{17} = s_j \qquad (11.23)$$

$$- x_{19} - x_{22} + p_3 = s_k \qquad (11.24)$$

$$x_{15} - x_{18} + x_{19} - x_{20} - x_{23} = s_l \qquad (11.25)$$

$$x_{22} - x_{24} - x_{26} = s_m \qquad (11.26)$$

$$x_{23} + x_{24} - x_{25} - x_{27} = s_n \qquad (11.27)$$

$$x_{20} + x_{21} + x_{25} - x_{28} = s_o \qquad (11.28)$$

$$x_{26} - x_{29} + x_{30} = s_p \qquad (11.29)$$

$$x_{27} + x_{28} - x_{30} - x_{31} - x_{32} - x_{33} - x_{34} + p_4 = s_q \qquad (11.30)$$

$$x_{29} + x_{31} - x_{35} = s_r \qquad (11.31)$$

$$x_{32} + x_{35} - x_{36} = s_s \qquad (11.32)$$

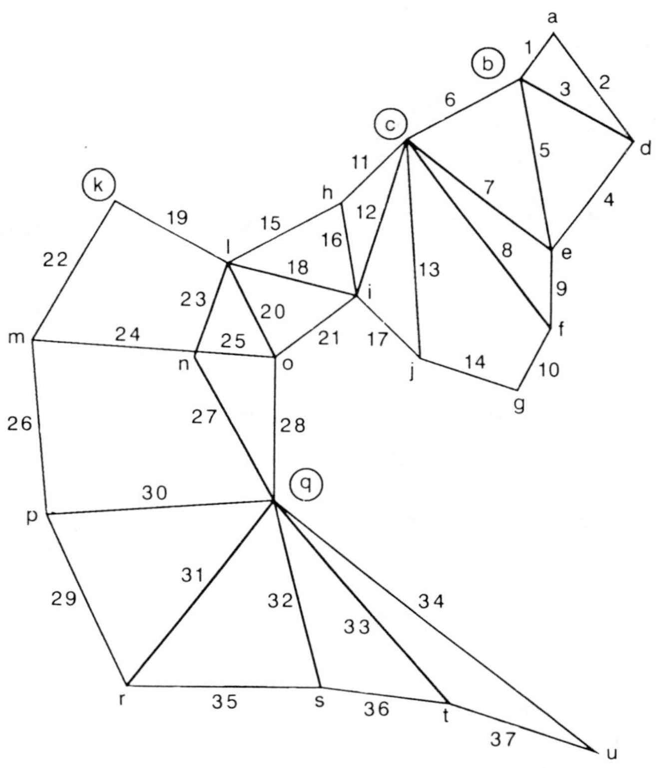

Figure 11.8 Conceptual network (○ indicates plant locations).

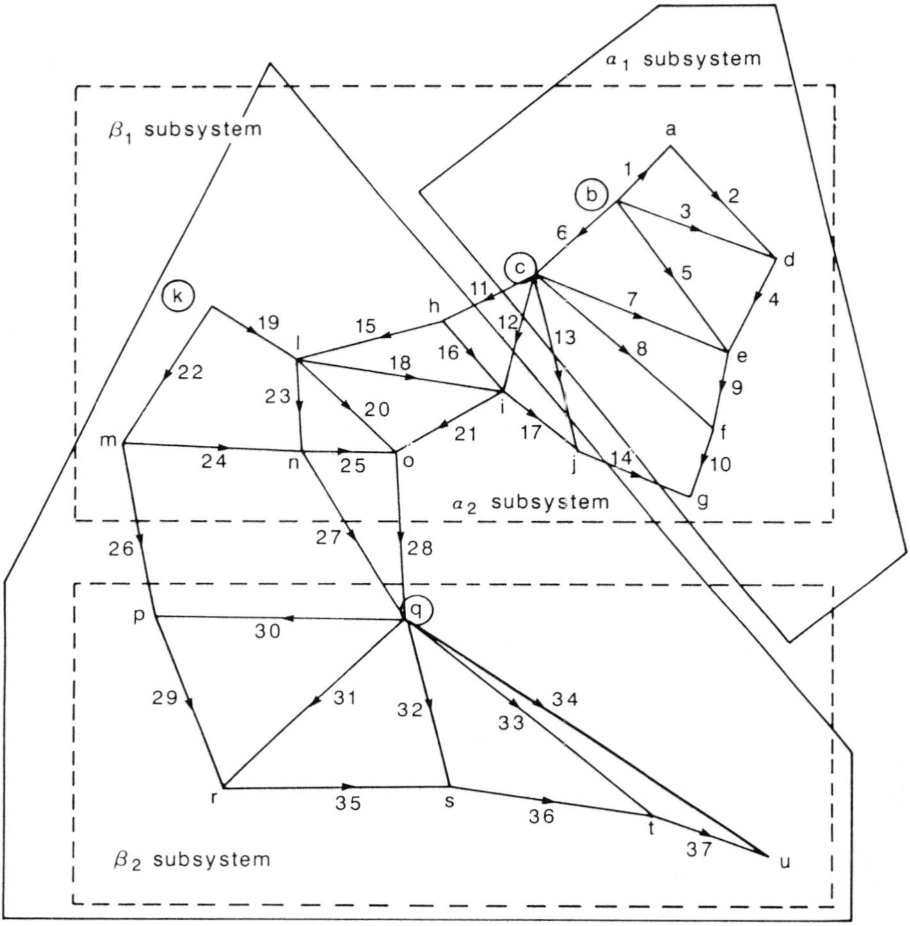

Figure 11.9 The network and the α and β decomposition. ○, places where plants are located; (—), α decomposition; (----), β decomposition.

$$x_{33} + x_{36} - x_{37} = s_t \tag{11.33}$$

$$x_{34} + x_{37} = s_u \tag{11.34}$$

$$p_{i,\min} \leqslant p_i \leqslant p_{i,\max} \qquad (i = 1, \ldots, 4) \tag{11.35}$$

where

F_1 = approximate total cost [(production cost) + (distribution and other costs)] ($\times\ 10^2$ dollars)
F_2 = approximate energy consumption (kiloliters, kl)
p_i = the amount of products produced at the ith plant ($\times\ 10^3$ units)

x_i = the amount of products distributed on the ith path of the network ($\times\ 10^3$ units)
s_i = the demand of ith market ($\times\ 10^3$ units)
$p_{i,\min}, p_{i,\max}$ = production limit and constraint of the ith plant ($\times\ 10^3$ units)

The α decomposition (plant managers' subproblems) and the β decomposition (area managers' subproblems) are as follows:

α Decomposition (plant decomposition)

α1 Subsystem (α1 plant managers' subproblem):

$$\min \begin{bmatrix} f_1^{\alpha 1} = \sum_{i=1}^{2} (a_i p_i^2 + b_i p_i + c_i) + \sum_{i=1}^{14} (d_i x_i^2 + e_i) \\ f_2^{\alpha 1} = \sum_{i=1}^{2} (f_i p_i + g_i) \end{bmatrix} \tag{11.36}$$

subject to

$$\left. \begin{aligned} x_1 - x_2 &= s_a \\ -x_1 - x_3 - x_5 - x_6 + p_1 &= s_b \\ x_6 - x_7 - x_8 - x_{11} - x_{12} - x_{13} + p_2 &= s_c \\ x_2 + x_3 - x_4 &= s_d \\ x_4 + x_5 + x_7 - x_9 &= s_e \\ x_8 + x_9 - x_{10} &= s_f \\ x_{10} + x_{14} &= s_g \\ x_{11} &= y_{\alpha 1} \\ x_{12} &= y_{\alpha 2} \\ x_{13} &= y_{\alpha 3} \\ x_{14} &= y_{\alpha 4} \\ p_{i,\max} \geqslant p_i \geqslant p_{i,\min} \qquad i &= 1, 2 \end{aligned} \right] \tag{11.37}$$

α2 Subsystem (α2 plant managers' subproblem):

$$\min \begin{bmatrix} f_1^{\alpha 2} = \sum_{i=3}^{4} (a_i p_i^2 + b_i p_i + c_i) + \sum_{i=15}^{37} (d_i x_i^2 + e_i) \\ f_2^{\alpha 2} = \sum_{i=3}^{4} (f_i p_i + g_i) \end{bmatrix} \tag{11.38}$$

subject to

$$\left.\begin{aligned}
-x_{15}-x_{16} &= s_h - y_{\alpha 1}\\
x_{16}-x_{17}+x_{18}-x_{21} &= s_i - y_{\alpha 2}\\
+x_{17} &= s_j - y_{\alpha 3} + y_{\alpha 4}\\
-x_{19}-x_{22}+p_3 &= s_k\\
x_{15}-x_{18}+x_{19}-x_{20}-x_{23} &= s_l\\
x_{22}-x_{24}-x_{26} &= s_m\\
x_{23}+x_{24}-x_{25}-x_{27} &= s_n\\
x_{20}+x_{21}+x_{25}-x_{28} &= s_o\\
x_{26}-x_{29}+x_{30} &= s_p\\
x_{27}+x_{28}-x_{30}-x_{31}-x_{32}-x_{33}-x_{34}+p_4 &= s_q\\
x_{29}+x_{31}-x_{35} &= s_r\\
x_{32}+x_{35}-x_{36} &= s_s\\
x_{33}+x_{36}-x_{37} &= s_t\\
x_{34}+x_{37} &= s_u\\
p_{i,\min} \leqslant p_i \leqslant p_{i,\max} \qquad i &= 3,\ 4
\end{aligned}\right] \tag{11.39}$$

***β* Decomposition [*area (market) decomposition*]**

β1 Subsystem (β1 area managers' subproblem):

$$\min \begin{bmatrix} f_1^{\beta 1} = \sum_{i=1}^{3} (a_i p_i^2 + b_i p_i + c_i) + \sum_{i=1}^{25} (d_i x_i^2 + e_i) \\ f_2^{\beta 1} = \sum_{i=1}^{3} (f_i p_i + g_i) \end{bmatrix} \tag{11.40}$$

subject to

$$\left.\begin{aligned}
x_1 - x_2 &= s_a\\
-x_1 - x_3 - x_5 - x_6 + p_1 &= s_b\\
x_6 - x_7 - x_8 - x_{11} - x_{12} - x_{13} + p_2 &= s_c\\
x_2 + x_3 - x_4 &= s_d\\
x_4 + x_5 - x_7 - x_9 &= s_e\\
x_8 + x_9 - x_{10} &= s_f\\
x_{10} + x_{14} &= s_g
\end{aligned}\right.$$

$$
\left.\begin{aligned}
x_{11} - x_{15} - x_{16} &= s_h \\
x_{12} + x_{16} - x_{17} + x_{18} - x_{21} &= s_i \\
x_{13} - x_{14} + x_{17} &= s_j \\
- x_{19} - x_{22} + p_3 &= s_k \\
x_{15} - x_{18} + x_{19} - x_{20} - x_{23} &= s_l \\
x_{22} - x_{24} &= s_m + y_{\beta 1} \\
x_{23} + x_{24} - x_{25} &= s_n + y_{\beta 2} \\
x_{20} + x_{21} + x_{25} &= s_o + y_{\beta 3} \\
p_{i,\min} \leqslant p_i \leqslant p_{i,\max} \qquad i &= 1,\ 2,\ 3
\end{aligned}\right] \tag{11.41}
$$

β2 Subsystem (β2 area managers' subproblem):

$$
\min \begin{bmatrix} f_1^{\beta 2} = (a_4 p_4^2 + b_4 p_4 + c_4) + \sum_{i=26}^{37} (d_i x_i^2 + e_i) \\ \\ f_2^{\beta 2} = (f_4 p_4 + g_4) \end{bmatrix} \tag{11.42}
$$

subject to

$$
\left.\begin{aligned}
x_{26} - x_{29} + x_{30} &= s_p \\
x_{27} + x_{28} - x_{30} - x_{31} - x_{32} - x_{33} - x_{34} + p_4 &= s_q \\
x_{29} + x_{31} - x_{35} &= s_r \\
x_{32} + x_{35} - x_{36} &= s_s \\
x_{33} + x_{36} - x_{37} &= s_t \\
x_{34} + x_{37} &= s_u \\
x_{26} &= y_{\beta 1} \\
x_{27} &= y_{\beta 2} \\
x_{28} &= y_{\beta 3} \\
p_{4,\min} \leqslant p_4 \leqslant p_{4,\max} &
\end{aligned}\right] \tag{11.43}
$$

Here x_i $(i = 1, \ldots, 37)$ is assumed to be positive if it flows according to the direction of the arrow shown in Fig. 11.9; negative x_i means that the amount x_i is distributed in the reverse direction of the arrow.

11.2.2.2 Solution procedure We will solve the problem using the top-down overlapping coordination approach. For the purpose of illustration, we assume the existence of a single mathematical decisionmaker who has the following utility function implicitly in mind:

$$U(F_1, F_2) = \mathbf{p}_0\mathbf{F} + \tfrac{1}{2}\,\mathbf{F}^T Q_0 \mathbf{F}$$

where $\mathbf{p}_0 = [p_1, p_2]$ and $\mathbf{F} = [F_1, F_2]^T$. Let

$$Q_0 = \begin{bmatrix} q_{11} & q_{12} \\ q_{21} & q_{22} \end{bmatrix} = \begin{bmatrix} q_{11} & 0 \\ 0 & q_{22} \end{bmatrix}$$

Then

$$\lambda_{12}(\mathbf{F}) = \frac{\partial U/\partial F_2}{\partial U/\partial F_1} = \frac{p_2 + q_{22}F_2}{p_1 + q_{11}F_1}$$

$$\boldsymbol{\lambda}(\mathbf{F}) = [1, \lambda_{12}(\mathbf{F})]^T$$

where λ_{12} has a unit 10^2 \$/kl, and

$$\mathbf{F} = [F_1, F_2]^T \qquad \mathbf{F} = \mathbf{F}^\alpha = \mathbf{F}^\beta$$

$$\boldsymbol{\lambda}(\mathbf{F}) = \boldsymbol{\lambda}^\alpha(\mathbf{F}^\alpha) = \boldsymbol{\lambda}^\beta(\mathbf{F}^3)$$

We will assume that the coefficients are defined as shown in Table 11.3. The

Table 11.3 Coefficients of the example problem

i	a_i	b_i	c_i	f_i	g_i	$P_{i,\min}$	$P_{i,\max}$
1	0.05	50	400	50	700	100	500
2	0.06	90	500	30	500	150	600
3	0.04	30	300	70	800	80	500
4	0.02	20	200	110	1000	80	600

i	d_i	e_i	i	d_i	e_i	i	s_i
1	0.01	30	21	0.03	50	a	20
2	0.05	70	22	0.07	100	b	50
3	0.03	50	23	0.04	60	c	70
4	0.02	40	24	0.03	50	d	50
5	0.06	80	25	0.05	70	e	200
6	0.08	60	26	0.09	110	f	150
7	0.03	50	27	0.11	130	g	50
8	0.07	100	28	0.09	110	h	70
9	0.01	20	29	0.03	50	i	50
10	0.01	30	30	0.16	180	j	50
11	0.02	40	31	0.18	200	k	20
12	0.01	30	32	0.11	130	l	170
13	0.04	60	33	0.08	100	m	50
14	0.02	40	34	0.1	120	n	50
15	0.03	50	35	0.07	80	o	50
16	0.01	30	36	0.15	170	p	80
17	0.01	30	37	0.04	60	q	80
18	0.05	70				r	80
19	0.03	50				s	70
20	0.04	60				t	70
						u	70

Table 11.4 Initial values

i	x_i (× 10^3 units)	i	x_i (× 10^3 units)	i	P_i (× 10^3 units)
1	20	21	0	1	320
2	0	22	50	2	440
3	50	23	50	3	340
4	0	24	0	4	450
5	200	25	0		
6	0	26	0		
7	0	27	0		
8	200	28	0		
9	0	29	0		
10	50	30	80		
11	70	31	80		
12	50	32	70		
13	50	33	70		
14	0	34	70		
15	0	35	0		
16	0	36	0		
17	0	37	0		
18	0				
19	270				
20	50				

following are also assumed to be given: $p_1 = 20$, $p_2 = 40$, $q_{11} = -7.707 \times 10^{-5}$, and $q_{22} = -10^4$.

Here, the coefficients have the following units:

a_i: 10^2 dollars per 10^6 units
b_i: 10^2 dollars per 10^3 units
c_i: 10^2 dollars
d_i: 10^2 dollars per 10^3 units
e_i: 10^2 dollars
f_i: kl per 10^3 units
g_i: kl
p_i: 10^3 units
s_i: 10^3 units

We are now ready to solve the problem. We will stop the calculation if the following conditions are satisfied:

$$|\lambda_{12}^{\beta(n)} - \lambda_{12}^{\alpha(n)}| \leq 10^{-2} \qquad \text{or} \qquad |\lambda_{12}^{\alpha(n+1)} - \lambda_{12}^{\beta(n)}| \leq 10^{-2}$$

and

$$\Delta_i = |F_i^{\beta(n)} - F_i^{\alpha(n)}| \leq 1.0 \qquad \text{or} \qquad \Delta_i = |F_i^{\alpha(n+1)} - F_i^{\beta(n)}| \leq 1.0$$

Note that F_i ($i = 1, 2$) has a large value; ($\cdot$) means the iteration number.

We solve the problem.

0. First, the initial feasible values given in Table 11.4 are arbitrarily assumed. Then

$$F_1^{(0)} = F_1^{\alpha(0)} = 115{,}056\ (\times\ 10^2 \text{ dollars})$$

$$F_2^{(0)} = F_2^{\alpha(0)} = 105{,}500\ (\text{kl})$$

Based on the initial information, the decisionmaker evaluates the trade-off. Here, having assumed a mathematical decisionmaker for illustrative purposes, we have

$$\lambda_{12}^{(0)} = \lambda_{12}^{\alpha(1)} = \frac{p_2 + q_{22}F_2^{\alpha}(0)}{p_1 + q_{11}F_1^{\alpha}(0)} = 2.66 \qquad (\times\ 10^2 \text{ dollars/kl})$$

1. Iteration 1
   a. From the initial feasible values,

$$\mathbf{y}_\alpha = [y_{\alpha1}, y_{\alpha2}, y_{\alpha3}, y_{\alpha4}]^T = [x_{11}, x_{12}, x_{13}, x_{14}]^T$$
$$= [70, 50, 50, 0]^T \qquad (\times\ 10^3 \text{ units})$$

Given the above trade-off $\lambda_{12}^{\alpha(1)}$, the trade-off $\lambda_{12}^{\alpha i(1)}$ in each αi subsystem is calculated using Eq. (10.58). Given the obtained values of trade-offs in

Table 11.5 Solutions for iteration 1, α decomposition, multiobjective case

Subsystem	i	x_i ($\times$ 10³ units)	i_s/P_i	i_s/P_i ($\times$ 10³ units)
$\alpha 1^a$	1	70.8	9	127.2
	2	50.8	10	50.0
	3	108.3	11 ($y_{\alpha1}$)	70.0
	4	109.1	12 ($y_{\alpha2}$)	50.0
	5	90.5	13 ($y_{\alpha3}$)	50.0
	6	20.0	14 ($y_{\alpha4}$)	0.0
	7	127.6	P_1	339.6
	8	72.8	P_2	420.4
$\alpha 2^b$	15	−21.8	28	36.4
	16	21.8	29	47.8
	17	0.0	30	31.9
	18	17.6	31	36.3
	19	317.9	32	62.2
	20	51.6	33	78.6
	21	39.3	34	65.0
	22	162.1	35	4.2
	23	57.0	36	−3.6
	24	16.2	37	5.0
	25	−4.5	—	—
	26	95.9	P_3	500.0
	27	27.7	P_4	290.0

[a] $f_1^{\alpha1} = 75346.9 \times 10^2$ (dollars); $f_2^{\alpha1} = 30792.5$ (kl).
[b] $f_1^{\alpha2} = 43085.9 \times 10^2$ (dollars); $f_2^{\alpha2} = 68700.0$ (kl); $F_1^{\alpha(1)} = 118432.8 \times 10^2$ (dollars); $F_2^{\alpha(1)} = 99492.5$ (kl).

Table 11.6 Solutions for iteration 1, β decomposition, multiobjective case

Subsystem	i	x_i ($\times 10^3$ units)	i/P_i	x_i/P_i ($\times 10^3$ units)
β1[a]	1	81.9	15	81.4
	2	61.9	16	−18.3
	3	130.6	17	17.5
	4	142.5	18	−51.5
	5	112.4	19	126.0
	6	59.9	20	17.9
	7	66.1	21	110.9
	8	45.0	22	110.2
	9	121.0	23	71.1
	10	16.0	24	−35.7
	11	133.1	25	−42.4
	12	248.3	P_1	434.8
	13	66.5	P_2	569.0
	14	34.0	P_3	256.2
β2[b]	26 ($y_{\beta 1}$)	95.9	33	78.7
	27 ($y_{\beta 2}$)	27.7	34	65.0
	28 ($y_{\beta 3}$)	36.4	35	4.2
	29	47.9	36	−3.6
	30	32.0	37	5.0
	31	36.4	—	—
	32	62.2	P_4	290.0

[a] $f_1^{\beta 1} = 120849.6 \times 10^2$ (dollars); $f_2^{\beta 1} = 58743.8$ (kl).
[b] $f_1^{\beta 2} = 11969.7 \times 10^2$ (dollars); $f_2^{\beta 2} = 32900.0$ (kl); $F_1^{\beta(1)} = 132819.3 \times 10^2$ (dollars); $F_2^{\beta(1)} = 91643.8$ (kl).

the subsystems, the α decomposition is solved in each subsystem. [In this case, $\lambda_{12}^{\alpha} = \lambda_{12}^{\alpha 1} = \lambda_{12}^{\alpha 2}$, using Eq. (10.58).] Then the solutions shown in Table 11.5 are obtained.

b. From the α decomposition, the β decomposition has

$$\mathbf{y}_\beta = [y_{\beta 1}, y_{\beta 2}, y_{\beta 3}]^T = [x_{26}, x_{27}, x_{28}]^T$$
$$= [95.9, 27.7, 36.4]^T \qquad (\times 10^3 \text{ units})$$

Based on the information, the decisionmaker evaluates the trade-off. Again having assumed a mathematical decisionmaker for illustrative purposes [see Haimes and Tarvainen (1980)], we have

$$\lambda_{12}^{\beta(1)} = \frac{p_2 + q_{22} F_2^{\alpha(1)}}{p_2 + q_{11} F_1^{\alpha(1)}} = 2.76 \qquad (\times 10^2 \text{ dollars/kl})$$

Given this trade-off $\lambda_{12}^{\beta(1)}$, the trade-off $\lambda_{12}^{\beta i}$ in each βi subsystem is obtained using Eq. (10.59). Given the obtained values of trade-offs in the subsystems, the β decomposition is solved in each subsystem. In this case, $\lambda_{12}^{\beta} = \lambda_{12}^{\beta 1} = \lambda_{12}^{\beta 2}$ using Eq. (10.59). The solutions shown in Table 11.6 are obtained.

2. Iteration 2
   a. From the β decomposition, the α decomposition has

$$\mathbf{y}_\alpha = [x_{11}, x_{12}, x_{13}, x_{14}]^T = [133.1, 248.3, 66.5, 34.0]^T \qquad (\times 10^3 \text{ units})$$

Based on the information from the β decomposition, the decisionmaker evaluates his or her trade-off and here, in this case, we have $\lambda_{12}^{\alpha(2)} = 3.15$ ($\times 10^2$ dollars/kl). The α decomposition is solved in each subsystem and the solutions shown in Table 11.7 are obtained.

   b. From the α decomposition, the β decomposition has

$$\mathbf{y}_\beta = [x_{26}, x_{27}, x_{28}]^T = [139.2, 94.8, 136.0]^T \qquad (\times 10^3 \text{ units})$$

Based on the information obtained in the α decomposition, the decisionmaker evaluates his or her trade-off and here, in this case, we have $\lambda_{12}^{\beta(2)} = 3.69$ ($\times 10^2$ dollars/kl).

The β decomposition is solved in each subsystem, and the solutions shown in Table 11.8 are obtained.

Table 11.7 Iteration 2, α decomposition, multiobjective case

Subsystem	i	x_i ($\times 10^3$ units)	i/P_i	x_i/P_i ($\times 10^3$ units)
α1[a]	1	78.2	9	115.5
	2	58.2	10	16.0
	3	123.3	11 ($y_{\alpha 1}$)	133.1
	4	131.5	12 ($y_{\alpha 2}$)	248.3
	5	105.4	13 ($y_{\alpha 3}$)	66.5
	6	50.3	14 ($y_{\alpha 4}$)	34.0
	7	78.6	P_1	407.2
	8	50.5	P_2	596.6
α2[b]	15	58.4	28	136.0
	16	4.7	29	73.6
	17	17.5	30	14.4
	18	−34.2	31	25.1
	19	263.9	32	52.9
	20	70.8	33	75.4
	21	151.2	34	63.1
	22	182.3	35	18.6
	23	115.8	36	1.5
	24	−6.9	37	6.9
	25	−36.0	—	—
	26	139.2	P_3	466.2
	27	94.8	P_4	80.0

[a] $f_1^{\alpha 1} = 108874.2 \times 10^2$ (dollars); $f_2^{\alpha 1} = 39458.0$ (kl).
[b] $f_1^{\alpha 2} = 39825.5 \times 10^2$ (dollars); $f_2^{\alpha 2} = 43233.4$ (kl); $F_1^{\alpha(2)} = 147799.7 \times 10^2$ (dollars); $F_2^{\alpha(2)} = 82691.4$ (kl).

Table 11.8 Iteration 2, β decomposition, multiobjective case

Subsystem	i	x_i (× 10^3 units)	i/P_i	x_i/P_i (× 10^3 units)
β1[a]	1	90.7	15	98.7
	2	70.7	16	−13.4
	3	148.2	17	−7.8
	4	168.9	18	−61.8
	5	130.4	19	188.3
	6	80.7	20	57.4
	7	45.8	21	179.5
	8	40.4	22	161.7
	9	145.2	23	121.3
	10	35.5	24	−27.4
	11	155.3	25	−51.0
	12	296.9	P_1	500.0
	13	72.3	P_2	600.0
	14	14.5	P_3	370.0
β2[b]	26 ($y_{\beta 1}$)	139.2	33	75.4
	27 ($y_{\beta 2}$)	94.8	34	63.1
	28 ($y_{\beta 3}$)	136.0	35	18.6
	29	73.6	36	1.5
	30	14.4	37	6.9
	31	25.1	—	—
	32	52.8	P_4	80.0

[a] $f_1^{\beta 1} = 142496.2 \times 10^2$ (dollars); $f_2^{\beta 1} = 70900$ (kl).
[b] $f_1^{\beta 2} = 9260.5 \times 10^2$ (dollars); $f_2^{\beta 2} = 9800$ (kl); $F_1^{\beta(2)} = 151{,}756.7 \times 10^2$ (dollars); $F_2^{\beta(2)} = 80{,}700$ (kl).

3. Iteration 3
   a. From the β decomposition, the α decomposition has

$$\mathbf{y}_\alpha = [x_{11}, x_{12}, x_{13}, x_{14}]^T$$
$$= [155.3, 296.9, 72.3, 14.5]^T \qquad (\times 10^3 \text{ units})$$

   Based on the information obtained in the β decomposition, the decisionmaker evaluates the trade-off, and here we have $\lambda_{12}^{\alpha(3)} = 3.84$ (× 10^2 dollars/kl). Given this trade-off, the α decomposition is solved in each subsystem, and the solutions shown in Table 11.9 are obtained.

   b. From the α decomposition, the β decomposition has

$$\mathbf{y}_\beta = [x_{26}, x_{27}, x_{28}]^T = [135.1, 93.6, 141.3]^T \qquad (\times 10^3 \text{ units})$$

   The decisionmaker also evaluates the trade-off, and here we have $\lambda_{12}^{\beta(3)} = 3.84$ (× 10^2 dollars/kl). The preference (trade-off) is converged. However, the β problem must be solved again to get a noninferior solution. The β decomposition is solved in each subsystem, and the solutions shown in Table 11.10 are obtained.

Table 11.9 Iteration 3, α decomposition, multiobjective case

Subsystem	i	x_i (× 10^3 units)	i/P_i	x_i/P_i (× 10^3 units)
$\alpha 1^a$	1	90.8	9	145.2
	2	70.8	10	35.5
	3	148.2	11 ($y_{\alpha 1}$)	155.3
	4	169.0	12 ($y_{\alpha 2}$)	296.9
	5	130.4	13 ($y_{\alpha 3}$)	72.3
	6	80.6	14 ($y_{\alpha 4}$)	14.5
	7	45.8	P_1	500.0
	8	40.4	P_2	600.0
$\alpha 2^b$	15	97.8	28	141.3
	16	−12.5	29	71.4
	17	−7.8	30	16.3
	18	−60.9	31	26.0
	19	190.1	32	53.6
	20	59.0	33	75.8
	21	181.3	34	63.2
	22	159.9	35	17.4
	23	119.8	36	1.0
	24	−25.2	37	6.8
	25	−49.0		
	26	135.1	P_3	370.0
	27	93.6	P_4	80.0

[a] $f_1^{\alpha 1} = 119{,}780.3 \times 10^2$ (dollars); $f_1^{\alpha 1} = 44{,}200$ (kl).
[b] $f_1^{\alpha 2} = 31{,}969.7 \times 10^2$ (dollars); $f_2^{\alpha 2} = 36{,}500$ (kl); $F_1^{\alpha(3)} = 151{,}750.0 \times 10^2$ (dollars); $F_2^{\alpha(3)} =$ 80,700 (kl).

The decisionmaker's preference (trade-off) has already been converged. That is,

$$|\lambda_{12}^{\beta(3)} - \lambda_{12}^{\alpha(3)}| \leq 10^{-2}$$

and

$$\Delta_1 = |F_1^{\beta(3)} - F_1^{\alpha(3)}| = 0$$

$$\Delta_2 = |F_2^{\beta(3)} - F_2^{\alpha(3)}| = 0$$

We stop the calculation.

To check the solution, we solve the problem without decomposition. That is, we solve the following problem with the given value of the trade-off $\lambda_{12}^* = 3.84$:

$$\min F_1(\mathbf{x}) + \lambda_{12}^* F_2(\mathbf{x})$$

subject to the same constraints, Eqs. (11.14), ..., (11.35), with the same initial values.

Then, the solutions shown in Table 11.11 are obtained. Those solutions are almost the same as the solutions obtained by the MHOC procedure.

In this example, a mathematical decisionmaker was assumed for illustrative purposes. In real decisionmaking, however, the human decisionmaker (the top manager, or the members of the board of directors) must interact each time that the managers of the α or the β decompositions respond, that is, whenever the α and the β decompositions are solved. In this example problem, we have assumed that the decisionmaker is only interested in overall performance, that is, the values of the overall objective functions. We solved the problem by the top-down decisionmaking approach. To find the preferences (the indifference trade-offs) each time, the decisionmaker needs to be helped by the computer and by his or her staff (systems analysts). To find the preferences efficiently, a computer graphics display may be utilized.

11.2.3 Single-Objective HOC Case

In this section, we will solve a single-objective HOC problem. To do this, we will ignore the second objective of the previous example; that is, we will not consider the energy consumption. The decisionmaker is assumed to be interested only in minimizing the cost. Here, the overall problem, the α subproblems, and the β subproblems are defined in exactly the same way, except that the second objective is ignored. The overall problem is therefore defined as follows:

Table 11.10 Iteration 3, β decomposition, multiobjective case

Subsystem	i	x_i ($\times 10^3$ units)	i/P_i	x_i/P_i ($\times 10^3$ units)
β1[a]	1	90.8	15	97.9
	2	70.8	16	−12.8
	3	148.1	17	−7.9
	4	168.9	18	−61.3
	5	130.4	19	189.7
	6	80.7	20	59.0
	7	45.9	21	180.8
	8	40.3	22	160.3
	9	145.3	23	119.8
	10	35.6	24	−24.8
	11	155.1	25	−48.5
	12	297.0	P_1	500.0
	13	72.3	P_2	600.0
	14	14.4	P_3	370.0
β2[b]	26 ($y_{\beta 1}$)	135.1	33	75.7
	27 ($y_{\beta 2}$)	93.6	34	63.3
	28 ($y_{\beta 3}$)	141.3	35	17.3
	29	71.2	36	1.0
	30	16.0	37	6.7
	31	26.1	—	—
	32	53.7	P_4	80.0

[a] $f_1^{\beta 1} = 142{,}463.4 \times 10^2$ (dollars); $f_2^{\beta 1} = 70{,}900$ (kl).

[b] $f_1^{\beta 2} = 9286.6 \times 10^2$ (dollars); $f_2^{\beta 2} = 9800$ (kl); $F_1^{\beta(3)} = 151{,}750.0 \times 10^2$ (dollars); $F_2^{\beta(3)} = 80{,}700$ (kl).

Table 11.11 Solutions of the problem solved without decompositions[a]

i	x_i	i	x_i or p_i
1	90.9	21	181.0
2	70.9	22	160.1
3	148.1	23	119.7
4	169.0	24	−24.8
5	130.3	25	−48.7
6	80.7	26	134.9
7	46.0	27	93.7
8	40.3	28	141.4
9	145.3	29	71.1
10	35.6	30	16.2
11	155.0	31	26.2
12	297.1	32	53.7
13	72.3	33	75.7
14	14.4	34	63.3
15	97.9	35	17.3
16	−12.9	36	1.0
17	−7.9	37	6.7
18	−61.1	P_1	500.0
19	189.9	P_2	600.0
20	59.1	P_3	370.0
		P_4	80.0

[a] $F_1 = 151{,}750.0 \times 10^2$ (dollars); $F_2 = 80{,}700.0$ (kl).

Table 11.12 Iteration 1, α decomposition, single-objective case

Subsystem	i	x_i ($\times 10^3$ units)	i/F_i	x_i/P_i ($\times 10^3$ units)
α1[a]	1	91.2	9	155.0
	2	71.2	10	50.0
	3	149.1	11 ($y_{\alpha 1}$)	70.0
	4	170.3	12 ($y_{\alpha 2}$)	50.0
	5	131.3	13 ($y_{\alpha 3}$)	50.0
	6	78.4	14 ($y_{\alpha 4}$)	0.0
	7	53.4	P_1	500.0
	8	45.0	P_2	260.0
α2[b]	15	4.9	28	−74.8
	16	−4.9	29	6.2
	17	0.0	30	60.3
	18	−3.9	31	54.7
	19	144.1	32	77.2
	20	−11.5	33	83.9
	21	−8.8	34	68.0
	22	63.7	35	−19.2
	23	−5.7	36	−11.9
	24	−12.2	37	2.0
	25	−4.6	P_3	227.8
	26	25.9	P_4	562.2
	27	−63.3		

[a] $f_{\alpha 1} = 70{,}381.3 \times 10^2$ (dollars).
[b] $f_{\alpha 2} = 33{,}812.4 \times 10^2$ (dollars); $f_\alpha^{(1)}$ $(= f_{\alpha 1} + f_{\alpha 2})$ = 104.193.7 $\times 10^2$ (dollars).

$$\min_{\mathbf{x},\mathbf{p}} f(\mathbf{x}, \mathbf{p}) = \sum_{i=1}^{4} (a_i p_i^2 + b_i p_i + c_i) + \sum_{i=1}^{37} (d_i x_i^2 + e_i)$$

subject to Eqs. (11.14)–(11.35).

The α and the β decompositions are also represented in the same way, except that the second objective is ignored; that is,

α Decomposition

α1 Subsystem:

$$\min f_{\alpha 1} = \sum_{i=1}^{2} (a_i p_i^2 + b_i p_i + c_i) + \sum_{i=1}^{14} (d_i x_i^2 + e_i)$$

subject to Eq. (11.37).

α2 Subsystem:

$$\min f_{\alpha 2} = \sum_{i=3}^{4} (a_i p_i^2 + b_i p_i + c_i) + \sum_{i=15}^{37} (d_i x_i^2 + e_i)$$

subject to Eq. (11.39).

Table 11.13 Iteration 1, β decomposition, single-objective case

Subsystem	i	x_i ($\times 10^3$ units)	i/P_i	x_i/P_i ($\times 10^3$ units)
β1[a]	1	72.2	15	99.3
	2	52.2	16	18.3
	3	110.0	17	105.0
	4	113.2	18	63.2
	5	93.3	19	333.3
	6	43.2	20	20.7
	7	71.4	21	−77.9
	8	41.8	22	114.8
	9	77.9	23	−19.9
	10	−30.4	24	39.0
	11	−11.0	25	32.4
	12	−4.3	P_1	369.6
	13	25.3	P_2	150.0
	14	80.4	P_3	468.2
β2[b]	26 ($y_{\beta 1}$)	25.9	33	83.9
	27 ($y_{\beta 2}$)	−63.3	34	68.0
	28 ($y_{\beta 3}$)	−74.8	35	−19.1
	29	6.2	36	−11.9
	30	60.3	37	2.0
	31	54.7		
	32	77.3	P_4	562.2

[a] $f_{\beta 1} = 72{,}659.7 \times 10^2$ (dollars).
[b] $f_{\beta 2} = 23{,}060.7 \times 10^2$ (dollars); $f_\beta^{(1)}$ $(= f_{\beta 1} + f_{\beta 2}) = 95{,}720.4 \times 10^2$ (dollars).

β Decomposition

β1 Subsystem:

$$\min f_{\beta 1} = \sum_{i=1}^{3} (a_i p_i^2 + b_i p_i + c_i) + \sum_{i=1}^{25} (d_i x_i^2 + e_i)$$

subject to Eq. (11.41).

β2 Subsystem:

$$\min f_{\beta 2} = (a_4 p_4^2 + b_4 p_4 + c_4) + \sum_{i=26}^{37} (d_i x_i^2 + e_i)$$

subject to Eq. (11.43).

We now assume the same initial values for **p** and **x** (see Table 11.12) and solve the single-objective HOC problem. We stop the calculation if the following condition is satisfied:

$$\Delta_1 = |f_\beta^{(n)} - f_\alpha^{(n)}| \leq 1.0 \qquad \text{or} \qquad \Delta_2 = |f_\alpha^{(n+1)} - f_\beta^{(n)}| \leq 1.0$$

(Note that f has a large value.)

Table 11.14 Iteration 2, α decomposition, single-objective case

Subsystem	i	x_i (× 10^3 units)	i/P_i	x_i/P_i (× 10^3 units)
α1[a]	1	72.2	9	77.9
	2	52.2	10	−30.4
	3	111.0	11 ($y_{\alpha 1}$)	−11.0
	4	113.2	12 ($y_{\alpha 2}$)	−4.3
	5	93.3	13 ($y_{\alpha 3}$)	25.3
	6	43.2	14 ($y_{\alpha 4}$)	80.4
	7	71.4	P_1	369.6
	8	41.7	P_2	150.0
α2[b]	15	−93.8	28	−103.8
	16	12.8	29	6.7
	17	105.0	30	60.0
	18	58.9	31	54.4
	19	305.7	32	77.1
	20	7.9	33	83.8
	21	−87.7	34	68.0
	22	104.7	35	−18.8
	23	−24.9	36	−11.7
	24	27.9	37	2.0
	25	26.0	P_3	430.4
	26	26.8	P_4	600.0
	27	−73.0		

[a] $f_{\alpha 1}$ = 43,749.1 × 10^2 (dollars).
[b] $f_{\alpha 2}$ = 50,968.6 × 10^2 (dollars); $f_\alpha^{(2)}$ (= $f_{\alpha 1} - f_{\alpha 2}$) = 94,717.7 × 10^2 (dollars).

Table 11.15 Iteration 2, β decomposition, single-objective case

Subsystem	i	x_i (× 10^3 units)	i/P_i	x_i/P_i (× 10^3 units)
β1[a]	1	70.0	15	−100.6
	2	50.0	16	16.2
	3	106.6	17	110.3
	4	106.6	18	63.5
	4	88.9	19	318.3
	6	32.2	20	10.1
	7	76.7	21	−92.5
	8	43.1	22	108.3
	9	72.1	23	−25.8
	10	−34.7	24	31.5
	11	−14.4	25	28.7
	12	−11.9	P_1	353.5
	13	24.5	P_2	150.0
	14	84.7	P_3	446.5
β2[b]	26 ($y_{\beta 1}$)	26.7	33	83.9
	27 ($y_{\beta 2}$)	−73.0	34	67.9
	28 ($y_{\beta 3}$)	−103.8	35	−18.8
	29	6.7	36	−11.8
	30	60.0	37	2.1
	31	54.4	P_4	600.0
	32	77.1		

[a] $f_{\beta 1} = 69{,}383.9 \times 10^2$ (dollars).
[b] $f_{\beta 2} = 25{,}292.6 \times 10^2$ (dollars); $f_{\beta}^{(2)}$ $(= f_{\beta 1} + f_{\beta 2}) = 94{,}676.5 \times 10^2$ (dollars).

1. Iteration 1.
   a. Assuming the initial feasible values shown in Table 11.4, $\mathbf{y}_\alpha$ is given as follows:

$$\mathbf{y}_\alpha = (y_{\alpha 1}, y_{\alpha 2}, y_{\alpha 3}, y_{\alpha 4})^T = (x_{11}, x_{12}, x_{13}, x_{14})^T$$
$$= (70, 50, 50, 0)^T \qquad (\times 10^3 \text{ units})$$

   The α problem is solved in each subsystem. The solutions shown in Table 11.12 are obtained.

   b. From the α decomposition, the β decomposition has

$$\mathbf{y}_\beta = (y_{\beta 1}, y_{\beta 2}, y_{\beta 3})^T = (x_{26}, x_{27}, x_{28})^T$$
$$= (25.9, -63.3, -74.8)^T \qquad (\times 10^3 \text{ units})$$

   The β decomposition is solved in each subsystem. The solutions shown in Table 11.13 are obtained.

2. Iteration 2.
   a. From the β decomposition, the α decomposition has

$$\mathbf{y}_\alpha = (-11.10, -4.3, 25.3, 80.4)^T \qquad (\times 10^3 \text{ units})$$

Table 11.16 Iteration 3, α decomposition, single-objective case

Subsystem	i	x_i (× 10^3 units)	i/P_i	x_i/P_i (× 10^3 units)
α1[a]	1	70.0	9	72.1
	2	50.0	10	−34.7
	3	106.7	11 ($y_{\alpha 1}$)	−14.4
	4	106.6	12 ($y_{\alpha 2}$)	−11.9
	5	88.9	13 ($y_{\alpha 3}$)	24.5
	6	37.9	14 ($y_{\alpha 4}$)	84.7
	7	76.6	P_1	353.5
	8	43.1	P_2	150.0
α2[b]	15	−100.4	28	−104.7
	16	16.1	29	7.1
	17	110.3	30	59.7
	18	63.4	31	54.2
	19	318.1	32	76.9
	20	9.8	33	83.8
	21	−92.7	34	67.9
	22	108.5	35	−18.6
	23	−25.5	36	−11.7
	24	31.0	37	2.1
	25	28.3	P_3	446.5
	26	27.5	P_4	600.0
	27	−72.8		

[a] $f_{\alpha 1} = 42{,}244.2 \times 10^2$ (dollars).
[b] $f_{\alpha 2} = 52{,}432.2 \times 10^2$ (dollars); $f_\alpha^{(3)}$ $(= f_{\alpha 1} + f_{\alpha 2}) = 94{,}676.4 \times 10^2$ (dollars).

Table 11.17 Solutions obtained without decompositions[a]

i	x_i	i	x_i or P_i
1	69.9	21	−92.8
2	49.9	22	108.4
3	106.7	23	−25.5
4	106.7	24	31.1
5	88.9	25	28.3
6	37.9	26	27.3
7	76.8	27	−72.7
8	43.1	28	−104.6
9	72.4	29	7.1
10	−34.4	30	59.8
11	−13.3	31	54.3
12	−13.5	32	76.9
13	24.7	33	83.8
14	84.4	34	67.9
15	−100.2	35	−18.6
16	16.8	36	−11.7
17	109.7	37	2.1
18	63.6	P_1	353.5
19	318.1	P_2	150.0
20	9.9	P_3	446.5
		P_4	600.0

[a] $f = 94{,}676.4 \times 10^2$ (dollars).

The α decomposition is solved in each subsystem. The solutions shown in Table 11.14 are obtained.

b. From the α decomposition, the β decomposition has

$$\mathbf{y}_\beta = (26.6, -73.1, -103.8)^T \qquad (\times 10^3 \text{ units})$$

The β decomposition is solved again in each subsystem. The solutions shown in Table 11.15 are obtained.

3. Iteration 3.

a. From the β decomposition, the α decomposition has

$$\mathbf{y}_\alpha = (-14.4, -11.9, 24.5, 84.7)^T \qquad (\times 10^3 \text{ units})$$

The α decomposition is solved in each subsystem. The solutions shown in Table 11.16 are obtained. Now, $\Delta = |f_\alpha^{(3)} - f_\beta^{(2)}| = 0.1 < 1.0$.

We stop the calculation. (Note that $\Delta f/f_\alpha^{(3)} = 1 \times 10^{-7}$.)

In order to check the solution, we solve the overall problem without decomposition for purposes of illustration. This will give almost the same solutions as those shown in Table 11.17. In this subsection, the single-objective problem has been solved. We modified the multiobjective problem by ignoring the second objective completely. Because of this, a better value for the first objective was obtained in the single-objective case than in the multiobjective case that was considered in the previous section.

REFERENCES

Davis, S. M., Trends in the Organization of Multinational Corporations, *Columbia J. World Business*, Summer 1976. [Also in *Organization*, ed. J. P. Kotter, L. A. Schlesinger, and V. Sathe, Richard D. Irwin, Inc., Homewood, Ill., 1979].

Davis, S. M., *Managing and Organizing Multinational Corporations*, Pergamon Press, New York, 1979.

Haimes, Y. Y., Hierarchical Holographic Modeling, *IEEE Trans. Syst., Man Cybern.*, vol. SMC-11, no. 9, pp. 606–617, 1981.

Haimes, Y. Y., and Macko, D., Hierarchical Structures in Water Resources Systems Management, *IEEE Trans. Syst. Man Cybern.*, vol. SMC-3, no. 4, pp. 396–402, 1973.

Haimes, Y. Y., and Tarvainen, K., Hierarchical-Multiobjective Framework for Large Scale Systems, in *Multicriteria Analysis in Practice*, ed. P. Nijkamp and J. Spronk, Gower Press, London, 1980.

Lawrence, P. R., Kolodny, H. F., and Davis, S. M., The Human Side of the Matrix, *Organization Dynamics*, Summer 1977. [Also in *Organization*, ed. J. P. Kotter, L. A. Schlessinger, and V. Sathe, Richard D. Irwin, Inc., Homewood, Ill., 1979.]

Macko, D., and Haimes, Y. Y., Overlapping Coordination of Hierarchical Structures, *IEEE Trans. Syst. Man Cybern.*, vol. SMC-8, no. 10, pp. 745–751, 1978.

Mendu, S. R., Haimes, Y. Y., and Macko, D., Computational Aspects of Overlapping Coordination Methodology for Linear Hierarchical Systems, *IEEE Trans. Syst. Man Cybern.*, vol. SMC-10, no. 2, pp. 68–78, 1980.

Shima, T., and Haimes, Y. Y., Convergence Properties of Hierarchical Overlapping Coordination, *IEEE Trans. Syst. Man Cybern.*, vol. SMC-14, no. 1, 1984.

Shima, T., Tarvainen, K., and Haimes, Y. Y., Multiobjective Hierarchical Overlapping Coordination, paper presented at the Fifth International Conference on Multiple Criterian Decision Making, Mons, Belgium, 1982.

Shima, T., Haimes, Y Y., Tarvainen, K., and Sung, K., Applications of Hierarchical Overlapping Coordination with Multiple Objectives, *Inform. Decision Technol.*, vol. 15, no. 1, pp. 33–57, 1989.

Sung, K., *Coordination of Overlapping Hierarchical Water Resources Systems*, Ph.D. dissertation, Case Western Reserve University, Cleveland, Ohio, 1978.

Vancil, R. F., What Kind of Management Control Do You Need? in *Organization*, ed. J. P. Kotter, L. A. Schlessinger, and V. Sathe, R. D. Irwin Inc., Homewood, Ill., 1979.

CHAPTER

TWELVE

HIERARCHICAL HOLOGRAPHIC MODELING

12.1 INTRODUCTION

The large-scale systems for which hierarchical optimization methods are developed usually have noncommensurable and conflicting objectives. The hierarchical multiobjective optimization (HMO) models developed by Tarvainen (1981) and Tarvainen and Haimes (1982) incorporate multiobjective functions, as appropriate, within each and every subsystem and level of the hierarchical structure (see Chap. 5). The HMO model has a unique decomposition, depending on the hierarchical structure of the system. However, in the real world, cases exist in which a multiplicity of constituencies, each with distinct and overlapping goals and perceptions, analyze, plan, and control the system. Each constituent prefers to decompose the whole system into several subsystems according to its own needs, study purposes, and data-base structures. Haimes and Macko (1973) and Macko and Haimes (1978) proposed the hierarchical overlapping coordination (HOC) scheme as a means of coordinating the different decompositions arising from pluriconstituencies in the single-objective case. The extensions of the HOC concept to accommodate models with multiple objectives are considered by Shima (1982) and Shima and Haimes (1984) (see Chaps. 9–11).

These HOC models, each under its own constituency, are composed of the same overall variables over constraints, and overall objective functions. Therefore, these models address only the overlapping actions (decisions) and objectives of the different constituencies, neglecting their own nonoverlapping actions and goals. Extending the HOC models to take into account the nonoverlapping (distinct) actions, objectives, and technical and/or input-output relationships motivates this chapter. These overlapping models are referred to as hierarchical holographic submodels (HHSs) (Haimes, 1981). They have both common and distinct sets of objective functions, constraints, and decision variables.

Bateson (1979) has shown that analysis of multiple descriptions—comparing information from two or more sources (i.e., HHSs)—yields a dual bonus: first, it provides two or more distinct views of the same system; second, analysis of the difference between the multiple views yields insights into a new dimension (i.e., information of a higher logical type).

In the real world, many good examples exist where hierarchical holographic modeling (HHM) has potential capability for modeling and consensus solution seeking. A good example is the problem of designing a bridge across a river. At least two groups are interested in this project. One group takes the role of the transportation agency, while the other group represents the environmental protection agency. The operation and maintenance control programs may be viewed as distinct decisions and/or actions under the transportation agency's jurisdiction. The natural resource and water quality preservation programs around the bridge construction area are planned and implemented by the environmental protection agency. The transportation agency and the environmental protection agency are interested in achieving equilibrium between the economic operation of the bridge and preservation of the environmental quality—an overlapping objective which motivates their involvement in designing the bridge capacity. Thus, the decisions and/or actions about the capacity of the bridge can be viewed as overlapping decision variables that are under both groups' interest. The constraints representing the operation and maintenance schedules are under the transportation agency's authority (distinct constraints), while the technical constraints on water quality control programs are viewed as distinct constraints belonging to the environmental protection agency; the constraints representing the technical aspects of the bridge capacity design can be viewed as overlapping constraints appearing in both groups' HHSs. Both groups have varying magnitudes of preference concerning the operation and capacity levels of the bridge. HHM establishes formal linkages, facilitates information exchanges between these two groups, and provides a way to analyze the differences between their preferences, thereby moving away from conflict toward harmony.

In this chapter, we discuss an algorithm that generates one subset of Pareto optimal solutions for two HHSs formulated by two paraindependent planning groups. In the next section we give mathematical formulations of the HHSs in detail and decompose them into submodels. The development of an algorithm to generate the Pareto optimal solutions of two HHSs with particular structured constraint sets is considered in Sect. 12.3.

12.2 MATHEMATICAL FORMULATION

Suppose there are only two independent agencies or groups involved in planning, managing, and controlling a large complex system, and these two have diversified and overlapping goals and aspirations. Call one group the α group and the other the β group. The HHS under the α group's control is denoted by α-HHS, while that of the β group is denoted by β-HHS.

The decision vectors are defined as follows:

$\mathbf{X}$ = the *distinct* decision vector that is completely under the α-group's control
$\mathbf{Z}$ = the *overlapping* decision vector that is common to the α and β groups, who may share either equally or unequally in authority and responsibility for their control and manipulation

The coupling constraints representing the input-output relationships between $\mathbf{X}$ and $\mathbf{Z}$ are assumed to be separable with respect to $\mathbf{X}$ and $\mathbf{Z}$:

$$\mathbf{G}(\mathbf{X}) + \mathbf{H}(\mathbf{Z}) \geqslant \mathbf{b} \tag{12.1}$$

where $\mathbf{b}$ is a vector of dimension l.

The constraints that are composed of only the distinct decision vector $\mathbf{X}$ are denoted by

$$\mathbf{A}(\mathbf{X}) \geqslant \mathbf{0} \tag{12.2}$$

which is a vector of dimension m.

The constraints that consist of only the overlapping decision vector $\mathbf{Z}$ and which also belong to both the α-HHS and the β-HHS are denoted by

$$\mathbf{D}(\mathbf{Z}) \geqslant \mathbf{0} \tag{12.3}$$

which is a vector of dimension n.

These overlapping constraints represent the technical and social relationship pertaining to the overlapping decision vector or the restrictions imposed by the overlapping physical structures of both groups. The multiobjective vector that represents the distinct objectives of the α group, which may be noncommensurable and conflicting, is denoted by

$$\mathbf{E}(\mathbf{X}) \tag{12.4}$$

which is to be minimized and has r objective components:

$$\mathbf{E}(\mathbf{X}) = [E_1(\mathbf{X}), \ldots, E_r(\mathbf{X})]^T$$

The multiobjective vector that represents the overlapping objectives of both groups is denoted by $\mathbf{O}(\mathbf{Z})$, which is a function of only the overlapping decision vector $\mathbf{Z}$ and has the dimension o:

$$\mathbf{O}(\mathbf{Z}) = [O_1(\mathbf{Z}), \ldots, O_o(\mathbf{Z})]^T \tag{12.5}$$

We assume that all the objective components in E and O are to be minimized. All the decision variables are assumed to be nonnegative.

The α-HHS is summarized as follows:

$$\min\ [\mathbf{E}(\mathbf{X}), \mathbf{O}(\mathbf{Z})]^T \tag{12.6}$$

subject to

$$\mathbf{G}(\mathbf{X}) + \mathbf{H}(\mathbf{Z}) \geqslant \mathbf{B} \qquad A(\mathbf{X}) \geqslant \mathbf{0} \qquad D(\mathbf{Z}) \geqslant \mathbf{0} \qquad \mathbf{X}, \mathbf{Z} \geqslant \mathbf{0}$$

The β-HHS of the β group is formulated in a similar fashion:

$$\min \, [\mathbf{F}(\mathbf{Y}), \mathbf{O}(\mathbf{Z})]^T \tag{12.7}$$

subject to

$$I(\mathbf{Y}) + J(\mathbf{Z}) \geqslant \mathbf{C} \qquad K(\mathbf{Y}) \geqslant \mathbf{0} \qquad D(\mathbf{Z}) \geqslant \mathbf{0} \qquad \mathbf{Y}, \mathbf{Z} \geqslant \mathbf{0}$$

where **Y** is the distinct decision vector, which is completely controlled by the β group. **Z** is already defined above.

$\mathbf{I}(\mathbf{Y}) + \mathbf{J}(\mathbf{Z}) \geqslant \mathbf{C}$ represents the coupling constraint vector of dimension u, reflecting the input-output relationships between the distinct decision vector **Y** and overlapping decision vector **Z**.

$\mathbf{K}(\mathbf{Y}) \geqslant \mathbf{0}$ is the distinct constraint vector of dimension v, which is formulated, restricted, and controlled by the β group. $F(\mathbf{Y})$ is the multiobjective vector of dimension w, which represents those objectives of the β group that do not overlap those of the α group. The multiobjective vectors **E**, **F**, and **O** are assumed to be nonlinear, while the constraint sets can be nonlinear or linear or both.

We will now try to decompose each HHS into two submodels, one with respect to the distinct decision vector and the other one with respect to the overlapping decision vector.

The α group and the β group can assign upper bound vectors, say **L** and **M** for **H**(**Z**) and **J**(**Z**), respectively.

This is done in order to decouple the coupling constraints with respect to the distinct decision vector and the overlapping decision vector. That is, Eq. (12.1) becomes

$$\mathbf{G}(\mathbf{X}) \geqslant \mathbf{B} - \mathbf{L} \qquad \mathbf{H}(\mathbf{Z}) = \mathbf{L} \tag{12.8}$$

The coupling constraining vector of the β-HHS, $I(\mathbf{Y}) + J(\mathbf{Z}) \geqslant \mathbf{C}$, is equivalent to

$$\mathbf{I}(\mathbf{Y}) \geqslant \mathbf{C} - \mathbf{M} \qquad \mathbf{J}(\mathbf{Z}) = \mathbf{M} \tag{12.9}$$

Here

$\mathbf{L} = [\mathbf{L}_1 \ \mathbf{L}_2 \ \dots \ \mathbf{L}_v]^T$ is specified by the α group.
$\mathbf{M} = [\mathbf{M}_1 \ \mathbf{M}_2 \ \dots \ \mathbf{M}_u]^T$ is specified by the β group.

Using the assigned values of **L**, the α-HHS summarized in Eq. (12.6) is equivalent to the following problem:

$$\min \, [E(\mathbf{X}), O(\mathbf{Z})]^T$$

subject to

$$\mathbf{G}(\mathbf{X}) \geqslant \mathbf{B} - \mathbf{L} \qquad \mathbf{H}(\mathbf{Z}) = \mathbf{L} \qquad \mathbf{A}(\mathbf{X}) \geqslant \mathbf{0} \qquad \mathbf{D}(\mathbf{Z}) \geqslant \mathbf{0} \qquad \mathbf{X}, \mathbf{Z} \geqslant \mathbf{0} \tag{12.10}$$

which can be decomposed into the following two submodels:

$$\min \mathbf{E}(\mathbf{X})$$

subject to

$$\mathbf{G}(\mathbf{X}) \geqslant \mathbf{B} - \mathbf{L} \qquad \mathbf{A}(\mathbf{X}) \geqslant \mathbf{0} \qquad \mathbf{X} \geqslant \mathbf{0} \tag{12.11}$$

and

$$\min \mathbf{O}(\mathbf{Z})$$

subject to

$$\mathbf{H}(\mathbf{Z}) = \mathbf{L} \qquad \mathbf{D}(\mathbf{Z}) \geqslant \mathbf{0} \qquad \mathbf{Z} \geqslant \mathbf{0} \tag{12.12}$$

In similar fashion, using the values of $\mathbf{M}$, the β-HHS [Eq. (12.7)] is equivalent to the following multiobjective problem:

$$\min [\mathbf{F}(\mathbf{Y}), \mathbf{O}(\mathbf{Z})]^T$$

subject to

$$\mathbf{I}(\mathbf{Y}) \geqslant \mathbf{C} - \mathbf{M} \qquad \mathbf{J}(\mathbf{Z}) = \mathbf{M} \qquad \mathbf{D}(\mathbf{Z}) \geqslant \mathbf{0} \qquad \mathbf{K}(\mathbf{Y}) \geqslant \mathbf{0} \qquad \mathbf{Y}, \mathbf{Z} \geqslant \mathbf{0} \tag{12.13}$$

which can be decomposed into the following two submodels:

$$\min \mathbf{F}(\mathbf{Y})$$

subject to

$$\mathbf{I}(\mathbf{Y}) \geqslant \mathbf{C} - \mathbf{M} \qquad K(\mathbf{Y}) \geqslant \mathbf{0} \qquad \mathbf{Y} \geqslant \mathbf{0} \tag{12.14}$$

and

$$\min \mathbf{O}(\mathbf{Z})$$

subject to

$$\mathbf{J}(\mathbf{Z}) = \mathbf{M} \qquad \mathbf{D}(\mathbf{Z}) \geqslant \mathbf{0} \qquad \mathbf{Z} \geqslant \mathbf{0} \tag{12.15}$$

The submodel Eq. (12.11) of the α-HHS Eq. (12.10) may be an aggregation of a number of submodels, all under the jurisdiction of the α group. Each submodel may represent a subsystem that is completely controlled by a local controller belonging to the α group. In other words, the submodel Eq. (12.11) can be viewed as an HMO problem. The decomposition into subsystems depends on (1) the structure of the system that the submodel represents, (2) the purpose of the study/analysis/design, (3) the objective of the α group, and (4) the structure of the data bases. Thus, there exists a decomposition for the α-HHS that could be different from that of β-HHS. For the Maumee River basin example given in Haimes (1981), the agriculture planning group (the α group) used a geographic decomposition, while the flood control group (the β group) used a hydrological decomposition. Let us call the decomposition under α group the α decomposition and that of β group the β decomposition.

We will assume that the multiobjective functions of the submodel Eq. (12.11)—which is termed the α-HMO submodel—are additively separable with respect to the decision variables of the decomposed subsystems. Similarly, the objective vector $O(\mathbf{Z})$ of submodel Eq. (12.12) is additively separable. Also, the constraint vectors $G(\mathbf{X})$, $A(\mathbf{X})$, $H(\mathbf{Z})$ and $D(\mathbf{Z})$ are separable with respect to the subsystem decision variables. Assume that there are α subsystems (sub-submodels) in the α-HMO submodel under its own α decomposition. There may exist input-output relationships among the subsystems of the α-HMO submodel. Let $\mathbf{U}$ be the vector of interaction variables, that is, input-output variables from some subsystems to other subsystems. We add the following constraint vector to the α-HMO submodel:

$$T\mathbf{X} = \mathbf{U}$$

where T is a transformation matrix which is block diagonal.

The addition of the constraints $T\mathbf{X} = \mathbf{U}$ means that the values of $\mathbf{U}$ are fixed. Likewise, let $\mathbf{R}$ be the interaction vector among the α subsystems of submodel Eq. (12.12), under the α decomposition. The constraint vector $N\mathbf{Z} = \mathbf{R}$ is added to submodel Eq. (12.12). N is a block diagonal transformation matrix. Then, the α-HMO submodel Eq. (12.11) and submodel Eq. (12.12) become:

$$\min \mathbf{E}(\mathbf{X}; \mathbf{U})$$

subject to

$$\mathbf{G}(\mathbf{X}; \mathbf{U}) \geqslant \mathbf{B} - \mathbf{L} \qquad \mathbf{A}(\mathbf{X}; \mathbf{U}) \geqslant \mathbf{0} \qquad T\mathbf{X} = \mathbf{U} \qquad \mathbf{X} \geqslant \mathbf{0} \qquad (12.16)$$

and

$$\min \mathbf{O}(\mathbf{Z}; \mathbf{R})$$

subject to

$$\mathbf{H}(\mathbf{Z}; \mathbf{R}) = \mathbf{L} \qquad \mathbf{D}(\mathbf{Z}; \mathbf{R}) \geqslant \mathbf{0} \qquad N\mathbf{Z} = \mathbf{R} \qquad \mathbf{Z} \geqslant \mathbf{0} \qquad (12.17)$$

The arguments appearing after the semicolon (;) in the objective functions and constraint functions are parameters, and

$$\begin{aligned}\mathbf{E}(\mathbf{X}; \mathbf{U}) &= [E_1(\mathbf{X}; \mathbf{U}), E_2(\mathbf{X}; \mathbf{U}), \ldots, E_r(\mathbf{X}; \mathbf{U})]^T \\ &= \left[\sum_{j=1}^{\alpha} E_{1,j}(\mathbf{X}_j; \mathbf{U}), \sum_{j=1}^{\alpha} E_{2,j}(\mathbf{X}_j; \mathbf{U}), \ldots, \sum_{j=1}^{\alpha} E_{r,j}(\mathbf{X}_j; \mathbf{U})\right]^T\end{aligned}$$

Under the α decomposition of the α group, the α_j-HMO sub-submodel of the α-HMO submodel Eq. (12.16) and the α_j sub-submodel of the submodel Eq. (12.17) can be written as follows (cf. Chap. 9):

α_j-HMO Sub-submodel of the α-HMO Submodel Eq. (12.16) ($j = 1, 2, \ldots, \alpha$):

$$\min [\bar{E}_j(\mathbf{X}_j; \mathbf{U})]$$

subject to

$$\mathbf{G}_j(\mathbf{X}_j; \mathbf{U}) \geqslant \mathbf{B}_j - \mathbf{L}_j \qquad A(\mathbf{X}_j; \mathbf{U}) \geqslant \mathbf{0} \qquad T_j\mathbf{X}_j = \mathbf{U}_j \qquad \mathbf{X}_j \geqslant \mathbf{0} \qquad (12.18)$$

where $\bar{\mathbf{E}}_j = (E_{1,j}, E_{2,j}, \ldots, E_{r,j})T$

α_j-Sub-submodel of Submodel Eq. (12.17) (j = 1, 2, ..., α):

$$\min\,[\bar{\mathbf{O}}_j(\mathbf{Z}_j;\,\mathbf{R})]$$

subject to

$$\mathbf{H}_j(\mathbf{Z}_j;\,\mathbf{R}) = \mathbf{L}_j \qquad \mathbf{D}(\mathbf{Z}_j;\,\mathbf{R}) \geqslant \mathbf{0} \qquad N_j\mathbf{Z}_j = \mathbf{R}_j \qquad \mathbf{Z}_j \geqslant \mathbf{0} \tag{12.19}$$

For the definition of $\bar{\mathbf{O}}_j$, compare the definition of $\bar{\mathbf{E}}_j$, above. In a similar fashion, the β-HMO submodel Eq. (12.14) has β subsystems under its own β decomposition. Let **V** and **S** be interaction vectors among the subsystems of the β-HMO submodel and the subsystems of submodel Eq. (12.15), respectively. We add the following constraint vector to the β-HMO submodel:

$$Q\mathbf{Y} = \mathbf{V}$$

Also, the constraint vector $P\mathbf{Z} = \mathbf{S}$ is added to the submodel Eq. (12.15), where Q and P are transformation matrices that are block diagonal with respect to the subsystem decision subvectors.

Also, we assume that all the multiobjective vectors and constraint vectors of the β-HMO submodel Eqs. (12.14) and (12.15) are separable with respect to the subsystem decision variables.

Then, the β-HMO submodel Eq. (12.14) and submodel Eq. (12.15) can be written in the following equivalent forms:

$$\min\,[\mathbf{F}(\mathbf{Y};\,\mathbf{V})]$$

subject to

$$\mathbf{I}(\mathbf{Y};\,\mathbf{V}) \geqslant \mathbf{C} - \mathbf{M} \qquad \mathbf{K}(\mathbf{Y};\,\mathbf{V}) \geqslant \mathbf{0} \qquad Q\mathbf{Y} = \mathbf{V} \qquad \mathbf{Y} \geqslant \mathbf{0} \tag{12.20}$$

$$\min\,[\mathbf{O}(\mathbf{Z};\,\mathbf{S})]$$

subject to

$$\mathbf{J}(\mathbf{Z};\,\mathbf{S}) = \mathbf{M} \qquad \mathbf{D}(\mathbf{Z};\,\mathbf{S}) \geqslant \mathbf{0} \qquad P\mathbf{Z} = \mathbf{S} \qquad \mathbf{Z} \geqslant \mathbf{0} \tag{12.21}$$

Under the β decomposition of the β group, the β_j-HMO sub-submodel of the β-HMO submodel Eq. (12.20) and the β_j sub-submodel of submodel Eq. (12.21) can be written as follows (the $\bar{\mathbf{F}}_j$'s are defined as $\bar{\mathbf{E}}_j$ above):

β_j-HMO Sub-submodel of the β-HMO Submodel Eq. (12.20) (j = 1, 2, ..., β):

$$\min\,[\bar{\mathbf{F}}_j(\mathbf{Y}_j;\,\mathbf{V})]$$

subject to

$$\mathbf{I}_j(\mathbf{Y};\,\mathbf{V}) \geqslant \mathbf{C}_j - \mathbf{M}_j \qquad \mathbf{K}_j(\mathbf{Y}_j;\,\mathbf{V}) \geqslant \mathbf{0} \qquad Q_j\mathbf{Y}_j = \mathbf{V}_j \qquad \mathbf{U}_j \geqslant \mathbf{0} \tag{12.22}$$

β_j-Sub-submodel of Submodel Eq. (12.21) (j = 1, 2, ..., β):

$$\min\,[\bar{\mathbf{O}}_j(\mathbf{Z}_j;\,\mathbf{S})]$$

subject to

$$\mathbf{J}_j(\mathbf{Z}, \mathbf{S}) = \mathbf{M}_j \qquad \mathbf{B}_j(\mathbf{Z}_j; \mathbf{S}) \geq \mathbf{0} \qquad P_j\mathbf{Z}_j = \mathbf{S}_j \qquad \mathbf{Z}_j \geq \mathbf{0} \tag{12.23}$$

We look at the scope of harmony and conflict between the α group and the β group with respect to the evaluation and recommendation of a policy for implementation. This task of policy evaluation and analysis will depend significantly on the attained values of both distinct and overlapping objective functions in both the α-HHS and the β-HHS, with respect to their own Pareto optimal solutions. The overlapping multiobjective vector **O** will have identical optimal values in both HHS's if the overlapping decision vector **Z** has identical optimal values in both the α-HHS and the β-HHS with respect to their own respective Pareto optimal solutions. This will not happen; that is, the α group and the β group will not get identical optimal values for **Z**, because the constraint structures of submodels Eqs. (12.12) and (12.15) (under α-HHS and β-HHS, respectively) are not identical and, consequently, neither are their decomposed sub-submodels Eqs. (12.19) and (12.23) identical. Unless there is a scheme to generate identical optimal values for the overlapping decision vector, the α group and the β group may not reach a compromise solution. What we need is a modeling scheme that can treat the α group and the β group as independent constituencies, provide flexibility in decomposing the α-HHS and the β-HHS into their own respective subsystems, based on each group's own needs and data base structure, and also reach toward identical optimal values for the overlapping decision vector. To do this, we will consider a new submodel of the following form:

$$\min\ [\mathbf{O}(\mathbf{Z})]$$

subject to

$$\mathbf{H}(\mathbf{Z}) = \mathbf{L} \qquad \mathbf{J}(\mathbf{Z}) = \mathbf{M} \qquad \mathbf{D}(\mathbf{Z}) \geq \mathbf{0} \qquad \mathbf{Z} \geq \mathbf{0} \tag{12.24}$$

This submodel, termed the overlapping submodel, is identical to submodel Eq. (12.12) under the α group except for the additional constraints of $\mathbf{J}(\mathbf{Z}) = \mathbf{M}$. Also, this submodel is identical to submodel (12.15) under the β group except for the additional constraints $\mathbf{H}(\mathbf{Z}) = \mathbf{L}$. In other words, by adopting the overlapping submodel in Eq. (12.24), the α group has all the characteristics of its own submodel Eq. (12.12) except for some additional restrictions on the overlapping decision vector. The same argument holds for the β group.

By applying the HOC concept to the overlapping submodel, the α group and the β group can be treated as independent agencies, which provides flexibility in decomposition. Under the HOC scheme (Macko and Haimes, 1978; Shima, 1982; Chap. 10), the α group can decompose submodel Eq. (12.24) into submodels according to its own decomposition (the α decomposition) and the β group can also decompose the same submodel into submodels according to its own decomposition (the β decomposition). We assume here that the constraint set of the overlapping submodel has special structures that fit the characteristics of an HOC model. That is, when the overlapping submodel is decomposed under the α decomposition, the resulting interaction variables must not appear as interaction variables under the β decomposition, and vice versa. In other words, the intersection between the set of interaction variables under the β decomposition and that set of the β decomposition is null (cf. Chap. 9).

We will now apply the HOC concept, decomposing submodel Eq. (12.24) into submodels under the α decomposition and the β decomposition. Under the α decomposition, the overlapping submodel can be equivalently represented by

$$\min\ [\mathbf{O}_\alpha(\mathbf{Z}_\alpha; \mathbf{R})]$$

subject to

$$\mathbf{H}_\alpha(\mathbf{Z}_\alpha; \mathbf{R}) = \mathbf{L}_\alpha \qquad \mathbf{J}_\alpha(\mathbf{Z}_\alpha; \mathbf{R}) = \mathbf{M}_\alpha \qquad \mathbf{D}_\alpha(\mathbf{Z}_\alpha; \mathbf{R}) \geq \mathbf{0}$$

$$N\mathbf{Z}_\alpha = \mathbf{R} \qquad \mathbf{Z}_\alpha \geq \mathbf{0} \tag{12.25}$$

where

$$\mathbf{Z}_\alpha = A_\alpha \mathbf{Z} \tag{12.26}$$

$$\mathbf{L}_\alpha = B_\alpha \mathbf{L} \tag{12.27}$$

$$\mathbf{M}_\alpha = C_\alpha \mathbf{M} \tag{12.28}$$

$$\mathbf{H}(\mathbf{Z}) = \mathbf{H}_\alpha(A_\alpha \mathbf{Z}) \tag{12.29}$$

$$\mathbf{J}(\mathbf{Z}) = \mathbf{J}_\alpha(A_\alpha \mathbf{Z}) \tag{12.30}$$

$$\mathbf{D}(\mathbf{Z}) = \mathbf{D}_\alpha(A_\alpha \mathbf{Z}) \tag{12.31}$$

$$\mathbf{O}(\mathbf{Z}) = \mathbf{O}_\alpha(A_\alpha \mathbf{Z}) \tag{12.32}$$

A_α, B_α, and C_α are nonsingular matrices that rearrange the components of the vectors **Z**, **L**, and **M**, respectively. The overlapping submodel is essentially unchanged, except that the variables and the constraints are rearranged. The submodel in Eq. (12.25) is termed the α-HOC submodel.

Under the α decomposition, the decision vector $\mathbf{Z}_\alpha$ and the right-hand-side vectors $\mathbf{L}_\alpha$ and $\mathbf{M}_\alpha$ are partitioned into α subvectors, $\mathbf{Z}_{\alpha j}$, $\mathbf{L}_{\alpha j}$, and $\mathbf{M}_{\alpha j}$ for the jth subsystem. **R** is a vector of interaction variables among the α subsystems under the α decomposition, which is partitioned into α subvectors $\mathbf{R}_j$ associated with the jth subsystem.

$$\mathbf{R} = N\mathbf{Z}_\alpha \tag{12.33}$$

that is,

$$\mathbf{R}_j = N_j \mathbf{Z}_{\alpha j} \qquad j = 1, 2, \ldots, \alpha \tag{12.34}$$

where N is a transformation matrix. Furthermore, we assume O_α, H_α, J_α, and D_α are separable with respect to the subsystem variables. By assigning feasible values to the interaction variables **R**, the α-HOC submodel Eq. (12.25) is decomposed into α independent sub-submodels:

$$\min\ [\mathbf{O}_{\alpha j}(\mathbf{Z}_{\alpha j}; \mathbf{R})]$$

subject to

$$\mathbf{H}_{\alpha j}(\mathbf{Z}_{\alpha j}; \mathbf{R}) = \mathbf{L}_{\alpha j} \qquad \mathbf{J}_{\alpha j}(\mathbf{Z}_{\alpha j}; \mathbf{R}) = \mathbf{M}_{\alpha j} \qquad \mathbf{D}_{\alpha j}(\mathbf{Z}_{\alpha j}; \mathbf{R}) \geqslant \mathbf{0}$$
$$N_j\mathbf{Z}_{\alpha j} = \mathbf{R}_j \qquad \mathbf{Z}_{\alpha j} \geqslant \mathbf{0} \qquad j = 1, 2, \ldots, \alpha \tag{12.35}$$

This is identical to subproblem Eq. (12.19) except for the constraint set $\mathbf{J}_j(\mathbf{Z}_{\alpha j}; \mathbf{R}) = \mathbf{M}_j$, which comes from the β-HHS. The addition of the constraints $N_j\mathbf{Z}_{\alpha j} = \mathbf{R}_j$ means that the values of $\mathbf{R}_j$ are fixed.

Similarly, under the β decomposition of the β group, the overlapping submodel can be equivalently represented by

$$\min\, [\mathbf{O}_\beta(\mathbf{Z}_\beta; \mathbf{S})]$$

subject to

$$\mathbf{J}_\beta(\mathbf{Z}_\beta; \mathbf{S}) = \mathbf{M}_\beta \quad \mathbf{H}_\beta(\mathbf{Z}_\beta; \mathbf{S}) = \mathbf{L}_\beta \quad \mathbf{D}_\beta(\mathbf{Z}_\beta; \mathbf{S}) \geqslant \mathbf{0} \quad P\mathbf{Z}_\beta = \mathbf{S} \quad \mathbf{Z}_\beta \geqslant \mathbf{0} \tag{12.36}$$

where

$$\mathbf{Z}_\beta = A_\beta\mathbf{Z} \tag{12.37}$$

$$\mathbf{L}_\beta = B_\beta\mathbf{L} \tag{12.38}$$

$$\mathbf{M}_\beta = C_\beta\mathbf{M} \tag{12.39}$$

$$\mathbf{H}(\mathbf{Z}) = \mathbf{H}_\beta(A_\beta\mathbf{Z}) \tag{12.40}$$

$$\mathbf{J}(\mathbf{Z}) = \mathbf{J}_\beta(A_\beta\mathbf{Z}) \tag{12.41}$$

$$\mathbf{D}(\mathbf{Z}) = \mathbf{D}_\beta(A_\beta\mathbf{Z}) \tag{12.42}$$

$$\mathbf{O}(\mathbf{Z}) = \mathbf{O}_\beta(A_\beta\mathbf{Z}) \tag{12.43}$$

A_β, B_β, and C_β are nonsingular matrices that rearrange the components of the vectors $\mathbf{Z}$, $\mathbf{L}$, and $\mathbf{M}$, respectively, under the β decomposition. Submodel Eq. (12.36) is termed the β-HOC submodel, which can be further decoupled into β independent subproblems by fixing the values of the interaction variables $\mathbf{S}$, under assumptions that are similar to those used for the β-HOC submodel (i.e., $\mathbf{O}_\beta$, $\mathbf{H}_\beta$, $\mathbf{J}_\beta$, and $\mathbf{D}_\beta$ are separable with respect to the subsystem variables). Then, the jth subproblem of the β-HOC submodel Eq. (12.36) is

$$\min\, [\mathbf{O}_{\beta j}(\mathbf{Z}_{\beta j}; \mathbf{S})]$$

subject to

$$\mathbf{J}_{\beta j}(\mathbf{Z}_{\beta j}; \mathbf{S}) = \mathbf{M}_{\beta j} \qquad \mathbf{H}_{\beta j}(\mathbf{Z}_{\beta j}; \mathbf{S}) = \mathbf{L}_{\beta j} \qquad \mathbf{D}_{\beta j}(\mathbf{Z}_{\beta j}; \mathbf{S}) \geqslant \mathbf{0}$$
$$P_j\mathbf{Z}_{\beta j} = \mathbf{S}_j \qquad \mathbf{Z}_{\beta j} \geqslant \mathbf{0} \qquad j = 1, 2, \ldots, \beta \tag{12.44}$$

which is identical to subproblem Eq. (12.23) except for the constraint set $\mathbf{H}_{\beta j}(\mathbf{Z}_{\beta j}; \mathbf{S}) = \mathbf{L}_{\beta j}$, which comes from the α-HHS.

We see by the relationships in Eqs. (12.29)–(12.32) and (12.40)–(12.43) that the overlapping submodel, the α-HOC submodel Eq. (12.25), and the β-HOC submodel Eq. (12.36) are equivalent to one another:

$$\begin{aligned}
\mathbf{O}(\mathbf{Z}) &= \mathbf{O}_\alpha(A_\alpha\mathbf{Z}, NA_\alpha\mathbf{Z}) = \mathbf{O}_\beta(A_\beta\mathbf{Z}, PA_\beta\mathbf{Z}) \\
\mathbf{J}(\mathbf{Z}) &= \mathbf{J}_\alpha(A_\alpha\mathbf{Z}, NA_\alpha\mathbf{Z}) = \mathbf{J}_\beta(A_\beta\mathbf{Z}, PA_\beta\mathbf{Z}) \\
\mathbf{H}(\mathbf{Z}) &= \mathbf{H}_\alpha(A_\alpha\mathbf{Z}, NA_\alpha\mathbf{Z}) = \mathbf{H}_\beta(A_\beta\mathbf{Z}, PA_\beta\mathbf{Z}) \\
\mathbf{D}(\mathbf{Z}) &= \mathbf{D}_\alpha(A_\alpha\mathbf{Z}, NA_\alpha\mathbf{Z}) = \mathbf{D}_\beta(A_\beta\mathbf{Z}, PA_\beta\mathbf{Z})
\end{aligned} \tag{12.45}$$

Furthermore, the decision vectors $\mathbf{Z}_\alpha$ and $\mathbf{Z}_\beta$, under the α decomposition and the β decomposition, respectively, are related by a nonsingular transformation:

$$\mathbf{Z}_\alpha = A_{\alpha\beta}\mathbf{Z}_\beta \tag{12.46}$$

or

$$\mathbf{Z}_\beta = A_{\beta\alpha}\mathbf{Z}_\alpha \tag{12.47}$$

where

$$A_{\alpha\beta} = A_\beta A_\alpha^{-1} \tag{12.48}$$

$$A_{\beta\alpha} = A_\alpha A_\beta^{-1} \tag{12.49}$$

Because of these relationships, that is, Eqs. (12.46)–(12.49), the values of the interaction **R** in the α-HOC submodel can be fixed by the optimal values of the decision vector in the β-HOC submodel and vice versa. This is the basic concept of the HOC scheme (see Macko and Haimes, 1978, and Chap. 9). Here the interaction vector **R** can be viewed as those components of the overlapping decision vector **Z** over which the β group has more authority, and, therefore, the β group exerts control over **R** in its own β-HOC submodel and communicates the computed optimal values of **R** to the α group. The α group, in turn, manipulates the values of the interaction vector **S** (over which it has more power) in its own α-HOC submodel and conveys the information on **S** to the β group. We can see that the HOC scheme provides flexibility in manipulating and regulating the values of the overlapping decision vector, according to the magnitude of both groups' authority and duty.

In summary, we have illustrated how the α-HHS Eq. (12.6) can be transformed into the α-HMO submodel Eq. (12.16) and the α-HOC submodel Eq. (12.25), which can be further partitioned into their own sub-submodels, given in Eqs. (12.18) and (12.35), respectively. The same argument goes for the β group. See Fig. 12.1 for a flow chart of these decompositions.

12.3 PARETO OPTIMAL SOLUTION GENERATION SCHEME

12.3.1 Basis of Derivation of Solution Scheme

This section outlines the reasons for adopting the particular methodology (given in Section 12.3.2) that will lead to solution of the α- and β-HHSs that were described in the previous subsection. The necessary conditions and lemma that will

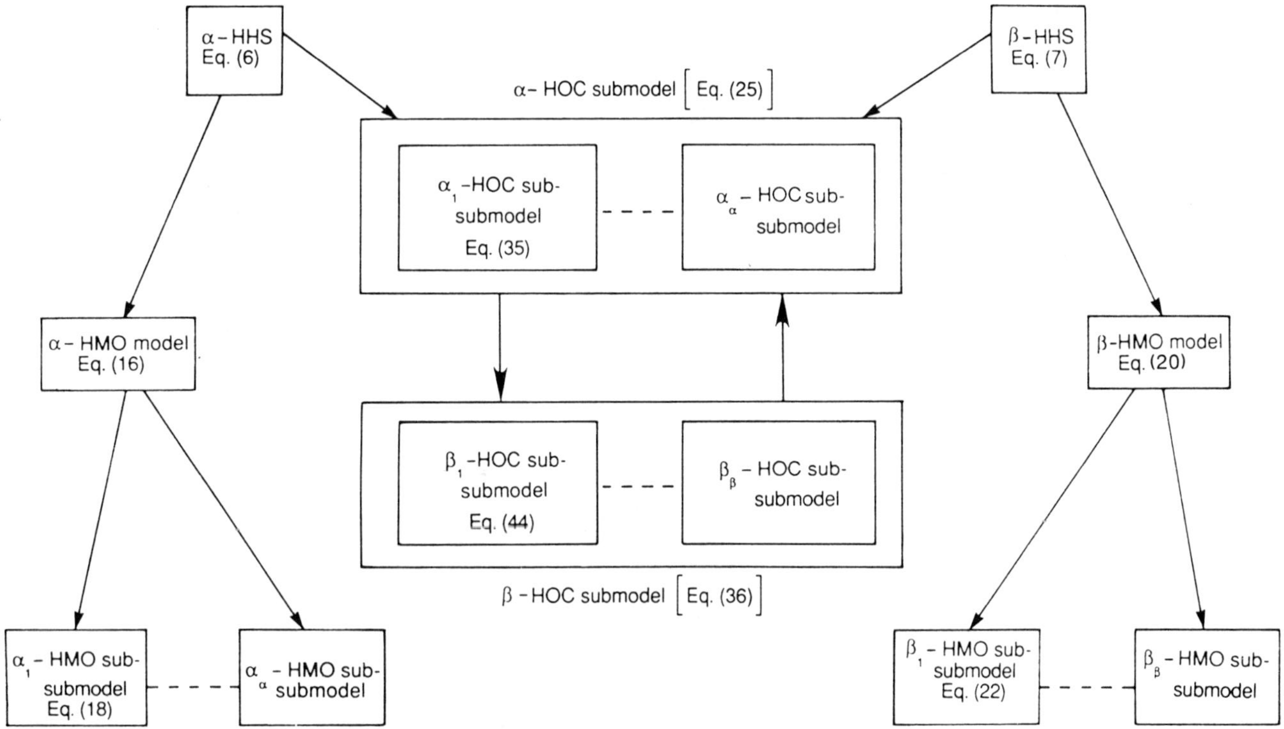

Figure 12.1 Transformation/decomposition of α- and β-HHSs into their own submodels.

lead to the Pareto optimal solutions and convergence aspects of the proposed algorithm can be found in Thadathil (1983) and Thadathil and Haimes (1984a, 1984b).

Suppose that the α group assigns feasible values to the interaction vectors **U** and **R** and the right-hand-side vector $\mathbf{L}_{\alpha j}$ $(j = 1, 2, \ldots, \alpha)$ and that the β group independently assigns feasible values to the **V**, **S**, and $\mathbf{M}_{\beta j}$, $j = 1, 2, \ldots, \beta$. The α and β analysts, by applying some multiobjective optimization technique, solve their own submodels [Eqs. (12.18), (12.19), and (12.22), (12.23), respectively] and update the interaction and right-hand-side vectors in a feasible fashion in order to generate Pareto optimal solution sets for their own HHS's. In the respective sets of Pareto optimal solutions, the α and β groups will most probably not have identical optimal values for the overlapping decision vector; consequently, they will have different optimum levels of the overlapping multiobjective vector $O(\mathbf{Z})$ with respect to each of their own Pareto optimal solutions. This happens because the nonidentical equality constraint sets $\mathbf{H}_{\alpha j}(\mathbf{Z}_{\alpha j}; \mathbf{R}) = \mathbf{L}_{\alpha j}$ and $\mathbf{J}_{\beta j}(\mathbf{Z}_{\beta j}; \mathbf{S}) = \mathbf{M}_{\beta j}$ are present in their own submodels—Eqs. (12.19) and (12.23), respectively. These different attained levels of the overlapping decision vector will prevent the two groups from reaching a compromise and recommending a policy.

To generate identical optimal values for the overlapping decision vector with respect to each Pareto optimal solution, the α group should adopt the α-HOC submodel Eq. (12.25) instead of the one in Eq. (12.17). These two submodels are identical except for the addition of the decoupled constraints $\mathbf{J}(\mathbf{Z}) = \mathbf{M}$ from the β-HHS. Similarly, the β group should adopt the β-HOC submodel Eq. (12.36) instead of Eq. (12.21); these two submodels are identical except for the constraints $\mathbf{H}(\mathbf{Z}) = \mathbf{L}$ decoupled from the α-HHS. As shown in Sect. 12.2, the α-HOC submodel is equivalent to the β-HOC submodel under the terminology of HOC. Here, under the HOC scheme, the α group can decompose the α-HOC submodel into submodels, based on its own aspects and solve the submodels at the fixed levels of the interaction variables, which are determined by the β group, while the β group can decompose the β-HOC submodel, based on its own decomposition and solve the submodels at the fixed levels of the interaction variables, which are specified by the α group. Because of the equivalence of the α- and the β-HOC submodels, the α and β groups should get identical optimal values for their own overlapping decision vectors with respect to each Pareto optimal solution generated by the HOC scheme for the α- and the β-HOC submodels.

An attempt is made to derive a scheme capable of generating a sufficient number of Pareto optimal solutions for the α-HHS and the β-HHS by

1. Solving the α-HMO submodel Eq. (12.16) and the β-HMO submodel Eq. (12.20) and updating the interaction variables and decoupled right-hand-side subvectors in a feasible fashion; this is done by the α and β analysts, respectively;
2. Solving the α-HOC submodel Eq. (12.25) and the β-HOC submodel Eq. (12.36) under the HOC scheme; this is done by the α and β analysts, respectively.

We will use the term *α-Pareto optimal solution* to describe the α-HHS's Pareto optimal solution worked out by the scheme that will be described in Sect.

12.3.2. This solution will have identical optimal values for the overlapping decision vector. Similarly, the β-HHS will have a β-Pareto optimal solution. For the α-HHS, the α-Pareto optimal solution is not Pareto optimal when the α-HHS is viewed as an independent model. Rather, the α-Pareto optimal solution can be justified as Pareto optimal when the α-HHS is visualized as a submodel of a hypothetical overall model that will be described shortly. The same argument holds for the β-Pareto optimal solution. Note that Pareto optimal solutions of the α-HHS and β-HHS do not have identical optimal values for **Z**.

The hypothetical overall model is the integrated overall model of the α-HMO submodel Eq. (12.11), the β-HMO submodel Eq. (12.14), and the overlapping submodel Eq. (12.24) and has the following form:

$$\min\ [\mathbf{E}(\mathbf{X}),\ \mathbf{F}(\mathbf{Y}),\ \mathbf{O}(\mathbf{Z})]^T$$

subject to

$$\begin{gathered} \mathbf{G}(\mathbf{X}) \geqslant \mathbf{B} - \mathbf{L} \quad \mathbf{A}(\mathbf{X}) \geqslant \mathbf{0} \quad \mathbf{I}(\mathbf{Y}) \geqslant \mathbf{C} - \mathbf{M} \\ \mathbf{K}(\mathbf{Y}) \geqslant \mathbf{0} \quad \mathbf{H}(\mathbf{Z}) = \mathbf{L} \quad \mathbf{J}(\mathbf{Z}) = \mathbf{M} \quad \mathbf{D}(\mathbf{Z}) \geqslant \mathbf{0} \quad \mathbf{X}, \mathbf{Y}, \mathbf{Z} \geqslant \mathbf{0} \end{gathered} \tag{12.50}$$

The integrated overall model Eq. (12.50) is not under control of either group. In other words, this model does not exist; rather, it is a hypothetical model formulated for deriving a solution scheme (Thadathil, 1983). We can argue that the above model looks exactly like an HMO model and wonder whether we might decompose model (12.50) into submodels (the α-HMO submodel, the β-HMO submodel, the overlapping submodel, and their own sub-submodels) and apply the HMO framework developed by Tarvainen (1981) (see Chap. 5). The answer is often negative. The overall model (12.50) possesses characteristics that may render the HMO scheme unsuitable for solving the model. The most pertinent characteristic of the overall model is that there are independent multiple decisionmakers—the α and the β decisionmakers—tied to their own limited missions and goals. These decisionmakers are accustomed to their own mission-oriented objectives and therefore they will likely find it impossible to estimate the overall indifference trade-off vector as a function of the overall multiobjective vector [(**E**(**X**), **F**(**Y**), **O**(**Z**)] since they have no idea what kind or magnitude of trade-off exists between the ith objective of **E**(**X**), $i = 1, 2, \ldots, r$, and the jth objective of **F**(**Y**), $j = 1, 2, \ldots, w$. In the HMO scheme, it is assumed there exists a single decisionmaker or a group with a unique utility function, and this person or group will estimate the overall indifference trade-off vector for selecting the preferred solution, whereas the scheme proposed in Sect. 12.3.2 uses a weighting approach to generate a sufficient number of Pareto optimal solutions from which the compromise solution is selected by the α and β decisionmakers.

The second distinct characteristic of the overall model is that the overlapping submodel can be decomposed in two different ways to suit the needs and data base structures of the α and β groups, whereas a typical HMO model has only one decomposition. The problem of multiple decompositions (the α and the β decomposition) is handled by the HOC scheme, which is integrated into the proposed scheme.

We will now write down the steps of the HOC scheme briefly (Macko and Haimes, 1978; Chap. 9). We make three assumptions regarding the multiobjective vectors and the constraint set.

Assumption 1 All the objective functions of **E(X)**, **E(Y)**, and **O(Z)** are continuously differentiable, convex, and nonlinear.

Assumption 2 All the constraint functions are continuously differentiable.

Assumption 3 The constraint set is compact and convex.

For the assigned levels of the right-hand-side subvectors $\mathbf{L}_{\alpha j}$ and $\mathbf{M}_{\alpha j}$ ($j = 1, 2, \ldots, \alpha$) and assigned weight vector $\mathbf{w}^o$, we transform submodels Eq. (12.35) of the α-HOC submodel into scaled optimization problems under the weighting approach [the $O_{i,\alpha j}$'s are defined as $\bar{E}_j$ in Eq. (12.18)]:

$$\min \sum_{i=1}^{o} \mathbf{w}_i^o O_{i,\alpha j}(\mathbf{Z}_{\alpha j}; \mathbf{R})$$

subject to

$$\begin{gathered} \mathbf{H}_{\alpha j}(\mathbf{Z}_{\alpha j}; \mathbf{R}) - \mathbf{L}_{\alpha j} = \mathbf{0} \qquad \mathbf{J}_{\alpha j}(\mathbf{Z}_{\alpha j}; \mathbf{R}) = \mathbf{M}\alpha_j \qquad \mathbf{D}_{\alpha j}(\mathbf{Z}_{\alpha j}; \mathbf{R}) \geqslant \mathbf{0} \\ N_j \mathbf{Z}_{\alpha j} = \mathbf{R}_j \qquad \mathbf{Z}_{\alpha j} \geqslant \mathbf{0} \end{gathered} \tag{12.51}$$

The scaled problems associated with the sub-submodels Eq. (12.44) of the β-HOC submodel are [the $O_{i,\beta j}$'s are defined as $\bar{E}_j$ in Eq. (12.18)]:

$$\min \sum_{i=1}^{o} \mathbf{w}_i^o O_{i\beta j}(\mathbf{Z}_{\beta j}; \mathbf{S})$$

subject to

$$\begin{gathered} \mathbf{J}_{\beta j}(\mathbf{Z}_{\beta j}; \mathbf{S}) - \mathbf{M}_{\beta j} = \mathbf{0} \qquad \mathbf{H}_{\beta j}(\mathbf{Z}_{\beta j}; \mathbf{S}) = \mathbf{L}_{\beta j} \qquad \mathbf{D}_{\beta j}(\mathbf{Z}_{\beta j}; \mathbf{S}) \geqslant \mathbf{0} \\ P_j \mathbf{Z}_{\beta j} = \mathbf{S}_j \qquad \mathbf{Z}_{\beta j} \geqslant \mathbf{0} \end{gathered} \tag{12.52}$$

12.3.1.1 HOC scheme for the α-HOC sub-submodels Eq. (12.51) and the β-HOC sub-submodels Eq. (12.52)

Step 1: Initial feasible values of $\mathbf{R}_j^{(1)}$ are estimated (guessed) by the β group. Based on these values, the α analyst solves the α-HOC sub-submodels (12.51), computes the solution $\mathbf{Z}_{\alpha j}^{(1)}$ $j = 1, 2, \ldots, \alpha$, and thereby fixes the values of **S** by $\mathbf{S}^{(1)} = PA_{\beta\alpha}\mathbf{Z}_{\alpha}^{(1)}$.

Step 2: Using the information of $\mathbf{S}^{(1)}$ from the α group, the β analyst computes $\mathbf{Z}_{\beta j}^{(1)}$, $j = 1, 2, \ldots, \beta$, from the β-HOC sub-submodels (12.52) and then fixes the values of **R** by $\mathbf{R}^{(2)} = NA_{\alpha\beta}\mathbf{Z}_{\beta}^{(1)}$.

Step 3: Based on the information of $\mathbf{R}^{(2)}$, the α analyst solves the α-HOC sub-submodels (12.51) again.

This process is repeated until convergence is reached. That is, the process is stopped when the following stopping criteria hold:

$$\left| \sum_{i=1}^{o} \sum_{j=1}^{\alpha} \mathbf{w}_i^o O_{i.\alpha j}(\mathbf{Z}_{\alpha j}^{(n+1)}; \mathbf{R}^{(n+1)}) - \sum_{i=1}^{o} \sum_{j=1}^{\alpha} \mathbf{w}_i^o O_{i.\alpha j}(\mathbf{Z}_{\alpha j}^{(n)}; \mathbf{R}^{(n)}) \right| \leq \epsilon \quad (12.53)$$

$$\left| \sum_{i=1}^{o} \sum_{j=1}^{\beta} \mathbf{w}_i^o O_{i.\beta j}(\mathbf{Z}_{\beta j}^{n+1}; \mathbf{S}^{(n+1)}) - \sum_{i=1}^{o} \sum_{j=1}^{\beta} \mathbf{w}_i^o O_{i.\beta j}(\mathbf{Z}_{\beta j}^{(n)}; \mathbf{S}^{(n)}) \right| \leq \epsilon \quad (12.54)$$

where ϵ is a prespecified, very small number.

12.3.2 Procedure to Generate Pareto Optimal Solutions

This section discusses how the α and β groups can solve their own HHSs to generate α- and β-Pareto optimal solutions, respectively.

Regarding the weights to generate Pareto optimal solutions, the α and the β analysts can confer and work out

1. A sequence $\mathbf{R}_\alpha$ of $\mathbf{w}^\alpha$ for the multiobjective vector $\mathbf{E}$
2. A sequence $\mathbf{R}_\beta$ of $\mathbf{w}^\beta$ for the multiobjective vector $\mathbf{F}$ (12.55)
3. A sequence $\mathbf{R}_o$ of $\mathbf{w}^o$ for the multiobjective vector $\mathbf{O}$

satisfying

$$\sum_{i=1}^{r} \mathbf{w}_i^{\alpha(l)} + \sum_{j=1}^{w} \mathbf{w}_j^{\beta(l)} + \sum_{k=1}^{o} \mathbf{w}_k^{o(l)} = 1$$

where

$\mathbf{w}_i^{\theta(l)}$ = the ith weight of the lth weight vector in $\mathbf{R}_\theta$
θ = α, β, o
l = $1, 2, \ldots, L$
L = total number of elements in $\mathbf{R}_\theta$

For example,

$$\mathbf{R}_\alpha = \{(\mathbf{w}_1^{\alpha(1)}, \mathbf{w}_2^{\alpha(1)}, \ldots, \mathbf{w}_r^{\alpha(1)}), \ldots, (\mathbf{w}_1^{\alpha(L)}, \mathbf{w}_2^{\alpha(L)}, \ldots, \mathbf{w}_r^{\alpha(L)})\}$$

12.3.2.1 The hierarchical holographic overlapping coordination (HHOC) algorithm We now describe a proposed algorithm, termed the hierarchical holographic overlapping coordination (HHOC) algorithm. This algorithm uses the HOC scheme as a subroutine.

Step 0: The α analyst and β analyst confer and assign values of weights in $\mathbf{R}_\alpha$, $\mathbf{R}_\beta$, and $\mathbf{R}_o$ [Eq. (12.55)]. Set $l = 1$ and $k = 1$.

Step 1: The α and β groups assign feasible values to $\mathbf{L}^{(k)}$ and $\mathbf{M}^{(k)}$, respectively. Start with the assigned weight vectors $\mathbf{w}^{\alpha(l)}$, $\mathbf{w}^{\beta(l)}$, and $\mathbf{w}^{o(l)}$ and perform steps 2 and 3.

Step 2: The α and β analysts perform the following steps independently. [The double parentheses (( )) mean that the contents belong to the β analyst].

a. Based on the fixed/updated levels of $\mathbf{L}^{(k)}$ (($\mathbf{M}^{(k)}$)), the α analyst ((β analyst)) solves the weighted sub-submodels of the α-HMO submodel Eq. (12.16) ((the β-HMO submodel Eq. (12.20))), at the initially fixed values of the interaction vector **U** ((**V**)), by using any nonlinear optimization technique such as the generalized reduced gradient (GRG) algorithm (Lasdon et al., 1982).
b. The α analyst ((β analyst)) checks whether the following stopping criteria (12.56) ((12.57)) are satisfied:

$$\left| \sum_{i=1}^{r} \mathbf{w}_i^{\alpha} \mathbf{E}_i(\mathbf{X}^{(k+1)}) - \sum_{i=1}^{r} \mathbf{w}_i^{\alpha} \mathbf{E}_i(\mathbf{X}^{(k)}) \right| < \epsilon_{\alpha} \tag{12.56}$$

$$\left| \sum_{i=1}^{w} \mathbf{w}_i^{\beta} \mathbf{F}_i(\mathbf{Y}^{(k+1)}) - \sum_{i=1}^{w} \mathbf{w}_i^{\beta} \mathbf{F}_i(\mathbf{Y}^{(k)}) \right| < \epsilon_{\beta} \tag{12.57}$$

where

ϵ_θ = a very small number, prespecified by the θ analyst (θ = α, β)
$\xi^{(k)}$ = optimal values of ξ attained at kth iteration (ξ = **X**, **Y**)

If these equations are satisfied, let the optimal solution of the weighted α-HMO submodel ((β-HMO submodel)) be $\mathbf{X}^{(k)}$ (($\mathbf{Y}^{(k)}$)) and go to step 4. Otherwise, go to step 2c.
c. The α analyst ((β analyst)) updates the interaction vector ***U*** ((**v**)) by a feasible method (see Chap. 3). The analysts then solve their own weighted sub-submodels. Go to step 2b.

Step 3:

a. Set $n = 1$. Based on the fixed or updated levels of $\mathbf{L}_{\alpha}^{(k)}$ and $\mathbf{M}_{\alpha}^{(k)}$, the α analyst solves the α-HOC sub-submodels (12.51) at the initial feasible values of $\mathbf{R}^{(0)}$ fixed by the β group, to compute the optimal solutions $\mathbf{Z}_{\alpha 1}^{(n-1)}$, $j = 1, 2, \ldots,$ α, and then fixes the values of **S** by $\mathbf{S}^{(n)} = PA_{\beta\alpha}\mathbf{Z}_{\alpha}^{(n-1)}$.
b. Based on the fixed or updated levels of $\mathbf{L}_{\alpha}^{(k)}$ and $\mathbf{M}_{\alpha}^{(k)}$, the β analyst computes the optimal solutions $\mathbf{Z}_{\beta i}^{(n)}$, $j = 1, \ldots,$ β, from the β-HOC sub-submodels (12.52), using the information of **S** from the α group (i.e., step 3a or 3d) and thereby fixes the values of **R** by $\mathbf{R}^{(n)} = NA_{\alpha\beta}\mathbf{Z}_{\beta}^{(n)}$.
c. Both analysts check whether their own stopping criteria, Eqs. (12.53) and (12.54), respectively, are satisfied. If yes, let the optimal solutions of the weighted α-HOC submodel and the weighted β-HOC submodel be $\mathbf{Z}_{\alpha}^{(k)}$ and $\mathbf{Z}_{\beta}^{(k)}$, respectively, and go to step 4. Otherwise, set $n = n + 1$ and go to step 3d.
d. Based on the fixed or updated levels of $\mathbf{L}_{\alpha}^{(k)}$ and $\mathbf{M}_{\beta}^{(k)}$ and using the information of **R** from the β group (i.e., step 3b), the α analyst computes the optimal solutions $\mathbf{Z}_{\alpha j}^{(n-1)}$, $j = 1, 2, \ldots,$ α, from the α-HOC sub-submodels (12.51) and fixes the values of **S** by $\mathbf{S}^{(n)} = PA_{\beta\alpha}\mathbf{Z}_{\alpha}^{(n-1)}$. Go to step 3b.

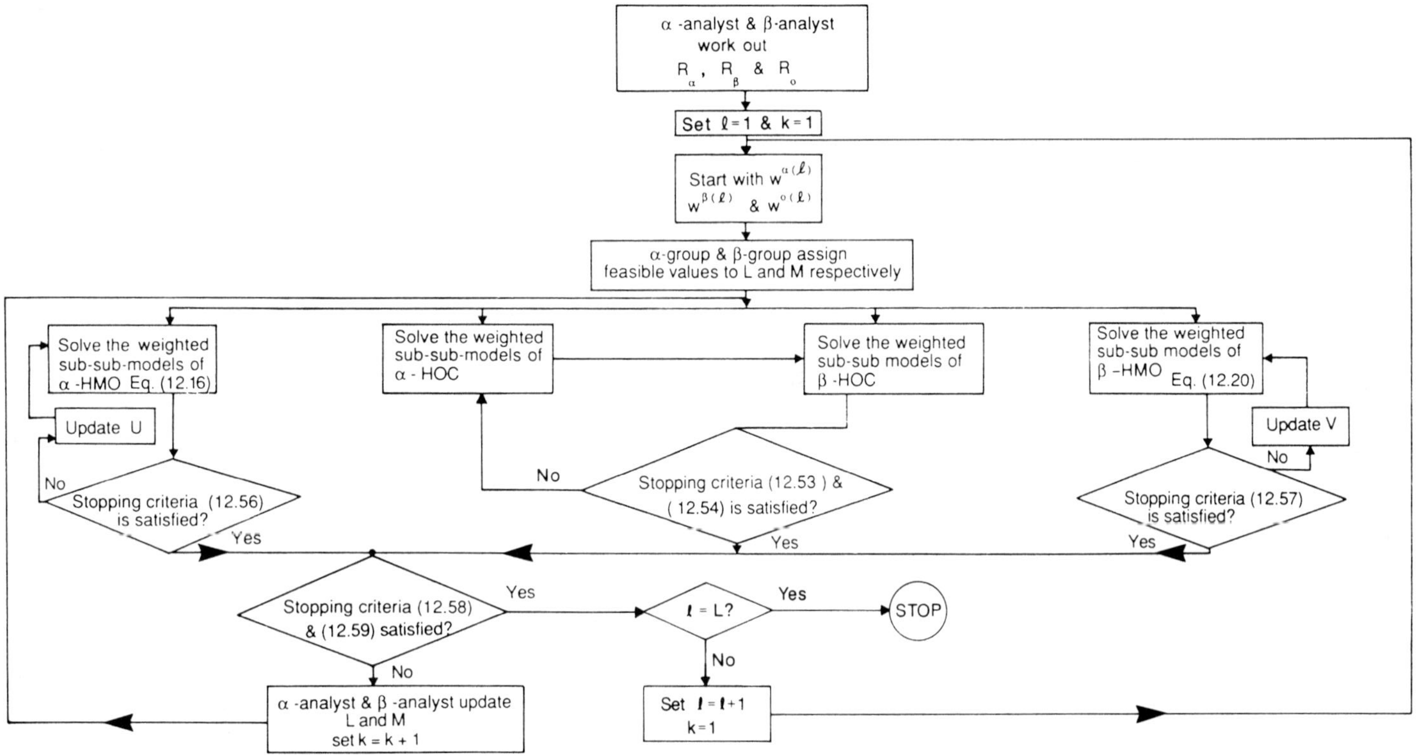

Figure 12.2 HHOC algorithm.

Step 4: The α and β analysts check whether the stopping criteria—Eqs. (12.58) and (12.59), respectively—hold:

$$\left| \sum_{i=1}^{r} \mathbf{w}_i^{\alpha} \mathbf{E}_i(\mathbf{X}^{(k+1)}) + \sum_{i=1}^{o} \mathbf{w}_i^{o} O_i(\mathbf{Z}_{\alpha}^{(k+1)}) - \sum_{i=1}^{r} \mathbf{w}_i^{\alpha} \mathbf{E}_i(\mathbf{X}^{(k)}) - \sum_{i=1}^{o} \mathbf{w}_i^{o} O_i(\mathbf{Z}_{\alpha}^{(k)}) \right| \leq \epsilon_{\alpha} \tag{12.58}$$

and

$$\left| \sum_{i=1}^{\beta} \mathbf{w}_i^{\beta} F_i(\mathbf{Y}^{(k+1)}) + \sum_{i=1}^{o} \mathbf{w}_i^{o} O_i(\mathbf{Z}_{\beta}^{(k+1)}) - \sum_{i=1}^{w} \mathbf{w}_i^{\beta} F_i(\mathbf{Y}^{(k)}) - \sum_{i=1}^{o} \mathbf{w}_i^{o} O_i(\mathbf{Z}_{\beta}^{(k)}) \right| \leq \epsilon_{\beta} \tag{12.59}$$

where

$\epsilon_{\alpha}, \epsilon_{\beta}$ = very small numbers prespecified by the α and β analysts, respectively

$\theta^{(k)}$ = optimal value of θ achieved at the kth iteration $\theta = \mathbf{X}, \mathbf{Y}, \mathbf{Z}$

If yes, the α-Pareto optimal solution $(\mathbf{X}^{(k)}, \mathbf{Z}_{\alpha}^{(k)})$ and the β-Pareto optimal solution $(\mathbf{Y}^{(k)}, \mathbf{Z}_{\beta}^{(k)})$ with respect to the assigned weight vectors $\mathbf{w}^{\alpha(l)}$, $\mathbf{w}^{\beta(l)}$, and $\mathbf{w}^{o(l)}$ have been arrived at. Go to step 5. Otherwise, the α and β analysts update the values of **L** and **M**, respectively, by a feasible method (see Chap. 3). Set $k = k + 1$ and perform steps 2 and 3.

Step 5: Check whether all the assigned weight values in $\mathbf{R}_{\alpha}$, $\mathbf{R}_{\beta}$, and $\mathbf{R}_{o}$ have been used (i.e., does $l = L$?). If yes, stop. Otherwise, set $l = l + 1$ and $k = 1$. Go to step 1.

A flow chart representing the HHOC algorithm is given in Fig. 12.2. This algorithm generates a set of Pareto optimal alternatives. Negotiation procedures are, then, needed for the decisionmakers to arrive at a compromise solution in the generated set. These procedures are not considered in this chapter; for one of the possible schemes, see Thadathil (1983).

REFERENCES

Bateson, G. *Mind and Nature*. Bantam, New York, 1979.

Haimes, Y. Y., Hierarchical Holographic Modeling, *IEEE Trans. Syst. Man Cybern.*, vol. SMC-11, no. 9, pp. 606–617, 1981.

Haimes, Y. Y., and Macko, D., Hierarchical Structures in Water Resources Systems Management, *IEEE Syst. Man Cybern.*, vol. SMC-3, no. 4, 1973.

Lasdon, L. S., Waren, A. D., and Ratner, M. W., *GRG2 User's Guide*, Department of Systems Engineering, Case Western Reserve University, Cleveland, Ohio 1982.

Macko, D., and Y. Y. Haimes, Overlapping Coordination of Hierarchical Structures, *IEEE Trans. Sys. Man Cybern.*, vol. SMC-8, no. 10, 1978.

Shima, T., *Hierarchical Overlapping Coordination of Large Scale Systems with Multiple Objectives*, Ph.D. dissertation, Case Western Reserve University, Cleveland, Ohio, May 1982.

Shima, T., and Haimes, Y. Y., The Convergence Properties of Hierarchical Overlapping Coordination, *IEEE Trans. Syst. Man Cybern.*, vol. 14, no. 1, 1984.

Tarvainen, K., *Hierarchical Multiobjective Optimization*, Ph.D. dissertation, Case Western Reserve University, Cleveland, Ohio, May 1981.

Tarvainen, K., and Haimes, Y. Y., Coordination of Hierarchical Multiobjective Systems: Theory and Methodology, *IEEE Trans. Syst. Man Cybern.*, vol. SMC-12, no. 6, 1982.

Thadathil, J., *Optimization Methods in Hierarchical Holographic Modeling*, Ph.D. dissertation, Case Western Reserve University, Cleveland, Ohio, August 1983.

Thadathil, J., and Haimes, Y. Y., Hierarchical Holographic Modeling: Coupling Resource Allocation and Compromise Solution, Technical report No. CLSS AOA #83-101, Systems Engineering Department, Case Western Reserve University, Cleveland, Ohio, 1984a.

Thadathil, J., and Haimes, Y. Y., Hierarchical Holographic Submodels with Particular Constraint Structures, Technical Report No. CLASS AOA #83-102, Systems Engineering Department, Case Western Reserve University, Cleveland, Ohio, 1984b.

AUTHOR INDEX

SUBJECT INDEX